COMPREHENSIVE CHEMICAL KINETICS

COMPREHENSIVE

Section 1.   THE PRACTICE AND THEORY OF KINETICS (3 volumes)

Section 2.   HOMOGENEOUS DECOMPOSITION AND ISOMERISATION REACTIONS (2 volumes)

Section 3.   INORGANIC REACTIONS (2 volumes)

Section 4.   ORGANIC REACTIONS (5 volumes)

Section 5.   POLYMERISATION REACTIONS (3 volumes)

Section 6.   OXIDATION AND COMBUSTION REACTIONS (2 volumes)

Section 7.   SELECTED ELEMENTARY REACTIONS (1 volume)

Section 8.   HETEROGENEOUS REACTIONS (4 volumes)

Section 9.   KINETICS AND CHEMICAL TECHNOLOGY (1 volume)

Section 10.   MODERN METHODS, THEORY AND DATA

CHEMICAL KINETICS

EDITED BY

N.J.B. GREEN

King's College London
London, England

VOLUME 41

OXOACIDITY: REACTIONS OF OXO-COMPOUNDS IN IONIC SOLVENTS

V.L. CHERGINETS

Institute for Scintillation Materials
Lenin Avenue, 60
Kharkov 61001, Ukraine

2005

ELSEVIER

AMSTERDAM – BOSTON – HEIDELBERG – LONDON – NEW YORK – OXFORD – PARIS
SAN DIEGO – SAN FRANCISCO – SINGAPORE – SYDNEY – TOKYO

ELSEVIER B.V.
Radarweg 29
P.O. Box 211, 1000 AE Amsterdam
The Netherlands

ELSEVIER Inc.
525 B Street, Suite 1900
San Diego, CA 92101-4495
USA

ELSEVIER Ltd
The Boulevard, Langford Lane
Kidlington, Oxford OX5 1GB
UK

ELSEVIER Ltd
84 Theobalds Road
London WC1X 8RR
UK

First edition 2005

Library of Congress Cataloging in Publication Data
A catalog record is available from the Library of Congress.

British Library Cataloguing in Publication Data
A catalogue record is available from the British Library.

ISBN: 0 444 51782 0
ISSN: 0069 8040 (Series)

♾ The paper used in this publication meets the requirements of ANSI/NISO Z39.48-1992 (Permanence of Paper). Printed in Great Britain

COMPREHENSIVE CHEMICAL KINETICS

ADVISORY BOARD

Volumes in the Series

Section 1.   THE PRACTICE AND THEORY OF KINETICS
             (3 volumes)

Volume 1   The Practice of Kinetics
Volume 2   The Theory of Kinetics
Volume 3   The Formation and Decay of Excited Species

Section 2.   HOMOGENEOUS DECOMPOSITION AND
             ISOMERISATION REACTIONS (2 volumes)

Volume 4   Decomposition of Inorganic and Organometallic Compounds
Volume 5   Decomposition and Isomerisation of Organic Compounds

Section 3.   INORGANIC REACTIONS (2 volumes)

Volume 6   Reactions of Non-metallic Inorganic Compounds
Volume 7   Reactions of Metallic Salts and Complexes, and Organometallic Compounds

Section 4.   ORGANIC REACTIONS (5 volumes)

Volume 8   Proton Transfer
Volume 9   Addition and Elimination Reactions of Aliphatic Compounds
Volume 10   Ester Formation and Hydrolysis and Related Reactions
Volume 12   Electrophilic Substitution at a Saturated Carbon Atom
Volume 13   Reactions of Aromatic Compounds

Section 5.   POLYMERISATION REACTIONS (3 volumes)

Volume 14   Degradation of Polymers
Volume 14A   Free-radical Polymerisation
Voulme 15   Non-radical Polymerisation

Section 6.   OXIDATION AND COMBUSTION REACTIONS
             (2 volumes)

Volume 16   Liquid-phase Oxidation
Volume 17   Gas-phase Combustion

Section 7.   SELECTED ELEMENTARY REACTIONS (1 volume)

Volume 18   Selected Elementary Reactions

Section 8.   HETEROGENEOUS REACTIONS (4 volumes)

Volume 19   Simple Processes at the Gas-Solid Interface
Volume 20   Complex Catalytic Processes
Volume 21   Reactions of Solids with Gases
Volume 22   Reactions in the Solid State

Section 9.   KINETICS AND CHEMICAL TECHNOLOGY (1 volume)

Volume 23   Kinetics and Chemical Technology

Section 10.   MODERN METHODS, THEORY, AND DATA

Volume 24   Modern Methods in Kinetics
Volume 25   Diffusion-limited Reactions
Volume 26   Electrode Kinetics: Principles and Methodology
Volume 27   Electrode Kinetics: Reactions
Volume 28   Reactions at the Liquid-Solid Interface
Volume 29   New Techniques for the Study of Electrodes and their Reactions
Volume 30   Electron Tunneling in Chemistry. Chemical Reactions over Large Distances
Volume 31   Mechanism and Kinetics of Addition Polymerizations
Volume 32   Kinetic Models of Catalytic Reactions
Volume 33   Catastrophe Theory
Volume 34   Modern Aspects of Diffusion-Controlled Reactions
Volume 35   Low-temperature Combustion and Autoignition
Volume 36   Photokinetics: Theoretical Fundamentals and Applications
Volume 37   Applications of Kinetic Modelling
Volume 38   Kinetics of Homogeneous Multistep Reactions
Volume 39   Unimolecular Kinetics, Part 1. The Reaction Step
Volume 40   Kinetics of Multistep Reactions, 2nd Edition

List of symbols

A	–	an acid, the generalized designation,
B	–	a base, the generalized designation,
β	–	the buffer number of solutions in melts $= (d\Delta b/dpO)$, the values are negative in studies of oxoacidity,
D	–	density, $kg\,m^{-3}$,
E, E_0	–	e.m.f., the standard e.m.f.,
ε_L	–	dielectric constant of a solvent 'L',
F	–	Faraday number, $96484{,}56\ Q\ mol^{-1}$,
ΔG_c	–	Gibbs energy of re-solvation, $kJ\ mol^{-1}$,
ΔG_x^f	–	Gibbs energy of component 'X' formation, $kJ\ mol^{-1}$,
K	–	equilibrium constant,
K	–	the pre-logarithmic coefficient of the Nernst equation, V,
L	–	a solvent; generalized designation,
l^+	–	the cation of solvent 'L'; generalized designation
l^-	–	the anion of solvent 'L'; generalized designation
m_x, m_x^0	–	molality and initial molality of component 'x' of a solvent, in $mol\,kg^{-1}$
M	–	molar mass,
μ_x, μ_x^0	–	chemical potential; standard chemical potential of component 'x',
N_A	–	the Avogadro number, $6.022 \times 10^{23}\ mol^{-1}$,
$\bar{n}$	–	the ligand number, the ratio of the base consumed in a titration of an acid to the initial molality of the acid, $\bar{n} = (m_{O^{2-}}^0 - 10^{-pO})/m_A^0$,
P_{MeO}	–	the solubility product of an oxide of composition MeO,
PO	–	the index (the negative logarithm) of the equilibrium molality of O^{2-},
pO^*	–	the index (the negative logarithm) of the initial molality of O^{2-},

Ω	–	the index (the negative logarithm) of the equilibrium molality of O^{2-} in units of the common oxoacidity function,
R	–	the universal gas constant, $8.314\,J\,mol\,K^{-1}$;
s_x	–	the molality of component 'x' in the saturated solution, $mol\,kg^{-1}$ of solvent,
Σs_{MeO}	–	the 'complete' solubility of oxide MeO in ionic solvent, *i.e.*, the sum of ionic and non-dissociated forms of the oxide in the saturated solution, $mol\,kg^{-1}$ of solvent,
w_I	–	weight of a substance, 'i', measured in g,
YSZ	–	yttria- stabilized zirconia ($0.9ZrO_2+0.1Y_2O_3$) – a material for solid electrolyte membranes possessing oxide ion conductivity,
Z	–	the number of electrons in the Nernst equation.

The subscripts 'N' and 'm' for the thermodynamic parameters mean that these are calculated using molar fractions and molalities, respectively.

Introduction

Theoretical studies and the application of chemical reactions within the matrix of various molecular and ionic solvents are among the most important trends in modern chemistry and engineering since the nature of the solvent significantly affects the technological processes running in it [1, 2]. Modern solution technologies are mainly based on the use of molecular liquids as solvents at room temperature. However, an increase in the application of high-temperature ionic melts as liquid media has been evident in recent decades. This is a consequence of a number of unique features which are characteristic of this class of solvents.

Molten salts are now used widely in science and engineering as convenient media (solvents) for performing a range of technological processes, including surface treatments (etching, fluxes), the electrolytic obtaining of metals and alloys, refining and the electrochemical deposition of a number of galvanic and diffusion coatings [3]. Ionic melts serve as solvents or starting materials for growing a number of widely used scintillation [CsI (Tl), NaI (Tl)] and acoustical (KCl) single crystals. For some decades, high-temperature ionic media have been attracting considerable attention as solvents for the treatment of nuclear materials and, especially importantly, nuclear wastes without producing sewage [4]. We would also note the recent growth of interest in room-temperature ionic liquids, owing to their potential use as reaction media for green chemical reactions [5, 6], electrolytes for high-current-density batteries, etc.

Ionic melts possess a number of features which compare favourably with the conventional room-temperature molecular solvents. The practically complete dissociation of ionic media into the constituent ions allows one to create high (up to $10 \, \mathrm{A \, cm^{-2}}$) current densities in electrolysis. The absence of oxidants similar to H^+ makes it possible to obtain final products which cannot be extracted from aqueous solvents (i.e. alkali- and alkaline-earth metals, sub-ions, etc.). From an ecological standpoint, molten ionic media are especially applicable as technological solvents because their employment does not cause accumulation of liquid wastes since cooling to room temperature transforms the ionic liquids into their solid state.

Processes which take place in ionic melts–solvents are considerably affected by impurities contained in the initial components of the melt or formed during the preparation (mainly by melting) of solid components of the solvents owing to high-temperature hydrolysis of the melts or their interactions with container materials (Al_2O_3, SiO_2, etc.) as well as active components of the atmosphere (O_2, CO_2, H_2O, etc.). The list of these impurities is long, and includes multivalent cations of transition metals and different complex anions (oxo- or halide anions). The effect of the mentioned admixtures on the processes in ionic melts is mainly dependent on the degree of their donor–acceptor interactions with constituent parts of ionic melts.

Oxide-containing impurities are the most widespread in molten salts and their effect on technological processes and the parameters of the products fabricated using ionic solvents is often negative. It consists in the fixation of reagents of acidic character (cations, polynuclear anions, acidic gases) that favours the formation of insoluble (precipitates or suspensions) or slightly dissociated products in the ionic melt. All these reasons cause retardation of the desired process owing to a decrease in the equilibrium concentrations of the starting acidic reagents, the appearance of inclusions of oxide particles in electrochemically deposited metals, additional absorption bands in optical single crystals, a reduction of their radiation resistance [7, 8], substantial corrosion of container materials, etc.

Quantitative investigations of the reactions of oxide ions and oxo-compounds in high-temperature ionic solvents are, therefore, of considerable scientific and applied importance. The interactions of such kinds are referred to as acid–base ones, according to Lewis. Since 1939, when Lux proposed a definition of acids as oxide ion acceptors and bases as donors of O^{2-}, such acid–base interactions came to be called "oxoacidity". The most general scheme of a Lux acid–base interaction is presented by the following equation:

$$A(\ )+ \ddot{:O}^{2-} \rightleftarrows B\left(\equiv A(:)\ddot{O}^{2-}\right), \tag{0.1}$$

and the "pO" index introduced by Lux serves as a measure of the basic (acidic) properties of a melt

$$pO \equiv -\log a_{O^{2-}} \approx -\log m_{O^{2-}}. \tag{0.2}$$

This index is similar to the pOH scale in aqueous solutions. Measurements of the pO index during the running of acid–base reactions of various kinds make it possible to calculate their thermodynamic characteristics and, to a first approximation, the equilibrium constants. Because there are no reliable experimental data on the activity coefficients of most substances in ionic solutions or theoretical substantiations of their estimations, the equilibrium constants are often obtained as concentration, but not thermodynamic parameters. Therefore, the pO indices discussed in the following sections of this book are the negative logarithms of O^{2-} concentrations. Also, in most cases, the studies concern diluted solutions of ionic solids or liquids in ionic liquids, i.e. both ingredients are substances with ionic bonding. The same nature of the bonding allows us to consider the formed solutions as close to ideal, but not infinitely diluted ones, as takes place in solutions of ionic substances in water or other molecular solvents at room-temperature.

The ability of high-temperature ionic solvents to dissolve various substances (e.g. metals or their oxides) is considered among their most important characteristics. In the case of scintillation, single crystals based on alkali-metal halides, some halides-dopants (TlI, CeX_3, etc.), are available as admixtures since they create levels in the forbidden zone which are responsible for the scintillation. In contrast, oxygen-containing admixtures are harmful since their existence in the crystals causes a considerable reduction in their light-yield, radiation resistance, etc.

Therefore, investigations of oxide solubilities in molten salts are of great interest for the following reasons. For some purposes (fluxes, etching, growth of oxide crystals and composites from ionic melts), it is necessary to choose a melt where the solubility of oxide materials should be as high as possible. If an ionic melt is used as a material for the growth of an optical single crystal (alkali- and alkaline-earth metal halides) then the requirements should be quite the opposite: the solubility of the oxygen-containing admixtures should be very low to provide a low quantity of additional absorption bands and disseminating centres in the crystals.

Investigations of oxide solubilities in high-temperature ionic solvents, especially those based on alkali metal halides, represent an important branch of oxoacidity studies, although they have not developed considerably during the past few decades. One of the most serious causes of the absence of essential progress in the studies of oxide solubility is the seemingly bad reproducibility of the reliable data on the solubility-product values of oxides obtained by different investigators—their scatter covers some orders of pP_{MeO}

depending on the experimental routine used. Other reasons, such as the synthesis of the initial metal halides, the thorough purification of ionic solvents from traces of oxide ions and the design of the apparatus for the experiments, complicate determination of the solubility characteristics in high-temperature ionic melts.

Examinations of Lux acid–base reactions of various kinds were very intensive in the 1960s and 1970s, but then this intensity decreased. The experimental routines existing at that time essentially limited the range of ionic solvents and reagents which could be studied successfully. For example, molten bromides and iodides as media for oxoacidity processes were practically unstudied because of the impossibility of using gas–oxygen electrodes that provided the supply of gaseous oxygen in the said media possessing strong reducing properties. However, for the next two decades, various kinds of membrane oxygen electrodes were thoroughly tested and introduced into wide practice and the methods of treatment of the results, and their interpretation, were extensively revised. This necessitates newer studies of oxoacidity, taking into account the modified scientific basis. Extensive basic and applied research on oxoacidity may result in the creation of principally new solution technologies based on the employment of high-temperature liquids.

Up to now, attempts have not been made to summarize the results of oxoacidity investigations and to acquaint a wide range of readers with achievements in oxoacidity and subsequent studies required for scientific and applied purposes. Where attempts have been made in books [3, 9], the body of data considered was scant and without deep scientific interpretation.

The present book is devoted to a systematization of investigations of oxoacidity and oxide solubility studies in ionic melts published up to 1 July 2004. The book brings together material that is quite widely dispersed, including work published in journals of the former Soviet Union—which remains rather inaccessible and is certainly unknown world-wide—into a single focused volume. To a significant extent the presented material is based on the author's own investigations that have been performed since 1987. Together with experimental data, a number of problems of the physical chemistry of solutions are considered, with some more or less substantial additions made by the author. Some of them are known well enough (acid–base theories), the others are known to a smaller number of scientists (e.g. the solubility of phases of different dispersity).

To my mind, the results and generalizations presented in this monograph will not only be useful for specialists who explore molten salts but will also be of interest to chemists studying the protic acidity of non-aqueous media, solvosystems and solubility phenomena. The text of the monograph is not complicated by terms and general conceptions which are not known to most readers familiar with physical chemistry and chemical thermodynamics. There are many experimental data which await understanding and explanation, and some material presented in this book is now waiting for its final treatment and understanding.

I would like to express my deep gratitude to all my experienced colleagues, Prof. I.N. Vjunnik, Prof. V.M. Koshkin, Prof. A.B. Blank and Prof. Yu.Ya. Fialkov, whose comments and advice promoted the preparation of this manuscript. I thank my colleagues and students from the Institute for Single Crystals and Institute for Scintillation Materials of National Academy of Sciences of Ukraine, O.V. Demirskaya, Dr. E.G. Kubrak (Khailova) and Dr. T.P. Rebrova (Boyarchuk), for their inestimable help in the obtaining and discussion of most of the experimental data on the oxoacidity of high-temperature ionic melts. Finally, I am very grateful to T.Yu. Zheltukhina for her kind help in the preparation of the English version of this monograph.

It should be stressed that our oxoacidity studies could not have been realized without the kind financial support of the International Science Foundation (Grant No. YSU 083062), the State Foundation of Basic Researches of Ukraine (Project No. 03.07/00138) and the National Academy of Sciences of Ukraine.

All comments and remarks will be accepted with appreciation.

VICTOR L. CHERGINETS

Contents

List of symbols   VIII

Introduction   XI

1 Homogeneous acid-base equilibria and acidity scales in ionic melts.   1

Part 1   Definitions of acids and bases   1

   1.1.1   Definitions of particles possessing acid or base properties   1
   1.1.2   Definitions of solvents system   5
   1.1.3   Hard and soft acids and bases (Pearson's concept)   11
   1.1.4   Generalized definition of solvent system. Solvents of kinds I and II.   17

Part 2   Studies of homogeneous acid-base reactions in ionic melts   33

   1.2.1   Features of high-temperature ionic solvents as media for
         Lux acid–base interactions   34
   1.2.2   Methods of investigations   37
   1.2.3   Ionic solvents based on alkali metal nitrates   48
   1.2.4   Molten alkali metal sulfates.   64
   1.2.5   Silicate melts   67
   1.2.6   KCl−NaCl equimolar mixture   69
         1.2.6.1   Oxocompounds of chromium(VI)   69
         1.2.6.2   Oxoacids of molybdenum(VI)   74
         1.2.6.3   Oxocompounds of tungsten(VI)   77
         1.2.6.4   Oxoacidic properties of phosphates   81
         1.2.6.5   Oxoacids of vanadium(V)   84
         1.2.6.6   Oxoacids of boron(III)   87
         1.2.6.7   Acidic properties of Ge(IV) and Nb(V) oxocompounds   89
   1.2.7   Molten KCl−LiCl (0.41:0.59) EUTECTIC   91
   1.2.8   Molten NaI   93
   1.2.9   Other alkaline-metal halides   98
   1.2.10 Conclusion   100

Part 3   Acid-base ranges in ionic melts. Estimation of relative acidic properties
        of ionic melts   103

   1.3.1   The oxobasicity index as a measure of relative oxoacidic properties of
         high-temperature ionic solvents   112

1.3.2   Oxoacidity scales for melts based on alkali- and alkaline-earth
        metal halides... 118
1.3.3   Conclusion 128

2   Oxygen electrodes in ionic melts. Oxide ion donors.................... 129

Part 4   Oxygen electrode reversibility in ionic melts 129
2.4.1   Potentiometric method of study of oxygen electrode reversibility........ 135
        2.4.1.1   Direct calibration 135
        2.4.1.2   Indirect calibration of oxygen electrodes 137
2.4.2   Experimental results 142
        2.4.2.1   Oxygen-containing melts 142
        2.4.2.2   Melts based on alkali metal halides 148
        2.4.2.3   Melts based on alkali- and alkaline-earth halides 175
        2.4.2.4   KCl–NaCl–NaF eutectic 176
2.4.3   Conclusions 178

Part 5   Investigations of dissociation of Lux bases in ionic melts 181
2.5.1   Reactions of ionic melts with gases of acidic or base character 182
        2.5.1.1   High-temperature hydrolysis of melts based on alkali metal
                  halides... 183
        2.5.1.2   Purification of halide ionic melts from oxide-ion admixtures .. 188
2.5.2   Behaviour of Lux bases in ionic melts 200
        2.5.2.1   Sodium peroxide, Na_2O_2 202
        2.5.2.2   Alkali metal carbonates, Me_2CO_3 206
        2.5.2.3   Alkali metal hydroxides, MeOH 217

3   Equilibria in "solid oxide–ionic melt" systems......................... 229

Part 6   Characteristics of oxide solubilities and methods of their
         determination 229
3.6.1   Parameters describing solubilities of solid substances in ionic solvents .. 229
3.6.2   Methods of oxide solubility determination 231
        3.6.2.1   Isothermal saturation method 231
        3.6.2.2   Potentiometric titration method 233
        3.6.2.3   Sequential addition method 257

Part 7   Regularities of oxide solubilities in melts based on alkali and
         alkaline-earth metal halides 265
3.7.1   Molten alkali-metal halides and their mixtures 267
        3.7.1.1   KCl–LiCl (0.41:0.59) eutectic mixture 267
        3.7.1.2   KCl–NaCl (0.50:0.50) equimolar mixture 275
        3.7.1.3   CsCl–KCl–NaCl (0.455:0.245:0.30) eutectic 293

3.7.1.4   CsBr–KBr (0.66:0.34) melt   305
3.7.1.5   Molten CsI, 700 °C   308
3.7.1.6   Molten potassium halides   316
3.7.1.7   Other solvents based on alkali-metal halides   320
3.7.2   Oxide solubilities in melts based on alkali- and alkaline-earth
         metal halides.......................................   326
3.7.3   Solubilities of alkali earth metal carbonates in KCl–NaCl eutectic   337
3.7.4   Conclusion   343

Afterword   347

References   351

Formula Index...   373

Subject Index...   379

Chapter 1

Homogeneous Acid–Base Equilibria and Acidity Scales in Ionic Melts

Part 1. Definitions of Acids and Bases

At present, various kinds of donor–acceptor interactions are considered to be acid–base ones, in the framework of an appropriate acid–base concept [10–12]. The last century witnessed the development of a few such definitions, which can be translated conditionally to definitions of "carriers" of acidic (basic) properties, those for a solvent system, and principles for the prediction of the behaviour of acid–base reactions in various media.

1.1.1 DEFINITIONS OF PARTICLES POSSESSING ACID OR BASE PROPERTIES

The Arrhenius definition, known as the theory of electrolytic dissociation (TED) appropriate for the description of aqueous solutions, was historically the first definition of this kind. It states that an acid is a substance prone to dissociation with the formation of protons, H^+:

$$HCl \rightleftharpoons H^+ + Cl^-, \tag{1.1.1}$$

whereas a base is a donor of hydroxide ions, OH^-:

$$NaOH \rightleftharpoons Na^+ + OH^-. \tag{1.1.2}$$

In the ionic form, an acid–base interaction (neutralization) may be written in the following manner:

$$H^+ + OH^- \rightleftharpoons H_2O. \tag{1.1.3}$$

Similar interactions in molecular solutions are characterized by the constant of the intrinsic acid–base auto-ionization, which is the important parameter describing acid–base properties of a solvent. For example, the constant of auto-ionization of water is given by the following product:

$$K_w \rightleftharpoons [H^+][OH^-] = 10^{-14}, \quad 298.15 \text{ K}. \tag{1.1.4}$$

The lower the value of this constant, the larger the differences in acidity indices (pH) between the standard solutions of strong acids and bases, that results in a wider acid–base range for the solvent. This refers not only to the acid–base equilibria in aqueous solutions but also applies to any donor–acceptor interaction in molecular solvents which are prone to heterolytic dissociation with the formation of "acidic" and "basic" particles, as provided by an appropriate definition of acids and bases. It follows from equations (1.1.3) and (1.1.4) that the Arrhenius definition can only be used for the description of acid–base interactions in aqueous solutions, since the reaction between the "acid" of solvent and the "base" of solvent can result in the formation only of the solvent molecules. In the case considered, this solvent is water.

A more general definition of acid–base equilibrium in molecular solvents was proposed independently by Brønsted and Lowry, who extended the term "base". According to them, the acid is a donor of protons, and they defined a base only as an acceptor of H^+. According to the Brønsted–Lowry definition, the dissociation of an acid results in the formation of a proton and a "conjugate" base:

$$A \rightleftharpoons B + H^+. \tag{1.1.5}$$

In molecular solvents, process (1.1.5) is superimposed on other acid–base reactions, namely the reaction with neutral solvent molecules or with other dissolved substances of acidic or basic character. The complete acid–base process is represented by the equation of type:

$$A_1 + B_2(L) \rightleftharpoons A_2(LH^+) + B_1. \tag{1.1.6}$$

Of course, protons cannot exist as free particles in molecular solvents: they should react with the solvent molecules, and such an interaction in aqueous solutions results in the formation of $H_5O_2^+$ ($H^+ \cdot 2H_2O$ or $[H_2O \cdots H \cdots OH_2]^+$) particles. Similarly, hydroxide ions in aqueous solutions exist as $H_3O_2^-$ ($OH_2 \cdots OH^-$). The stronger the acid A_1 compared to the acid A_2, the more complete is the shift of reaction (1.1.6) to the right. The Brønsted–Lowry definition thus extends the term "acid–base interaction" to reactions accompanied with proton transfer from one base to another. For example, according to this definition, hydrolysis is a kind of acid–base interaction

$$HCO_3^- + H_2O \rightleftharpoons CO_2 \cdot H_2O + OH^- \tag{1.1.7}$$

with the participation of two conjugate pairs consisting of an acid and a base which differ by one proton:

$$HCO_3^- + H^+ \rightleftharpoons H_2O, \tag{1.1.8}$$

$$H_2O - H^+ \rightleftharpoons OH^-. \tag{1.1.9}$$

An obvious advantage of the Brønsted–Lowry definition compared with TED is the fact that acid–base interaction reaches an equilibrium between two conjugate pairs in a solvent. When this definition was formulated it considered only different protolytic media as appropriate solvents, whose dissociation process can be described by the following scheme:

$$HL \rightleftharpoons H^+ + L^-. \tag{1.1.10}$$

Nevertheless, it was used later for the description of the reactions of proton exchange in aprotic media, since this definition has no restrictions concerning the chemical composition of the solvent particles, or more precisely, their tendency to dissociate with the formation of a solvated proton.

It should be noted that the terms "acid" and "base" in the Brønsted–Lowry theory refer only to the function of the given substance in a protolytic reaction. The same substance under certain conditions may react as an acid, and in other cases it may demonstrate basic properties. Also, the relative strength of acids and bases is considerably affected by solvation of the reagents and reaction products.

Lewis [13] proposed the electronic theory of acids and bases and gave a more general definition of acids and bases in its framework. Acids are determined as acceptors of an electron pair, and its donors are classified as bases. The principal scheme of the Lewis acid–base interaction is described by the following equation:

$$A(\ ) + : B \rightleftharpoons A(:)B. \tag{1.1.11}$$

According to this equation, both the definitions considered above concerning proton acidity are merely particular cases of the Lewis definition. Indeed, a proton is an electron-deficient particle and an acceptor of an electron pair, hence it is referred to as an acid and its acceptors are bases.

Besides protons, other electron-deficient particles can be classified as Lewis acids, regardless of their charge (Ni^{2+}, BF_3, etc.).

The most general definition of acids and bases was formulated by Usanovitch [14, 15] who proposed the definition of acids as participants in a chemical reaction delivering cations (a proton being among these) or fixing an anion (including a single electron). Thus, bases are donors of anions (electron) or acceptors of cation (proton). However, this definition has not become generally accepted, mainly because of its excessive universality—since most chemical reactions may be classified as acid–base ones in the framework of the Usanovitch definition. For example, the following redox reaction will be an acid–base reaction:

$$Fe^{3+} + Cu^+ \rightleftharpoons Fe^{2+} + Cu^{2+\cdot} \tag{1.1.12}$$

here, Fe^{3+} (an electron-donor) is an acid, and Cu^+ (the acceptor of an electron) is a base.

Usually, definitions of acids and bases are necessary for the classification of different kinds of chemical reactions, i.e. for dividing them into acid–base and "other" ones. For example, the Brønsted–Lowry definition divides reactions into acid–base, which are characterized by redistribution of protons, and other ones. The wider Lewis definition makes the division of reactions into acid–base and redox ones, meaning that in the former case there is redistribution of electron density on account of electron pairs, and the latter case concerns reactions with the transfer of single electrons. Since chemistry concerns just the redistribution of electrons of external shells, we can classify all chemical reactions as acid–base ones by the Usanovitch definition.

Nevertheless, there are some positive moments in the Usanovitch definition. So, acids and bases are classified on the basis of their reactions with ions (anions and cations), which are Lewis bases and acids. Therefore, this feature should be added to the Lewis definition to make the most general definition of acids and bases [16]: acids are either acceptors of an undivided electron pair (or anion) or donors of a cation, and bases are either donors of an electron pair (or anion) or acceptors of a cation.

However, let us consider the following reaction, which is an acid–base one (according to the Lux–Flood principle):

$$S_2O_7^{2-} + O^{2-} \rightleftharpoons 2SO_4^{2-}. \tag{1.1.13}$$

According to the above definitions, both reagents on the left-hand side of reaction (1.1.13) are acceptors of an anion, and therefore, should be treated as acids. However, this is not the case, and molten pyrosulfates are known to be very aggressive acidic media. The Lux definition provides one possible way to remove this seeming contradiction. Let us recollect that in 1939 Lux proposed a definition where acids were determined as acceptors of oxide ions, whereas bases were considered to be oxide-ion donors [17]. So, the principal Lux–Flood acid–base interaction can be described by equation (0.1). Later, some studies were performed by Flood *et al.* on molten potassium pyrosulfate [18–20]. The branch of studies which concerns the reactions of oxide ion exchange in molten salts is now called "oxoacidity".

It should be noted that there are some curious interpretations in the literature concerning oxoacidity, which distort to a greater or lesser degree the original definition. For example, Delimarskii and Barchuk [3] give as "the Lux definition" where a base is a donor of oxide or halide ions. Moreover, at least in the Russian translation of a book [21], "the Lux–Flood definition" is where an acid is a donor of oxide ions and a base is an acceptor of O^{2-}. Certainly, these "definitions" are erroneous, as may be seen by reading the original Lux paper [17].

1.1.2 DEFINITIONS OF SOLVENTS SYSTEM

Liquids of various natures (molecular, ionizing, ionic) have been mentioned in the Introduction as often serving as media for performing various technological processes, which can be considered as acid–base ones according to the appropriate acid–base definition. In such cases, the said processes are subject to the solvent action that shifts the equilibrium and favours or retards the formation of desirable products. The degree of this action is dependent on the acidic or basic properties of the solvent, and the same compound may act as an acid or a base in different solvents depending on their acid–base properties. For example, the well-known acetic acid, CH_3COOH, shows relatively weak acidic properties in aqueous solutions:

$$CH_3COOH + H_2O \rightleftharpoons H_3O^+ + CH_3COO^-, \qquad (1.1.14)$$

since the solvent (water) is not referred to acidic ones. Going to more basic solvents, such as liquid ammonia, results in considerable strengthening of the acidity of CH_3COOH. The increase of the solvent acidity (e.g. anhydrous sulfuric acid) forces acetic acid to demonstrate basic properties as is seen from the following example:

$$CH_3COOH + H_2SO_4 \rightleftharpoons CH_3COOH_2^+ + HSO_4^-. \qquad (1.1.15)$$

This necessitates the development of definitions for solvent systems, which should divide substances into acids and bases depending on their action on solvents and the particles formed by the intrinsic self-dissociation process of the said solvent. The best-known definition of such a kind is called the "definition for solvents system" or "solvosystem concept".

Historically, the first was the principle of division into acids and bases of substances dissolved in ionizing solvents, depending on their interactions with cations and anions formed as products of auto-ionization. This definition was given by Franklin in 1923, and is known as the solvosystem concept [16]. The process of auto-ionization of an ionizing solvent can be written as follows:

$$L \rightleftharpoons 1^+ + 1^-, \qquad (1.1.16)$$

and the corresponding expression of the mass action law is

$$K'_L = \frac{[l^+][l^-]}{[L]} \qquad (1.1.17)$$

If the degree of auto-ionization is negligible (and, as a rule, it is so) then the initial concentration of solvent molecules is practically equal to their equilibrium concentration, therefore, the term in the denominator can be entered in the constant that gives the following final form:

$$K_L = K'_L[L] = [l^+][l^-]. \qquad (1.1.18)$$

The constant K_L is called "the ionic product of the solvent"; and it is just the value that causes the difference in acidity (basicity) parameter units between solutions of strong acids and bases, whereas the pK_L value itself defines the width of the acid–base range of an ionizing solvent. Let us proceed with consideration of Franklin's definition. So, a substance whose addition leads to increase of the concentration of l^+ particles in the solution (it means automatically that concentration of l^- decreases) is referred to acids; on the contrary, if addition of a substance causes reduction of l^+ concentration as compared with its concentration in pure solvent, it should be classified among bases.

To illustrate the above items, let us consider phosphorus oxychloride, $POCl_3$ [22] as an example of ionizing solvents. The auto-ionization process of $POCl_3$ may be described by the following equation:

$$POCl_3 \rightleftharpoons POCl_2^+ + Cl^-. \tag{1.1.19}$$

Dissolution of phosphorus pentachloride, PCl_5, in this solvent will cause an increase of $POCl_2^+$ concentration that allows us to refer PCl_5 to acids in the said solvent:

$$PCl_5 + POCl_3 \rightleftharpoons POCl_2^+(\text{acid}) + PCl_6^-. \tag{1.1.20}$$

Addition of NH_4Cl to $POCl_3$ will result in an increase of anion particles formed by the autodissociation of the solvent:

$$NH_4Cl \rightleftharpoons NH_4^+ + Cl^-(\text{base}), \tag{1.1.21}$$

hence, NH_4Cl shows basic properties in $POCl_3$.

As is seen from equation (1.1.16), the solvosystem concept is appropriate for describing acid–base interactions in any molecular solvent with a relatively small ionic product and, consequently, slight auto-ionization. Also, it may be used for the description of interactions in covalent melts; mercury, zinc and aluminium halides should be mentioned among these. In relation to the terms "acid" and "base", this definition is more common than those formulated by Arrhenius or Brønsted and Lowry, although there are charge limitations on "acid" and "base": an acid of solvent is a cation particle whereas a "base" one is an anion.

In the case where an acid (base) dissolved in the solvent is stronger than the acid (base) formed by the solvent auto-ionization, the phenomenon of levelling of acidic (basic) properties is observed. It consists in complete solvolysis of the dissolved acid or base with the formation of an equivalent quantity of the acid or the base of the solvent. For example, in aqueous medium the following process takes place:

$$HClO_4 + H_2O \rightarrow H_3^+O + ClO_4^-, \tag{1.1.22}$$

Let us consider examples of non-aqueous solvents working in complete agreement with the solvosystem concept. In this respect, antimony(III) chloride and bromide have been studied. According to results of Refs. [23, 24]

their dissociation occurs as follows:

$$SbCl_3 \rightleftharpoons SbCl_2^+ + SbCl_4^-, \tag{1.1.23}$$

$$SbBr_3 \rightleftharpoons SbBr_2^+ + Br^-. \tag{1.1.24}$$

Liquid antimony bromide possesses a relatively slight intrinsic electro-conductivity, of the order of $1 \times 10^{-5} \Omega^{-1}$ cm [23]. As follows from equation (1.1.24), the solutions of ionic bromides such as $(CH_3)_4NBr$, NH_4Br, KBr, TlBr should show basic properties. This is really so, and three former bases were found to dissociate completely up to a concentration of 10^{-2} mol kg^{-1}. On the contrary, covalent acceptor bromides such as $AlBr_3$ and some Lewis acids of antimony $[Sb(CH_3C_6H_4SO_3)_3$ and adducts of $SbCl_3$ of SbX_2AlCl_4, SbX_2FeCl_4 [24]] were found to be acids, since their addition led to an increase of the SbX_2^+ concentration in the liquid antimony halides.

Whitney *et al.* reported ClF_3 and BF_5 to be promising liquid media for fluorination and the synthesis of complex fluorides based on alkali metal salts [25]. These solvents were prone to auto-ionization according to equations

$$ClF_3 \rightleftharpoons ClF_2^+ + ClF_4^- \tag{1.1.25}$$

and

$$BrF_5 \rightleftharpoons BrF_4^+ + BrF_6^-. \tag{1.1.26}$$

Both the solvents are characterized by relatively wide temperature ranges of their liquid state: -76.3 to $11.75\,°C$ and -61.4 to $40.4\,°C$, respectively. Studies of the reactions of metal fluorides which show basic properties in these solvents, show that their degree of their transformation into the corresponding complex fluorides

$$MeF + ClF_3 \rightleftharpoons MeClF_4, \tag{1.1.27}$$

$$MeF + BrF_5 \rightleftharpoons MeBrF_6, \tag{1.1.28}$$

increases with the atomic number of the alkali metals, i.e. and with a decrease in the "hardness" of the cation acids in the sequence, $Na^+-K^+-Rb^+-Cs^+$, that causes weakening of the "Me–F" bond in the sequence given, and facilitates formation of the halide complexes. In principle, this fact may be explained by an increase in the solubilities of the fluorides in the fluoride solvents.

ClF$_3$ demonstrates lower acidic properties compared to BrF$_5$ that can be seen both from the reaction of these compounds with one another:

$$ClF_3 + BrF_5 \rightleftharpoons ClF_2^+ + BrF_6^-,$$ (1.1.29)

and from the behaviour of ClF$_3$ in anhydrous HF as a superacid:

$$ClF_3 + HF \rightleftharpoons ClF_2^+ + HF_2^-.$$ (1.1.30)

The lowered acidity of ClF$_3$ (compared with BrF$_5$) explains the fact that there are no reactions of ClF$_3$ with LiF and with NaF, whereas BrF$_5$ reacts appreciably with the latter fluoride. As for halogen pentafluorides, their acidity increases in the sequence ClF$_5$–BrF$_5$–IF$_5$. This conclusion can be made on the basis of the stability of the corresponding ammonium salts (NH$_4$HalF$_6$).

However, the cases considered are the exceptions rather than the rule. Practical use of the solvosystem concept is laboured since for the systems similar to equation (1.1.19) identification of the ions formed by auto-ionization and measurement of their equilibrium concentrations are very difficult. Also, the systems mentioned have low dielectric constants that create additional obstacles for the investigations; the first being the incomplete dissociation and formation of ionic associates even in diluted solutions. For example, it is known that in liquid sulfur dioxide the following acid–base interaction takes place [26]:

$$Cs_2SO_3 + SOCl_2 \rightleftharpoons 2CsCl \downarrow + SO_2,$$ (1.1.31)

which supposedly gives evidence for the auto-ionization of SO$_2$ with the formation of doubly charged anions and cations at low temperatures (!) according to the equation:

$$2SO_2 \rightleftharpoons SO_3^{2-} + SO^{2+}.$$ (1.1.32)

Nevertheless, the isotopic exchange examinations of liquid SO$_2$ with the use of ^{18}O [27] showed that the equilibrium (1.1.32) did not take place. From the law of mass action, it means that in equation (1.1.32) the degree of ionization in liquid SO$_2$ is extremely low, and the isotopic exchange cannot be revealed even by isotopic techniques. High-temperature melts–solvents differ from those described above, first by their considerably higher degree of auto-ionization: in such media, ions but not molecules are the main components. For this reason the use of the solvosystem concept to characterize acid–base equilibria as being dependent on an auto-ionization process in the solvent fails—apart from some covalent halides of Cd, Zn, Hg. Indeed, small additions of an acceptor of chloride ions (e.g. $LnCl_3$) to molten KCl cannot change appreciably the equilibrium concentration of Cl^-, which is just the parameter that serves as a measure of the melt basicity according to the solvosystem concept. For this reason, pCl^- ($\equiv -\log[Cl^-]$) in similar systems will be practically unchanged and insensitive to small concentrations of acids and bases unless these additions lead to appreciable changes of the melt composition. The same applies to the second ion, K^+. So, this melt possesses unchangeable acid–base properties in relation to K^+ and Cl^-. From the viewpoint of the solvosystem concept, only equilibria having low dissociation constant values, not only auto-ionization, may be considered as "intrinsic" acid–base equilibria of a solvent. In this case, the concentrations of acids and bases can be varied over a wide range. Such an approach will be more fruitful if it is used to describe donor–acceptor equilibria in ionic media as acid–base ones.

The choice of "acid" or "base" for solvent is simplified appreciably for melts containing complex ions (as a rule, they are anions), which are prone to the acid–base dissociation. Dissociation of this ion is assumed as the intrinsic acid–base equilibrium of a melt of such kind. In this case, the simpler eliminated anion will be considered as the "base" of the solvent and the coordinationally unsaturated residue will be the acidic particle of the solvent. Naturally, the division of particles formed by the auto-dissociation into acids and bases is made on the basis of the Lewis definition [13]: an acid is the acceptor of an electron pair and a base is the donor of this electron pair. Ionic melts based on complex halides of gallium(III) [28], aluminium(III) [29] and boron(III) [30, 31] may serve as examples of successful application of the above approach. The electron-deficient covalent halide (e.g. $AlCl_3$, BF_3) in these melts is the solvent acid, and the corresponding halide ion is the base of the solvents:

$$2AlCl_4^- \rightleftharpoons Al_2Cl_7^- \text{ (acid)} + Cl^- \text{ (base)}, \qquad (1.1.33)$$

$$BF_4^- \rightleftharpoons BF_3\text{(acid)} + F^- \text{(base)}. \qquad (1.1.34)$$

The dissociation constants of the complex halide ions, which are the acid–base auto-dissociation constants of the corresponding solvents, are relatively small. Therefore, the equilibrium concentrations of the acid and the base of the solvent will be affected considerably by dissolved substances of acidic and basic character. The existence of several acids of the solvent is possible in melts of such kind, e.g. it follows from equation (1.1.33) that in the molten $NaAlCl_4$ we may assume the existence of two acids of the solvent— $AlCl_3$ and $Al_2Cl_7^-$. The simultaneous existence of two intrinsic acid–base equilibria of the solvent leads to considerable complication in the interpretation of the experimental results on the acid–base equilibria in these media [32]. Ionic melts consisting of salts of oxygen-containing acids (sulfates, carbonates, nitrates) may serve as another case of melts with complex anions. Dissociation of complex anions in similar melts results in the formation of oxide ion as a basic particle, and the division of substances into acids and bases may be performed according to the Lux–Flood definition [17]. Beside oxygen-containing melts, this definition is widely used for the description of the reactions between donors and acceptors of oxide ions in oxygen-free melts, although these reactions fail to be considered in the framework of the solvosystem concept because the base (O^{2-}) cannot be formed as a product of dissociation of the solvent constituents. Nevertheless, the effect of molten background on Lux–Flood acid–base interactions is appreciable, and is dependent on the strength of interactions between cations and anions of the melts from the one side with the base and acid, respectively.

To conclude this section, it should be emphasized that the solvosystem concept is restricted to self-ionizing solvents: it is, therefore, not suitable for completely describing ionic solvents and those which have no ionizing ability.

1.1.3 HARD AND SOFT ACIDS AND BASES (PEARSON'S CONCEPT)

In spite of the definitions of acids and bases there are some developments of acid–base theory which cannot be called "an acid–base definition" since there are no new definitions of the "acid" and "base" terms. These works concern peculiarities of the running of different acid–base interactions. The "hard and soft acids and bases" (HSAB) concept introduced by Pearson is the best known of them [33–35]. The original classification by Pearson considers metal cations of subgroups "a" of Periodic Table as "hard" acids, whereas F^- and Cl^- anions and ammonia are referred to as "hard" bases. As for the "soft" reagents, it should be mentioned that cations of subgroups "b" of the Periodic Table are referred to as intermediate and soft acids. Anions such as CN^-, I^-, and phosphines are classified among soft bases. According to Pearson, the hardest acid is the proton (H^+) and the softest one is CH_3Hg^+; the hardest bases are OH^- and F^-, the softest ones are I^- and H^-.

The generalized (modern) interpretation of the HSAB principle is as follows: small particles of weak polarization and high electronegativity are referred to as "hard" reagents; "soft" reagents are relatively large particles possessing stronger polarization and weaker electronegativity. Hard acids are hardly reduced, and hard bases are hardly oxidized, whereas soft acids can be easily reduced and soft bases are easily oxidized. According to Pearson's HSAB principle, acid–base interactions result in preferential formation of the adducts "hard acid–hard base" and "soft acid–soft base". From this, it follows that any acid or base may be classified among hard or soft acids and bases depending on its ability to form adducts with generally accepted "hard" and "soft" bases and acids.

The shift of an acid–base interaction towards the formation of the "hard–hard" and "soft–soft" adducts may be explained easily by the fact that the interaction between orbitals possessing close energies is more thermodynamically favourable. Such interactions lead to the preferential formation of adducts with ionic ("hard" acids and bases, electrostatic interaction), or covalent ("soft" acids and bases) characters of chemical bonding. Table 1.1.1 gives a list of some HSAB, whose properties will be discussed in the subsequent sections. It is very interesting for us that the attribution of acids and bases to hard- and soft-groups on the basis of their ionic radii is incorrect. Indeed, the cations Mg^{2+} and Ni^{2+} having the same ionic radius (0.074 nm) do not belong to the same class of "acids", although this is true for Cd^{2+} and Pb^{2+} ions whose radii are 0.099 and 0.126 nm, respectively.

Owing to its relative simplicity, Pearson's HSAB principle is often used to predict the run of different acid–base interactions on the background of media

TABLE 1.1.1
Hard and soft acids and bases

Acids	Bases
Hard	
H^+, Li^+, Na^+, K^+, Cs^+, Mg^{2+}, Ca^{2+}, Sr^{2+}, Ba^{2+}, CO_2	O^{2-}, NO_3^-, CO_3^{2-}, SO_4^{2-}, F^-, Cl^-, OH^-
Intermediate	
Fe^{2+}, Co^{2+}, Ni^{2+}, Cu^{2+}, Zn^{2+}, Pb^{2+}	NO_2^-, Br^-, SO_3^{2-}
Soft	
Cu^+, Cd^{2+}, Ag^+, Cu^+	I^-, H^-, CN^-

of different kinds. However, in order to obtain correct results, one should not only take into account both the class the acids and bases belong to but also their strength in the given solvent. If there are no considerable distinctions in the strength of acids or bases, then the use of Pearson's principle is fruitful enough for estimations: in the opposite case, many exclusions from this rule arise. For example, although aqueous solutions of mercury(II) chloride are known to interact with potassium iodide, and this interaction runs according to the predictions of the HSAB concept:

$$2KI + HgCl_2 \rightarrow KCl + HgI_2, \tag{1.1.35}$$

where the initial compounds KI and $HgCl_2$ are "hard−soft" and "soft−hard" ones, respectively, and the products are "hard−hard" (KCl) and "soft−soft" (HgI_2) compounds. The same solutions react with KOH solutions formed by the more "hard−hard" acid and base:

$$2KOH + HgCl_2 \rightarrow 2KCl + HgO + H_2O. \tag{1.1.36}$$

One of the fields where Pearson's HSAB principle can be used successfully is the prediction of the direction of donor−acceptor interactions in ionic melts, which are prone to dissociation with the formation of halide complexes which, in their turn, break down into halide ions and electron-deficient molecular halide or its adduct to the melt constituents. In particular, we analysed the effect of different ions belonging to hard and soft acids on the value of the acid−base product in melts based on complex "aluminium + alkali metal halides", "gallium + alkali metal halides" [36, 37]. The complex anion in the former kind of melts can dissociate according to equation (1.1.33). The melts based on aluminium halides, with compositions $AlCl_3−MCl$ (M = Li, Na, K, Cs) with $AlCl_3$ content exceeding 50 mol%, were investigated in the 175−400 °C temperature range [29, 38, 39]. These studies were performed with an electrode, reversible with respect to chloride ions. The pK values for the Na^+-containing melts were found to change from 7.1 to 5.0 with elevation of the melt temperature from 175 to 400 °C [39]. For the sequence of lithium-→sodium-→ caesium-containing halide melts at 400 °C, the pK values were changed in the manner: 3.8 (Li), 5.0 (Na), 5.8 (K), 7.4 (Cs) [29]. This fact may be explained by the higher stability of $Li^+−Cl^-$ complexes as compared with $Na^+−Cl^-$ and $Cs^+−Cl^-$ ones, since, although all the complexes are formed by the same hard base (chloride ion), the "hardness" of the cation acids reduces in the sequence $Li^+ - Na^+ - Cs^+$.

Investigations of $KGaX_4$ ionic melts demonstrate [28] that the acid–base product of the solvents increases from $KGaCl_4$ to $KGaI_4$ according to the following equations:

$$2GaCl_4^- \rightleftharpoons Ga_2Cl_7^- (GaCl_3 \cdot GaCl_4^-) + Cl^-, \quad pK = 4.25 \pm 0.05, \tag{1.1.37}$$

$$GaI_4^- \rightleftharpoons GaI_3 + I^-, \quad pK = 2.6 \pm 0.05. \tag{1.1.38}$$

Here, the base in the co-ordination shell around Ga^{3+} changes from a hard base (Cl^-) to a soft base (I^-). Since the Ga^{3+} ion is a hard acid, the stability of the halide complexes reduces in the sequence $GaCl_4^- \rightarrow GaI_4^-$.

Pearson's principle simplifies the treatment of the results of investigations performed by Lambertin *et al.* [40]. The authors of this work examined the effect of the addition of fluoride ion on the equilibrium between different oxidation states of americium in the molten KCl–LiCl eutectic at 743 K. The increase of fluoride ion concentration was shown to result in a shift to the right of the following reaction:

$$3Am^{2+} \rightleftharpoons 2Am^{3+} + Am^0. \tag{1.1.39}$$

This result is explained by the fact that addition of a hard base (fluoride ions) to the chloride melt containing a weaker hard base (chloride ions) favours stabilization of the halide complexes formed by a harder acid (in our case it is Am^{3+}) as compared with a weaker acid (Am^{2+}). This causes the shift of the equilibrium (1.1.39) to the right, and the disproportionation reaction runs.

We have found an interesting result for the system, which is classified among the ideal ones. The interactions between silver chloride and potassium halides:

$$AgCl + KX \rightleftharpoons AgX + KCl, \quad \Delta G_X \tag{1.1.40}$$

were studied by Tchernov *et al.* [41]. This reaction is very slow in aqueous solutions and proceeds according to the mechanism "solid–liquid–solid" that lasts for several hundred hours [42]. The Gibbs energies for these systems are -0.903 and -2.431 kJ mol^{-1} (at 298 K) and -0.809 and -2.524 kJ mol^{-1} (at 900 K) for X = Br and I, respectively.

From the viewpoint of the HSAB concept, the softer acid, Ag^+ shows the affinity to softer (as compared with Cl^-) bases such as Br^- and I^-. This causes the high yield of AgBr and AgI achieving 83–92%, and the exchange reaction leads to the formation of silver bromide and iodide as commercially important products with sufficient yield.

One of the most interesting applications of the HSAB concept consists in the prediction of the stability of the complexes formed owing to interaction of alkali metal halides with rare-earth metal halides. These systems are of great interest for the materials science of scintillation materials: the said complex halides are now considered among the most promising scintillation detectors and sensors. Besides, the Li- and Gd-based materials are especially convenient as effective detectors of thermal neutrons. The compositions and stability of the formed compounds depend considerably on the kind of acids and bases from which the compound is formed. So, Li^+ cation is one of the hardest cation acids, and, therefore, the formation of stable complex halides of Li and lanthanides according to reaction:

$$MeX + LnX_3 \rightleftharpoons MeLnX_4 \qquad (1.1.41)$$

is hardly possible. Indeed, the phase diagrams show that for the fluoride materials the lanthanides from La to Sm do not form any complex fluoride compounds with lithium, since the corresponding fluorides are too weak as hard acids compared with Li^+. The lanthanides from Eu to Ho possess smaller ionic radii than the cations of metals belonging to the "La–Sm" group: therefore, they are referred to as harder acids, and their interaction with lithium fluoride results in the formation of substances of composition $LiLnF_4$ which are stable in the solid phase but subject to the incongruent melting. The following "lanthanide compression" (decrease of ionic radius) leads to the formation of strong complex fluoride compounds of lanthanides from Er to Lu, which exist in the solid state and remain relatively stable after melting.

If we substitute lithium by sodium cations, which are softer acids, the stability of the fluoride complexes increases and the first members of the lanthanide row react with NaF to form compounds $NaLnF_4$, which decompose during melting. It is observed that in the formation of $Na_5Ln_9F_{32}$ in the $NaF–LnF_3$ systems, their stability increases with the atomic number of the element. If the Nd compounds are subjected to incongruent melting, the gadolinium derivatives possess clear melting points, some above 1000 °C. This is true also for the other lanthanides, whose cation radii are lower than that of Gd^{3+}.

Fluorides of potassium, rubidium and caesium form compounds of composition Me_3LnX_6, which are stable in the solid and liquid state. In the phase diagrams mentioned there are some other compounds which, however, are unstable and decompose in the solid state during temperature elevation (excluding LaF_3, where the adducts of $MeLaF_4$ are stable enough).

The change of halide ion results in weaker acidic properties for $LnCl_3$ as compared with LnF_3. This means that equilibrium (1.1.41) with the participation of alkali metal halide should be shifted to the left as compared with the fluoride complexes. That is, lithium chloride does not react with chlorides of the rare-earth elements with the formation of any compounds; the binary phase diagrams are characterized by one simple eutectic. The same situation is observed for the binary diagrams for lithium- and rare-earth bromides.

As for the complex compounds formed by other alkali metal halides, whose cations possess appreciably softer acidic properties, the situation is as follows. The $NaCl–LaCl_3$ is characterized by a simple eutectic, which means the absence of formation of any compounds. However, in the case of the $GdCl_3–NaCl$ binary system the formation of incongruently melted Na_3GdCl_6 and Na_3DyCl_6 compounds is observed [43]. On going from $DyCl_3$ to $LuCl_3$ the Na_3LnX_6 compounds become more stable. Concerning the complex compounds formed by K, Rb, Cs chlorides, it should be mentioned that potassium chloride interacts with rare-earth chlorides forming K_3LnCl_6 and KLn_3X_{10} compounds, which melt congruently. This is the case for caesium and rubidium chlorides. Bromide and iodide compounds resemble the corresponding complex chlorides: in the case of the sodium-based salts there is no formation of stable compounds, whereas K, Rb and Cs halides each form only one stable compound of composition Me_3LnCl_6. As well as the mentioned composition, in some cases $MeLn_2X_7$ compounds can be obtained: the doping of these compounds with Ce^{3+} results in the formation of materials which are referred to as the best scintillation materials for detecting gamma-rays (La- and Lu-based halide compounds) and neutrons (Gd- and Li-based materials).

Therefore, the stability of halide complexes formed by cations of different charges can be predicted successfully by using the HSAB concept: its possibilities for the estimation of interactions between hard and soft bases belonging to different valence groups are more straightforward.

1.1.4 GENERALIZED DEFINITION OF SOLVENT SYSTEM. SOLVENTS OF KINDS
I AND II

The preconditions of another approach to the treatment of acid–base concepts can be found in the "classic" solvosystem concept described above. Careful reading of this concept shows that the acid–base definition connects the terms "acid" and "base" only with the process of auto-dissociation of a molecular solvent or of one capable of ionization. Nevertheless, it is obvious that acid–base interactions can occur in those solvents, which are not able to form acid and base owing to a dissociation process. Aprotic solvents may serve as a typical example of solvents of such a kind; another case of solvents incapable of the acid–base auto-dissociation takes place if we consider the Lux acid–base equilibria in molten oxygen-free media. Therefore, in relation to any given acid or base there exist two kinds of solvents of differing auto-dissociation ability with the formation of the said acid or base [36, 37, 44, 45].

Acid–base processes in molten ionic media are hardly described in the framework of the classic solvosystem concept. The reason for the seeming principal distinction of ionic melts from room-temperature molecular solvents consists in the limitations contained in the solvosystem concept. The main reason is that the division of substances into acids and bases is performed in relation to their reactions with the products of a molecular solvent auto-dissociation and the degree of this auto-dissociation, in the pure solvent is negligible. On the contrary, an ionic melt is a case of a completely ionized solvent, and this is the fact that does not allow one to apply the solvosystem concept fruitfully for studies of acid–base interactions in this kind of liquid media.

In order to generalize the definition of the solvent system for the case of ionic media, we shall analyse Franklin's definition. First, it should be noted that the term "auto-ionization" in this definition should be substituted by "auto-dissociation" or "intrinsic acid–base equilibrium of the solvent", as a more common case of heterolytic break down of the constituent particles of a liquid. Indeed, for molecular solvents or those which are slightly ionized at room temperature, the terms "auto-ionization" and "intrinsic acid–base equilibrium of the solvent", relate to the same process, whereas for ionic liquids they differ considerably. For example, although sodium nitrate ($NaNO_3$) is subject to practically complete ionization in the liquid state, by the equation:

$$NaNO_3 \rightleftharpoons Na^+ + NO_3^-, \tag{1.1.42}$$

(according to what we have said, this process is called auto-ionization), its intrinsic acid–base product:

$$K_{NO_3^-} \rightleftharpoons [NO_2^+]\cdot[O^{2-}], \tag{1.1.43}$$

which is calculated according to the following equilibrium:

$$NO_3^- \rightleftharpoons NO_2^+ + O^{2-}, \tag{1.1.44}$$

is too small, and this fact provides the relatively wide range of variation of acidic (basic) properties of solutions in this solvent.

The next remark concerns the observation that the terms "acid" and "base" cannot be attributed exclusively to cation and anion particles formed in the process of intrinsic acid–base dissociation of the solvent. To illustrate this distinction let us compare features of the acid–base dissociation of nitrate and sulfate melts as backgrounds for the Lux acid–base reactions. The former melt corresponds to the case where NO_2^+ cation is an acid and O^{2-} anion is a base (equation (1.1.44)). Now let us consider a sulfate melt. Its intrinsic acid–base dissociation equilibrium is given by the following equation:

$$2SO_4^{2-} \rightleftharpoons S_2O_7^{2-} + O^{2-}, \tag{1.1.45}$$

from which we can see that all the reaction participants are anions. In both cases, anions, but not uncharged particles of the solvents, are the initial substances undergoing the breakdown into acids and bases.

Nevertheless, the behaviour of oxide-containing melts as a kind of background for acid–base reactions can be described in a manner similar to the solvosystem concept. However, for this purpose the solvosystem concept should be generalized in some ways. As mentioned above, the main object in the basis of Franklin's solvosystem concept is a molecular solvent prone to auto-ionization, and this process results in the formation of small concentrations of cations and anions of the solvent. Water, glycol, different spirits and some other room-temperature liquids are to be mentioned as typical examples of such solvents.

However, one can imagine a solvent that is not prone to ionization; which possesses its intrinsic acid–base equilibrium involving acid and base of solvent, without breakdown to ionic particles. An example of such a solvent and its acid–base dissociation process may be expressed by the equation

$$H_3N(:)BH_3 \rightleftharpoons H_3N : + (\ )BH_3. \qquad (1.1.46)$$

If we use the classic solvosystem concept to describe acid–base reactions in the ionic melt of $NaBF_4$, completely dissociated into the constituent ions Na^+ and BF_4^-, we shall have considerable difficulties, since the concentrations of Na^+ and BF_4^- cannot be varied over a wide range without essential changes of the solvent composition. On the other hand, equilibrium (1.1.34) at 420 °C is characterized by a pK value of 1.8 [30, 31], and if we believe it to be the acid–base dissociation of the melt–solvent then the electron-deficient particle BF_3 will demonstrate acidic properties in the melt, whereas fluoride ion F^- will be the solvent base.

The above speculations allow us to determine the statements of the classic solvosystem concept which should be generalized for the case of non-dissociated and ionic solvents:

- the acid and the base of a solvent may be formed not only owing to the auto-ionization equilibrium but also by any process of heterolytic dissociation of its constituent parts—both molecules and ions;
- the interpretation of the terms "acid" and "base", should be widened. Cation and anion particles of a solvent are not the only cases of the acid and base in the solvent; any Lewis acid may be the acid of the solvent, and similarly for the base of the solvent;
- in relation to the given definition of acids and bases solvents should be divided into two kinds—solvents of the first kind, and solvents of the second kind, taking into account the existence of intrinsic acid–base equilibria in these media. A solvent is referred to the first kind by considering an intrinsic acid–base equilibrium running in it. In contrast, if a solvent is unable to dissociate with the formation of the given acid and base (for example, O^{2-} is not a product of dissociation of sodium chloride), it belongs to solvents of the second kind. Such a solvent will affect acid–base processes in its background owing to the fixation of acid or base by its basic or acidic particles, respectively.

Moreover, it should be emphasized that the classic solvosystem concept considers the auto-ionization process to be primary, whereas substances dissolved are referred to as acids and bases by the formation of their adducts with cations and anions of the solvent.

However, the practice of investigations of acid–base equilibria consists in studies of the reactions with the transfer of definite particles (acid or base) in different solvents, irrespective of the fact whether they are prone to dissociation with the formation of given acid (base), or not. In order to give the most general formulation of the solvosystem concept we should divide the solvent in relation to the initially chosen definition of acids and bases (definition of acidic and basic particles).

Thus, the generalized definition for solvent system developed in our works [37, 44, 45] can be formulated as follows:

- the intrinsic acid–base dissociation (auto-dissociation) of a solvent is any (excluding redox) reaction of heterolytic dissociation of its constituent parts (molecules or ions), which results in the formation of electron-deficient particle–acid and electron-donor particle–base;
- a solvent is referred to the solvents of the first kind in the case of the formation of the corresponding acid (base) as a result of its intrinsic dissociation processes; in the other case it is referred to solvents of the second kind. (This item means that, in the common case, some auto-dissociation processes may exist simultaneously in a solvent.)

The auto-dissociation processes taking place in the solvents of the first kind can be described by the following basic scheme:

$$L \rightleftharpoons \sum_i l_i,$$ \hfill (1.1.47)

and the equilibrium

$$L \rightleftharpoons L$$ \hfill (1.1.48)

is a possible form of equation (1.1.47) and

$$l_i \rightleftharpoons A_{L,i} \cdot l_j + B_{L,i} \cdot l_k,$$ \hfill (1.1.49)

where l_i, l_j and l_k are the constituent parts of the solvent L which may be the same and A_l and B_l are the acid of the solvent and the base of the solvent; there are no limitations on the charge of these particles. As immediately follows from equations (1.1.47) and (1.1.49), there is a possibility of the simultaneous existence of some different acid–base equilibria in the same solvent.

The processes of interaction of a solvent of the first kind with an acid or a base added into it are called "solvolysis". They result in the formation of solvo-acid (solvo-base) and solvated conjugate base (acid). These interactions may be represented by the following equations:

$$A_1 + L \rightleftharpoons B_1(A_1B_{l,i}) \cdot l_k + A_{l,i} \cdot l_j \text{ (solvo-acid)}, \tag{1.1.50}$$

$$B_1 + L \rightleftharpoons A_1(B_1A_{l,i}) \cdot l_j + B_{l,i} \cdot l_k \text{ (solvo-base)}, \tag{1.1.51}$$

where A_1 and B_1 are the conjugate pair, acid and base. The system of equations presented shows that acidic properties of acids in solvents of the first kind are partially or completely levelled to the properties of $A_{l,i}l_j$, as for the base, it is obvious that its basic properties are levelled to those of $B_{l,i}l_k$.

In particular, for the conjugate acid–base pair A_1/B_1, which is located on the acidity scale over the acid–base range of the solvent L of the first kind (see Fig. 1.1.1), the complete transformation into conjugate acid with the formation of the equivalent concentration of the base of the solvent will be observed. It should be added that the acid formed would possess no acidic properties in the said solvent. Similarly, the conjugate pair A_2/B_2 is completely transformed into the conjugate base, which shows no basicity in the solvent. Hence, the acid–base ranges in solvents of the first kind are limited on two sides.

Water may serve as an example of a solvent of the first kind, in relation to protic acidity. Its dissociation, which occurs according to equation (1.1.3), leads to the formation of protons (H^+) or, more correctly, $H_5O_2^+$. Both pH and

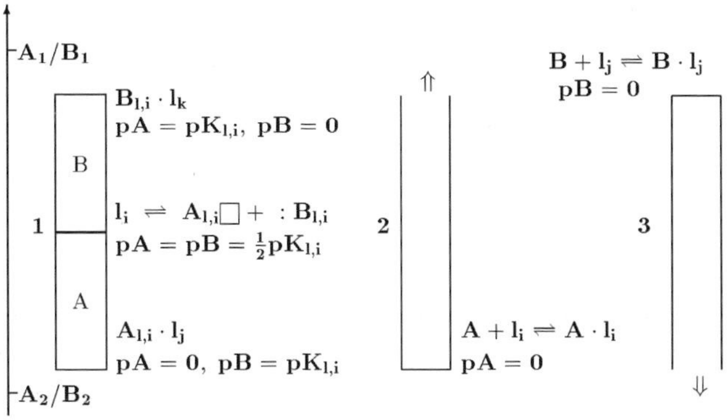

Fig. 1.1.1. Acid–base ranges in solvents of different kinds: (1) a solvent of the first kind; (2) a solvent of the second kind with defined acidic particle; (3) a solvent of the second kind with defined basic particle.

pOH indices describe definitively the solution acidity. The neutral point of the solvent is located at pH $=$ pOH $= 7$. Those solutions having pH values less than 7 are acidic ones, whereas values of pH > 7 are characteristic of basic solutions. The phenomenon of the levelling of acidic and basic properties described in detail may be illustrated by the following examples:

$$HClO_4 + H_2O \rightarrow ClO_4^- + H_3O^+, \qquad (1.1.52)$$

$$H^- + H_2O \rightarrow H_2 \uparrow + OH^- (H_2O). \qquad (1.1.53)$$

It follows from reactions (1.1.52) and (1.1.53) that there are no acids stronger than H_3O^+ and no bases stronger than solvated OH^- in aqueous solutions.

Ionic solvents of the first kind can be illustrated exhaustively by nitrate melts such as potassium nitrate KNO_3 with the melting point at $336\,°C$, or its low-melting equimolar mixture with sodium nitrate $KNO_3–NaNO_3$ having the melting point near $220\,°C$. The intrinsic acid–base equilibrium in these solvents can be expressed by equation (1.1.44). It is seen that oxide ion is formed as one of the products of NO_3^- dissociation, and just that allows us to attribute these solvents to those of the first kind. Acidic and basic solutions are defined in relation to the neutrality point, whose position is $pO = 1/2pK$. The magnitude of the intrinsic acid–base product of the nitrate melts has been estimated more than once. Zambonin presented the values in the molten equimolar mixture $KNO_3–NaNO_3$ to be 10^{-18}(molalities) [46], although Kust and Duke calculated these values as 2.7×10^{-26} at $250\,°C$ and 2.7×10^{-24} at $300\,°C$ [47]. The phenomenon of the levelling of acidic properties in the mentioned nitrate melts may be illustrated by the following examples:

$$2CrO_3 + NO_3^- \rightarrow Cr_2O_7^{2-} + NO_2^+, \qquad (1.1.54)$$

$$O^{2-} + nNa^+ \rightarrow Na_nO^{n-2}. \qquad (1.1.55)$$

The latter equation (1.1.55) is written taking into account the fact that the strength of a base is levelled to the basic properties of the adduct of O^{2-} to the most acidic cation of the ionic solvent (in $KNO_3–NaNO_3$ equimolar mixture this cation is Na^+ since it possesses a smaller radius than K^+).

Consequently, solvents of the first kind may be characterized by the limit of the levelling of acidic properties, equal to $pA = pK_a$, by the neutrality point located at $pA = pK_a + 1/2\ pK_s$, where K_s is the constant of the intrinsic acid–base dissociation, and by the limit of the strength of bases

$pA = pK_a + pK_s$. As for a solvent of the first kind, the measure of strength of acid (base) is the degree of its interaction with the solvent leading to the formation of solvated acid (base) of the solvent. Since the equilibrium concentration of the acid (base) of solvent cannot exceed the initial concentration of the added monobasic acid (monoacidic base), for this case the following relationships will be correct:

$$c_A^0 \geq c_{A_{l,i}} \quad \text{and} \quad c_B^0 \geq c_{B_{l,i}}.$$

Now let us analyse acid–base processes, which occur in solvents of the second kind, considering acid A as the "carrier" of acidic properties (the speculations concerning the "carrier" of basicity B are similar). The dissociation process of solvent L runs according to equation (1.1.47). If we add an acid (A_1) to the solvent the following interaction will take place:

$$A_1 + l_i \rightleftharpoons A_1 \cdot l_i. \tag{1.1.56}$$

From this equation, it is obvious that one of the constituent parts of the solvent, namely l_i, which fixes the acid, possesses basic properties. Additions of a base to the solvent will result in a similar reaction with the participation of constituent parts of acidic character. However, addition of a base into the pure solvent L does not lead to a change of acid A, since there are no particles of such a kind in the pure solvent.

As follows from Fig. 1.1.1, in the case of solvent of the second kind the acid–base pair A_2/B_2 is completely transformed into the conjugate base B_2, whereas the acid–base pair A_1/B_1 is not subjected to solvolysis.

As is seen, the acidic properties of strong acids are levelled to those of Al_i. Although the properties of bases are changed because of the reaction with the constituent parts of the solvent, they cannot be levelled to the properties of certain common base B. Theoretically, there are no limitations on the strength of bases in such a solvent of the second kind.

Hence, the solvents of the second kind are characterized by the limit of the levelling of acidic properties: for the standard solutions it is defined by the relationship $pA = pK_a$; the term "neutrality point" is theoretically senseless since there is no acid A in the pure solvent. An experimentally determined value of the neutrality point in the real solvent, and the pA index are dependent only on the content and nature of admixtures, which remain in the solvent after its purification.

The strength of bases cannot be expressed via the strength of a certain common base since the latter is not produced in the solution after the addition of another base. In simpler words, the role of bases in solvents of the second kind consists only in the fixation of acidic particles and increasing the pA of the solution, whereas their addition to the pure solvent does not result in the formation of the solvent's base. Therefore, acidity scales in solvents of the second kind are limited by the levelling of properties of the "carrier" of acidic (or basic) properties; there are no limitations on the opposite side of the acidity scale.

The above-considered features of the levelling of acidic and basic properties in solvents of different kinds allow us to conclude that the position of each particular solvent in the generalized scale of acidity is dependent on the properties, at least, of one conjugate acid–base pair, which acid or base is generated as one of the products of solvent auto-dissociation. The position of the acid–base range of a solvent in the generalized scale of acidity defines which acids and bases can exist in this solvent and which ones should be subject to solvolysis. Therefore, determination of the width of the acid–base range of a solvent, and its position in the generalized acidity scale, are important practical tasks for the chemistry of acids and bases. Now, for the case of protic acidity this question has been examined widely and the majority of solvents are classified in this respect. The methods of estimation of relative acidic properties for protic and aprotic room-temperature solvents are sufficiently developed, and they are well known to investigators in the field. We shall now consider these methods, briefly.

The Hammett method is based on the ability of the studied solution to give up a proton (acid) to the indicator base, whose basic strength is assumed to be independent of the solvent properties other than its acidity, i.e. the dielectric constant, structure, etc. Under these conditions the ratio of the concentrations of conjugate acid and base in the solution will be the measure of its common acidity, H. Usually, there is a subscript near "H" which shows the type of the conjugated base: the subscript "0" denotes uncharged bases (e.g. R_3N:), whereas " $-$ " means that for acidity estimation an anionic base (similar to $RCOO^-$, RO^-) was used. Most of the bases used are common acid–base indicators. So, for estimations of the generally accepted H_0 function, such indicators are used: methyl yellow, aminoazobenzene, anthraquinone, methyl orange, etc. As mentioned above, the non-thermodynamic assumption is made that the pK value of any indicator used is independent of the dielectric constant of the solvent or solution studied. To investigate acid–base properties of the solutions differing considerably in acidity a set of indicator bases possessing various basicities is used, i.e. for constants of proton addition:

$$B + H^+ \rightleftharpoons BH^+(A), \quad K_{Ind}. \tag{1.1.57}$$

After adding a base, colorimetric determination is performed of the concentrations of the base B and the conjugate acid BH^+ in the studied solution. The concentrations found allow one to estimate the acidity of a solution according to the following equation:

$$H_0 = pK_{Ind} + \log \frac{C_B}{C_{BH^+}}, \tag{1.1.58}$$

where pK_{Ind} is the value of pK determined in aqueous solutions. If we express pK_{Ind} by taking into account the values of the activity coefficients and the primary medium effect, we obtain the following equation:

$$pK_{Ind} = -\log a_{H^+} - \log \frac{C_B}{C_{BH^+}} + \log \frac{\gamma_B}{\gamma_{BH^+}} + \log \frac{\gamma_{0,B}}{\gamma_{0,BH^+}}. \tag{1.1.59}$$

If we consider the studied solution to be infinitely diluted we can neglect the term connected with the activity coefficients of the charged particles in the solution and obtain finally that the Hammett function is dependent only on the proton activity and the primary medium effects for the conjugate indicator pair:

$$H_0 = -\log a_{H^+} + \log \frac{\gamma_{0,B}}{\gamma_{0,BH^+}}. \tag{1.1.60}$$

The summand

$$\log \frac{\gamma_{0,B}}{\gamma_{0,BH^+}} = \log \gamma_{0,B} - \log \gamma_{0,BH^+} \tag{1.1.61}$$

is the difference between the primary medium effects of the base B and the conjugate acid BH^+. A zero value of this difference for an acid–base pair serves to substantiate its use for common acidity measurements in different solvents.

All that is said above concerns the Hammett acidity function, H_0.

Several years ago, Bates and Schwarzenbach introduced a similar acidity function H_- [48]. All the assumptions which substantiate the use of this function are similar to those for H_0. In this case, the relationship between

H_- and proton activities in non-aqueous media is presented as

$$H_- = -\log a_{H^+} + \log \frac{\gamma_{0,A^-}}{\gamma_{0,HA}}. \tag{1.1.62}$$

The correctness of the obtained values of H_0 and H_- is dependent on the validity of the original assumption that the solvent change does not result in a changing ratio of primary medium effects of the conjugate indicator base and acid. Naturally, it is hardly possible that these ratios of primary medium effects will be the same for different acid–base pairs, even when they belong to the same charge type (anionic or uncharged), in liquid media with appreciably different dielectric constants. Indeed, the values of H_0 and H_- for the same non-aqueous solution were shown to differ appreciably—i.e. the dependences of H_0 and H_- on the solvent composition in water–ethanol mixtures differ appreciably. All this means that the Hammett function in both forms is not the direct measure of proton activity in non-aqueous media; it is interesting as an express estimation of the relative acidities of solvents possessing strongly acidic properties (superacids). The use of the Hammett function for this purpose gives a visual representation for a solvent of such a kind. These magnitudes for the standard solutions are often cited in the literature instead of the corresponding primary effects of a proton.

Now let us return to the approaches connected with the estimation of the primary medium effect for protons, $\log \gamma_{0,H^+}$, that are used for obtaining quantitative information on the acidity of pure protolytic or aprotic solvents relative to the standard solution of a strong acid in water. From the thermodynamics, these are known to be a measure of the Gibbs free energy of proton transfer from the standard solution in water to the one in a non-aqueous solvent (M). This parameter is connected with the energy of proton re-solvation in the following way:

$$\log \gamma_{0,H^+} = \frac{\Delta G_{cH^+(H_2O)} - \Delta G_{cH^+(M)}}{2.3RT} = \frac{\Delta(\Delta G_{cH^+})}{2.3RT}. \tag{1.1.63}$$

According to Izmailov [49] the acidity of non-aqueous solution characterized by the pH_M index in the given medium is described by the function of common acidity pA_M, as

$$pA_M = pH_M + \log \gamma_{0,H^+}. \tag{1.1.64}$$

By comparing the Hammett acidity function with the function of common acidity we obtain

$$H_0 = \text{pH} - \log \gamma_{0,\text{H}^+} + \log \frac{\gamma_{0,\text{BH}^+}}{\gamma_{0,\text{B}}} = pA_\text{M} + \log \frac{\gamma_{0,\text{BH}^+}}{\gamma_{0,\text{B}}}. \quad (1.1.65)$$

This expression means that the values of pA_M and the H_0 values obtained for the standard solutions should coincide in the case when the primary medium effects for acid and base are the same, i.e. $\log \gamma_{0,\text{BH}^+} - \log \gamma_{0,\text{B}} = 0$. Practical investigations of proton acidity in different solvents gives the evidence that this is not usually the case [50].

Izmailov et $al.$ developed methods for the estimation of the primary medium effect of protons [51], which permit determination of the relative acidic properties of different solvents and solutions. The essence of their proposition consisted in the division of the primary medium effect of a separate ion into two terms

$$\log \gamma_{0,\text{H}^+} = 2 \log \gamma_0^\text{bas} + \log \gamma_0^\text{el}, \quad (1.1.66)$$

where the first term $2 \log \gamma_0^\text{bas}$ is a measure of the work of re-solvation

$$\text{MH}^+ + \text{H}_2\text{O} \rightleftharpoons \text{M} + \text{H}_3\text{O}^+, \quad \text{K}_\text{r}, \quad (1.1.67)$$

which is dependent on the affinity to the proton of water and the given non-aqueous solvent. Using the activities, this can be expressed as

$$\log \gamma_0^\text{bas} = \frac{1}{2} \text{K}_\text{r}^* + \log \frac{a^*_{\text{H}_3\text{O}^+}}{a^*_{\text{MH}^+}}, \quad (1.1.68)$$

where the activities $a^*_{\text{H}_3\text{O}^+}$ and $a^*_{\text{MH}^+}$ are related to an infinitely diluted solution in water and in the non-aqueous solution, respectively. The second term on the right-hand side of equation (1.1.66) is the average energy of solvation of lyonium cations and the anion of the acid, which is dependent on the dielectric constants of both liquids

$$\log \gamma_0^\text{el} = \frac{e^2 N_A}{9.2RT} \sum \frac{1}{r_i} \left(\frac{1}{\varepsilon_\text{M}} - \frac{1}{\varepsilon_{\text{H}_2\text{O}}} \right) + \frac{\Delta G_\text{c}}{2.3RT}. \quad (1.1.69)$$

Taking into account the known values of the constants of proton exchange

$$\text{K}_\text{r} = \log \frac{a^*_\text{M} a^*_{\text{H}_3\text{O}^+}}{a^*_{\text{H}_2\text{O}} a^*_{\text{MH}^+}}, \quad (1.1.70)$$

and the average-ionic primary medium effects $\log \gamma_0^{\pm}$, Izmailov *et al.* calculated the electrostatic terms (equation (1.1.69)) and then, according to equation (1.1.66) they estimated the values of the primary medium effects for the solvents such as ethanol, propanol, ammonia and formic acid. This determination, together with the data on acid–base products (pK_s), of the said non-aqueous solvents resulted in the construction of the common acidity scale, which would be reported in Ref. [51].

An example presented in Fig. 1.1.2 gives a visual presentation of the relative acidities and acid–base ranges of the most important non-aqueous protic and aprotic solvents. The positions of many known acid–base pairs are also presented on this scale. This scale illustrates the above-discussed phenomenon of the levelling of acidic and basic properties that can be seen from the following example concerning the behaviour of acetic acid as a solvent and as a dissolved substance in other solvents. There are three kinds of particles in the pure acetic

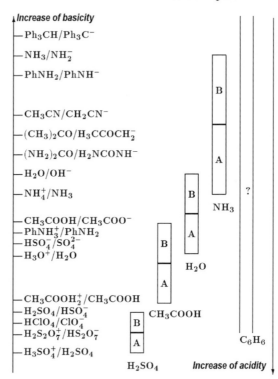

Fig. 1.1.2. The schematic common-acidity scale of protic and aprotic solvents constructed from the data of Refs. [16, 52] (where A designates acidic regions and, ? designates basic ones).

acid: the pure acid CH_3COOH, $CH_3COOH_2^+$ cations and CH_3COO^- anions. They coexist in appreciable concentrations in this solvent. If we consider sulfuric acid, we can see that there is complete transformation of CH_3COOH into the protonated form $CH_3COOH_2^+$ owing to the reaction (1.1.15). So, acetic acid in this medium is a strong base without any acidic properties.

In aqueous solutions, the existence of two forms of the acid, namely non-dissociated acid and acetate ion, is possible: this is described by equilibrium (1.1.14). Also, in basic solvents, such as liquid ammonia NH_3, we observe the complete transformation of acetic acid into the conjugate base

$$CH_3COOH + NH_3 \rightleftharpoons CH_3COO^- + NH_4^+, \tag{1.1.71}$$

which possesses no acidic properties in this medium.

Aprotic solvents (benzene, DMSO, CCl_4) belong to the solvents of the second kind, that makes the terms "acidic" and "basic" medium, conditional, whereas in water, acetic, and sulfuric acids and ammonia, there are clear boundaries between these media, which is located in the middle of the acid–base range.

It should be noted that Jolly and Hallada [52] did not show the limit of the levelling of acidic properties in benzene, although this limit should be observed. Indeed, benzene is well known to add a proton in anhydrous sulfuric acid, and this process is widely used in organic synthesis. Hence, the acid–base pair $C_6H_7^+/C_6H_6$ should be placed somewhat above the $H_3SO_4^+/H_2SO_4$ pair in the said common scale of acidity.

Therefore, the generalized solvosystem concept proposed above is applicable to the description of protic and aprotic solvents, although just for this case this does not allow us to obtain basically new results. As for the case of ionic melts, it is more important, since the theoretical basis of acid–base interactions in this kind of solvents is not developed so extensively. Nevertheless, there are many results on oxoacidity obtained in ionic solvents of the first and the second kinds, which have been treated in a proper way.

For ionic melts, we should mention that solvents of the second kind have been studied more intensively than those belonging to the first kind. Also, the treatment of the results former media is simpler. According to the data obtained on equilibrium constants, the general oxoacidity scale may be presented to a sufficient accuracy in the following manner (Fig. 1.1.3). The extremely wide range of liquid state of ionic liquids presented there forces us to picture solvents which cannot coexist at the same temperature in the same scale, but a similar situation is observed for protic solvents (liquid ammonia, water). So, the destruction of nitrate melts runs at temperatures near

500–550 °C whereas the minimum of the melting-point plot for the KCl–NaCl binary system is achieved at the component ratio of 1:1 and corresponds to a temperature near 658 °C. Naturally, such a considerable difference in temperature should cause appreciable changes in the acidic/basic properties of both substances arranged in one conjugate pair. Therefore, we shall consider further in Part 3 the general oxoacidity scales corresponding to the definite temperatures: 600, 700 and 800 °C.

The principal general acid–base scale presented in Fig. 1.1.3 shows that the oxygen-containing ionic melts, which belong to the solvents of the first kind, as a rule, are unavailable for the determination of the constants of oxide ion addition. This is owing to the volatility or thermal instability of the solvent acids, and the acidic properties of most strong acids are levelled in these media. The constant of the intrinsic acid–base dissociation of oxygen-containing ionic melts is one of the most important parameters of these liquids: it makes it possible to separate the regions of acidic and basic solutions. This equilibrium complicates the acid–base processes taking place in such solvents.

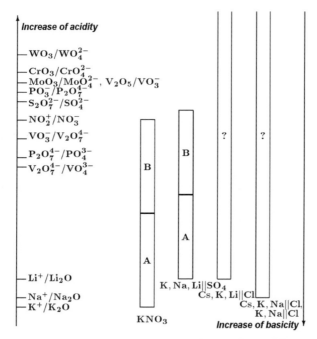

Fig. 1.1.3. The general oxoacidity scale of ionic melts. The positions of the conjugated acid–base pair corresponds to the KCl–NaCl eutectic at 700 °C.

Concerning molten alkali-metal halides (which are referred to the solvents of the second kind) it should be emphasized that there is no levelling of acidic properties in them, and, therefore, it is possible to determine the relative strength of acids by measurements in these media. Nevertheless, the properties of strong bases are levelled to those of the oxide formed by the most acidic constituent cation of the melt. As a rule, it is assumed that this oxide is formed by the alkali metal cation of the smallest radius (i.e. Li^+ in the KCl–LiCl eutectic, and Na^+ in the KCl–NaCl equimolar mixture).

Hence, on the basis of the generalized solvosystem concept it is possible to formulate the main distinctive features of different kinds of high-temperature ionic melts as media for acid–base interactions by the Lux–Flood method.

Oxygen-containing ionic melts, media of the first kind.

– These ionic solvents are characterized by the existence of the intrinsic acid–base dissociation, which possesses a constant acid–base product, $pK_{1,I}$, which is described by an equation similar to equation (1.1.43).
– They are distinguished by the levelling of both acidic and basic properties in these melts, owing to the solvolysis reactions. The properties of acids are levelled by the formation of an equivalent quantity of acid of the solvent that makes all the strong acids indistinguishable. The properties of strong bases are limited by the basicity of the oxide ion adduct to the most acidic constituent cation of the ionic solvent. Only pK values for weak acids and bases may be determined without appreciable distortions.
– Acid–base processes are complicated since they are affected, at least, by one intrinsic acid–base equilibrium of the melt. The upper limit of acidity caused by volatility or thermal decomposition of concentrated solutions of the solvent acids is often observed.
– The regions of acidic and basic solutions are separated clearly by the neutrality point, which is located at $pO_l = 1/2pK_{l,i}$.

Oxygen-less melts, media of the second kind.

– The constituent particles of these melts are not susceptible to the acid–base dissociation, therefore, there are no acids of solvent, and the solvent base is the adduct of oxide ions to constituent cations of the melt.

- They are characterized by the levelling of basic properties of the strongest bases, since they are transformed into an equivalent quantity of the solvent base. There are no limitations to the strength of acids, which makes it possible to compare the relative acidic properties of all the acids which are stable in these melts at a definite temperature.
- Their acidic and basic regions are determined conditionally with respect to the initial concentration of oxide ions in the melts, which, in its turn, is dependent on the concentration of oxygen admixtures in the "pure" solvent and, subsequently, on the quality of the melt purification from oxygen-containing impurities.

Therefore, melts–solvents of the first kind are of interest in the following scientific aspects: determination of the acid–base product of the ionic solvent and estimation of the upper limit of acidity of these solvents (such as nitrates, sulfates). The decrease of stability of the solvent acid can be used for the stepwise decomposition of acidic solutions of cations and synthesis of complex oxide compounds and composites by co-precipitation [53–56]. It is possible to obtain complex oxides containing alkali metals by precipitation of multivalent metal oxides with the alkali metal oxide as a strong Lux base, as was reported by Hong et al., who used $0.59LiNO_3–0.41LiOH$ mixed melt to obtain electrochemically active lithium cobaltate, $LiCoO_2$ [57].

The melts–solvent of the second kind, and molten alkali metal halides are used widely for practical purposes (electrochemical deposition of metals [58], alloys, crystal growth, etc.). Investigations of the acidic properties of these solvents allow one to estimate the stability of various dissolved oxygen-containing impurities, the solubility of constructional materials (alundum, yttria-stabilized zirconia, etc.) and to control the mentioned processes to take account of changes of the acidity (basicity) of the used melt–solvent. Naturally, knowledge of the acidic properties of "pure" ionic liquids makes it possible to choose the optimal ionic solvent for performing necessary technological processes on the basis of accurate quantitative parameters.

Therefore, investigations of acid–base reactions in molten salts meant for the determination of peculiarities of ionic melt composition on different acid–base equilibria are of importance both from the theoretical and applied viewpoints.

Part 2. Studies of Homogeneous Acid–Base Reactions in Ionic Melts

Nowadays, the following aspects of oxoacidity investigations in ionic melts are being developed [59]:

- Investigations of homogeneous acid–base reactions; among these the reactions of acidic oxides and coordinationally unsaturated oxoanions with bases should be mentioned.
- Studies of the processes of dissolution and dissociation of oxide materials in ionic solvents, and the effect of surface properties of oxide powders on their equilibrium solubility.
- Examinations of the interactions of ionic halide melts–solvents with gaseous reagents of acidic or basic character, which include high-temperature hydrolysis of molten alkali-metal halides, their purification from oxygen-containing admixtures, and studies of the dissociation of strong and weak Lux bases (such as CO_3^{2-}, OH^-, O_2^{2-}) in melts of different acidity.
- These investigations are inseparably linked with some directions, which can hardly be considered to be immediately related to oxoacidity studies. These are connected with the application of measurement methods (kinetic, potentiometric, chronopotentiometry, etc.) and accompanying problems (electrode reversibility), and with the treatment and explanation of the obtained results.

Although investigations of oxoacidity in ionic melts have been performed since 1939 [17], up to the present there are still no known attempts to collect and systematize all the obtained results. Some fragmentary data on oxoacidity may be found in monographs devoted to other aspects of molten salt chemistry. In particular, some monographs by Delimarsky [3], e.g. contain data on the oxoacidity studies made at the Institute of General and Inorganic Chemistry (NAS of Ukraine, Kiev) by Shapoval *et al.*: unfortunately, there are no comparative considerations of results obtained by other scientists working on this problem. The first attempt to consider homogeneous acid–base equilibria in molten salts was performed in our review published in 1997 [59], but subsequent years have brought many new experimental data. We shall now consider these studies using the solvosystem concept proposed in Part 1 to unite the scattered literature data for some systems.

1.2.1 FEATURES OF HIGH-TEMPERATURE IONIC SOLVENTS AS MEDIA FOR LUX ACID–BASE INTERACTIONS

All high-temperature ionic liquids used as media for performing acid–base reactions of different kinds may be divided into two groups: oxide-containing and oxygen-less melts, depending on whether or not the constitutional oxygen is present in them (i.e. entering into the composition of main components of the melt). The former melts belong to the solvents of the first kind, whereas oxygen-less melts are the solvents of the second kind. The peculiarities of running acid–base interactions in solvents of both kinds are defined by the chemical stability of the formed acid and base of the solvent. We shall now consider some of them [37, 60, 61].

Pure oxygen-less melts contain no oxide ions in any form, and, therefore, such pure melts cannot serve as donors of O^{2-}. The melts, which are solvents of the second kind, can affect acid–base interaction on their background in two manners: by fixation of oxide ions entering in the melt and by solvation of the conjugate acid or base. However, the ionic solvents of the second kind, used in practice for different measurements and applied purposes, contain admixtures of oxide-ion donors, which are formed in the melt from initial admixtures of oxo-anions such as SO_4^{2-}, CO_3^{2-} or OH^-. The second way of appearance of oxide ion admixtures in molten media is characteristic of the melts based on alkali metal halides: the process of high-temperature hydrolysis of the said halide melts results in the formation of hydroxide ions and, after their dissociation, of oxide ions:

$$MeX + H_2O \rightarrow MeOH(\equiv Me^+ + OH^-) + HX\uparrow. \qquad (1.2.1)$$

The presence of the oxide-containing admixtures in oxygen-less melts inevitably leads to their dissociation with the formation of oxide ions. For the admixtures mentioned, these processes are described in ionic form by the following equations:

$$SO_4^{2-} \rightleftharpoons SO_3\uparrow + O^{2-}, \qquad (1.2.2)$$

$$CO_3^{2-} \rightleftharpoons CO_2 \uparrow + O^{2-}, \tag{1.2.3}$$

$$2OH^- \rightleftharpoons H_2O \uparrow + O^{2-}. \tag{1.2.4}$$

Since the concentrations of all these admixtures depend on the purity of the salts used for preparation of the melts, they are not constant. This means that at each specific melt–solvent pO, the values of the initial solvent fluctuate about a certain magnitude. For example, pure CsI and KCl–NaCl melts are characterized by pO $\approx$ 4, whereas for NaI and KCl–LiCl melts these values are lower and vary between 3 and 3.5. For comparison, at 298 K pure water as a solvent is characterized by the pH value equal to 7, and this value remains constant after achieving a certain degree of purity of H_2O.

The degrees of the oxygen admixture dissociations differ substantially. For example, the addition of even relatively high concentrations of sulfate ions does not result in appreciable changes of O^{2-} concentration in the melt, whereas the effect of addition of CO_3^{2-} and OH^- on pO values in melts is more considerable. The investigations made on various ionic melts and over wide temperature ranges (from 500 to 1100 K) allow one to arrange the above-discussed oxoanions in the sequence $SO_4^{2-} \rightarrow CO_3^{2-} \rightarrow OH^-$ of increasing basic properties.

A preliminary thorough purification of oxygen-less ionic melts from oxygen-containing admixtures may provide the pO values varying within the limits of 3–4.5. This is the main cause of errors affecting calculations of thermodynamic parameters, especially at small concentrations of reagents in the melts, and it is impossible to eliminate this influence in any way. So, for alkali metal halide melts the use of the strongest halogenating agents does not result in essential reduction of the oxide ion concentration, since the action of such reagents favours the transformation of oxide ion into weaker bases such as CO_3^{2-}, and the latter admixture is hardly destroyed in the melt. Historically, earlier studies of oxoacidity in molten salts were performed without detailed analysis of the melts' composition, and the pO of "pure" oxygen-less melts was often used as a constant of the given solvent [62–65]. Proceeding from this assumption, the authors of the mentioned papers constructed scales of acidity of some oxygen-less melts and attributed acid–base ranges of a certain length to these melts. As follows from Part 1, the acid–base ranges in the ionic solvents of the second kind are half-open, and, therefore, their length cannot be estimated by a constant value.

The Lux acid−base interactions in oxygen-containing ionic melts have been studied more extensively than those in the media of the second kind. The complications are caused, first by the superimposition of the acid−base equilibrium of the melt−solvents themselves (autodissociation), which, according to what we have said in Part 1, belongs to the solvents of the first kind.

As the first-kind solvents, oxygen-containing ionic melts contain not only oxide ions but also the solvent acid as products of the intrinsic acid−base dissociation. This leads to the two-sided limitation of the acid−base range that gives rise to different specific features of these solvents, such as the levelling of properties of acids, narrowing of the acid−base range with rising temperature, etc. Mainly, the behaviour of the oxoacidic acid−base pairs in these solvents is similar to that of the Brønsted−Lowry acid−base pairs in aqueous solutions.

However, oxygen-containing ionic melts possess some features originating in relatively high temperatures of the liquid state. In some cases, the acid of the solvent is volatile or thermally unstable [17, 66] and, being formed as a result of the Lux acid−base process, it is gradually removed from the melt. This leads to a loss of the acid, and the pO of such an acidic solution approaches the value corresponding to the neutral point of the solvent. Therefore, besides the levelling of acidic properties, the oxygen-containing melts are often characterized by the upper limit of acidity defined either by the solubility of the formed gaseous acid in the solvent or by the decomposition temperature of this acid.

Thus, the highest oxides or chromium(VI) and molybdenum(VI) were established to destroy molten KNO_3 with subsequent evolution of NO_2: termination of this destruction resulted in stabilization of the pO values of the melt [67]. This value determined by the method of potentiometric titration was found to reduce as the temperature of the melt was elevated considerably. Such a behaviour of the acidic solution in the nitrate melt was explained by the formation of nitrogen pentoxide, N_2O_5, or nitronium cation, NO_2^+, in the acidic solution, together with the corresponding oxoanions of Cr(VI) and Mo(VI). Since the added acids are very strong, their reaction with the melt

$$A + NO_3^- \rightarrow B + NO_2^+, \tag{1.2.5}$$

is shifted completely to the right, and the strength of such acids is determined by the acidic properties of NO_2^+. The thermal stability of the latter cation is relatively small, and when its concentration exceeds a certain magnitude the reaction of nitronium with nitrate ions

$$NO_2^+ + NO_3^- \rightarrow 2NO_2 \uparrow + \tfrac{1}{2}O_2 \uparrow, \qquad (1.2.6)$$

becomes intense enough.

The so-called kinetic methods for the determination of nitrate-melt acidities are based on this reaction: the rate of NO_2 evolution from the melts is dependent only on their acidity. Further, in the process of reaction (1.2.6), the concentration of NO_2^+ cation decreases; this causes a fall in the rate of nitrogen dioxide evolution. Finally, the latter becomes so slow that a series of sequential measurements of pO demonstrates that this parameter is constant, i.e. the system passes to a quasi-equilibrium state. The pO value measured under these conditions is the upper limit of the melt acidity, which is independent of the composition of the added strong acid. All the above-said allows us to explain the thermal dependence of the upper limit of acidity for nitrate melts observed in [67]: the rise of the melt temperature results not only in the decrease of cation NO_2^+ stability but also in the increase of the rate of reaction (1.2.6).

Similar to the discussed nitrate melts, molten sulfates are characterized by an upper limit of acidity that is seemingly defined by both the limited solubility of SO_3 in these melts at temperatures near 800 °C and its instability under these conditions. This conclusion may be obtained from the data of Ref. [17]: there was the shift of pO (e.m.f.) of the acidic solution towards the neutrality point of the melt. This process was found by Lux to be caused by evaporation of SO_3 from the equimolar $K_2SO_4-Na_2SO_4$ mixture and by its possible decomposition according to the following scheme:

$$SO_3 \uparrow \rightarrow SO_2 \uparrow + \tfrac{1}{2}O_2 \uparrow. \qquad (1.2.7)$$

1.2.2 METHODS OF INVESTIGATIONS

The Lux acid–base equilibria and the acidic properties of ionic melts are examined using different methods, which can be divided into equilibrium and kinetic ones. The indicator method is one of the simplest equilibrium methods described in the literature. The essence of this method consists in Ref. [68]. A conventional acid–base indicator is added to melts at temperatures near 200 °C to indicate the equivalence point of an acid–base titration. The behaviour of 11 acid–base indicators in the molten equimolar mixture KNO_3–$LiNO_3$ at 210 °C, and in a KSCN melt at 200 °C was investigated in the following manner: the acid–base neutralization of potassium pyrosulfate with sodium hydroxide and peroxide, as well as the reverse processes, were examined. The transition in the solution colours was found to be caused by the transformation of the protonated form of an acid–base indicator (designated as HInd) into the deprotonated one (Ind$^-$) in the basic solutions

$$2HInd + O^{2-} \rightleftharpoons 2Ind^- + H_2O. \tag{1.2.8}$$

This colour transition in the nitrate melt was observed only in the case of phenolphthalein (from colourless to purple): other organic indicators seemed to be oxidized by the melt. In molten KSCN, which does not belong to oxidizing media, the colour transitions took place for other indicators used (methyl red, thymolphthalein, etc.). The use of organic acid–base indicators for oxoacidity investigations is limited, mainly, by their thermal instability, and ability to oxidize easily at elevated temperatures. As a rule, since the usual ionic melts are anhydrous media, the reverse transformations of the acid–base indicators in the protonated form are hardly possible. Probably, the corresponding indicator anhydro-acids are formed in the reverse transition. Such an observation may be confirmed by the following fact. Brough et al. [68] noted that titration of basic solutions with $K_2S_2O_7$ led to a colour transition from purple to yellow. The obvious drawback of this method is the scant information about the acid–base process studied: one can obtain only information on the acid–base reaction stoichiometry, whereas quantitative colorimetric studies have not yet been reported. Possibly, these studies are accompanied by instrumental difficulties, making the method of colorimetric indicator useless for molten salts.

Inorganic indicators are more readily available for studies of oxoacidity in ionic melts, since there are, in principle, no limitations of their thermal stability. In contrast, the majority of organic compounds are known to decompose at temperatures lower than 400 °C. Among inorganic substances, soft and intermediate cation acids are the most promising. Duffy and Ingram [69–73] successfully used the so-called metallo-indicator method to investigate the acidity of some oxygen-containing and oxygen-less melts. Some cations of heavy metals (Tl^+, Pb^{2+}, Bi^{3+}) introduced into oxide glasses were found by the authors to cause the appearance of an absorption maximum in the UV region [69–73]. This maximum arises owing to a $^1S_0-^3P_1$ transition and its location depends considerably on the anion neighbourhood of the metallo-indicator cation. There is a correlation between the relative shift of the absorption maximum and the concentration of Na_2O in molten glasses of the $Na_2O-B_2O_3$ system. Based on their studies, Duffy and Ingram introduced the parameter Λ, called the "optical basicity", which characterizes the relative oxoacidic properties of ionic melts, and is calculated from the following formula:

$$\Lambda_{Me^{n+}} = \frac{k_1 - \nu}{k_2}. \tag{1.2.9}$$

where $\Lambda_{Me^{n+}}$ is the "optical basicity" determined by the shift of the absorption maximum of Me^{n+}, ν (cm^{-1}) the frequency corresponding to the absorption maximum of the solution studied, k_1 the frequency of the maximum in the most acidic melt used as a standard: in this melt the metallo-indicator cation is believed to be completely unfixed to the oxide complexes, and this value is the same both for chloride melts and for oxyanion glasses. As for the k_2 coefficient, this is a constant which is specific for each metallo-indicator cation and anion of the melt studied. The studies of other melts resulted in similar linear dependencies differing in the numerical values in the denominator and the numerator of equation (1.2.9) the coefficients for chloride- and oxygen-containing melts are presented in Table 1.2.1. It can be seen that the denominator values in the chloride melts are appreciably higher than those in oxyanion melts, owing to stronger fixation of the metallo-indicator cations in the oxo-complexes. These values are dependent on the oxoacidic properties of the metal cations and increase with its acidity. The oxoacidic properties of the metallo-indicators are known to strengthen in the sequence $Tl^+-Pb^{2+}-Bi^{3+}$, the corresponding changes of k_2 values being 14,800–24,500–25,800.

TABLE 1.2.1
Parameters of equation (1.2.9) for estimations of optical basicities in different ionic melts

Cation	Chloride melts		Oxyanion glasses	
	k_1	k_2	k_1	k_2
Pb(II)	60,700	24,500	60,700	31,000
Bi(III)	56,000	25,800	56,000	28,800
Tl(I)	55,300	14,800	55,300	18,300

It is of interest to compare the values of optical basicity of melts and superacids obtained in the work mentioned, with the corresponding Hammett functions: this comparison is presented in Fig. 1.2.1. There are some deviations of the values obtained with the use of Bi(III) metallo-indicator from those obtained using Tl(I) and Pb(II). As seen from this figure, the acidic melts ($NaPO_3$ and K_2SO_4–$ZnSO_4$) are considerably more basic than the usual superacids, and there is an obvious correlation between H_0 and Λ. Nevertheless, the following estimations of optical basicities of such acidic melts as $NaCl$–$AlCl_3$ (67% of $AlCl_3$) and KCl–$LiCl$ (59% of $LiCl$) show that the basicities of the said melts are somewhat overestimated. Indeed, the Λ values of these melts are located within values of 0.865–0.867, whereas the latter melt possesses Λ values near 1 at 440 °C and 1.028 at 640 °C. This means, first, that the oxoacidic properties of chloride melts decrease with temperature (weakening the electrostatic attraction of the constituent particles as compared with the kinetic energy of the ions), but this decrease is very slight. The other conclusion concerns the higher basicity of the chloride melts as compared with the oxyanion ones. The sequence of the Li^+–Zn^{2+}–Al^{3+} cations is characterized by strengthening of the oxoacidic properties, and, indeed, the Al^{3+}-containing chloride melt has higher acidity than the Zn^{2+}-containing one ($ZnCl_2$); this melt, in its turn is more acidic than molten $LiCl$. The oxoacidic properties of the zinc halides were found to decrease

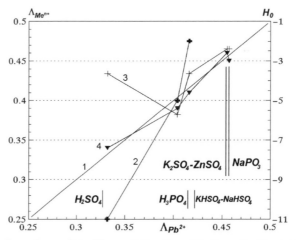

Fig. 1.2.1. The dependence of the Hammett functions (2), the optical basicities Λ, obtained with the use of Bi(III) (3) and Tl(I) metallo-indicators of some melts and superacids from the optical basicity obtained with the use of Pb(II) metallo-indicator (1).

proceeding from chloride to the bromide melt, although the HSAB concept predicted the inverse order for the melt acidity in this sequence. However, by comparing the optical basicity (Λ) of the $ZnCl_2$ melt and that of K_2SO_4–$ZnSO_4$ it can be found that the melt formed is appreciably more basic. This shows that the proposed approach is usable for the comparison of ionic melts belonging to the same class of melts (halide or oxyanion ones), but its application to melts belonging to different groups is not so fruitful.

For most of the studied glasses the Λ values were shown to rise together with the increase of their determined basicity, mainly by the ratio of the concentration of alkali- or alkaline-earth metal oxide to that of the acidic oxides (such as B_2O_3, SiO_2, Al_2O_3) in the glasses. The acidity scales for borate and some melts containing other acidic oxides are presented in Fig. 1.2.2. It follows from this figure that a reduction of the molar fraction of B_2O_3 in the melt from 1 to 0.8 leads to an increase in Λ, and that this increase is gradual, whereas in the vicinity of $N_{B_2O_3} = 0.8$ the melt-basicity increases sharply, and the dependence becomes similar to the usual potentiometric titration curve. It should be noted that at molar fractions of B_2O_3 exceeding 80% the

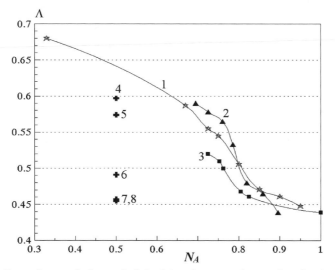

Fig. 1.2.2. Dependence of the optical basicity $\Lambda_{Pb(II)}$ against molar fraction of acidic oxide (B_2O_3) in some oxyanion melts: 1, Na_2O–B_2O_3; 2, K_2O–B_2O_3; 3, Li_2O–B_2O_3; 4, Na_2O–SiO_2 (0.3:0.7); 5, Na_2O–CaO–SiO_2 (0.2:0.1:0.7); 6, K_2O–Al_2O_3–B_2O_3 (0.154:0.154:0.692); 7, Na_2O–P_2O_5 (0.5:0.5); 8, K_2SO_4–$ZnSO_4$ (0.5:0.5). Scale constructed from the data of Ref. [71].

Λ-composition curves for all the studied oxides of alkali metals are practically coincident, and their acidic properties are comparable with those of $NaPO_3$ melt (Fig. 1.2.2, point 7). At the same time, at lower molar fractions of B_2O_3 the Λ values differ appreciably, and increase in the following sequence of cations: $Li^+ - Na^+ - K^+$.

A possible reason for the observed facts consists in the formation of oxo-complexes of metallo-indicator ion. Indeed, pure B_2O_3 is a molecular liquid and one cannot expect appreciable fixation of the metallo-indicator ions in oxo complexes. Increase of the concentration of Me_2O in the melt results in an increase of polyborate anions in the melts, and the observed "titration-like" form of the dependence may be connected with the process of titration of the metallo-indicator ion with a definite borate anion $(B_4O_7^{2-}, BO_2^-,$ etc.). Subsequent increase of alkali metal oxide concentration does not lead to corresponding drops in Λ, although the "absolute" equilibrium concentration of metallo-indicator in the melt decreases gradually. This explains the equality of Λ values measured in $NaPO_3$ and B_2O_3, although the latter substance is a weaker acid than the former. In the former case, the melt is characterized by a relatively strong acidity and a weaker complexation ability in comparison with multi-charged oxoanions. In the latter one, we have a melt which is prone to ionization and, naturally, possesses weak complexation ability.

As for alkali metal phosphate melts, they possess elevated complexation ability only in the case of the existence of multi-charged $P_2O_7^{4-}$ and PO_3^- ions in the melts, which favours the formation of compounds characterized by the following compositions: $M^IM^{II}PO_4$ (Pb^{II}), $M_3^IM^{III}(PO_4)_2$ and $M^IM^{III}P_2O_7$ (Bi^{III}), etc. [74].

As shown by the measurements performed using the metallo-indicators in silicate glasses of compositions SiO_2–CaO, SiO_2–MgO, SiO_2–Al_2O_3, calcium oxide possesses the strongest basic properties, whereas aluminium oxide is the weakest oxo-base of those studied. The rise in the basic properties of the oxides strengthens the effect of its concentration on the optical basicity of the melts.

As well as being used for molten alkali halides and glasses, the metallo-indicator method as described above has been used to investigate the basic properties of metallurgical slags [73], and a scale of relative "optical basicity" of metallurgical slags as against CaO as the standard, was constructed. Despite clear qualitative estimations of the relative basicity of molten ionic media provided by the metallo-indicator method (since this method is similar to the Hammett method discussed above) it was not developed to the level allowing it to give quantitative estimations of the basicity of molten salts. Similarly, optical investigations were performed using quenched samples, i.e. at temperatures quite different from the experimental ones, which brought additional errors to the obtained values of the relative basicity of the melts.

The investigations of $Na_2O-B_2O_3-Al_2O_3$, $Na_2O-B_2O_3-SiO_2$ and $Na_2O-SiO_2-Al_2O_3$ melts, performed by the method of X-ray fluorescent spectroscopy are reported in Ref. [75]. The relative acidity of acidic oxides was shown to decrease in the following sequence: $B_2O_3 > SiO_2 > Al_2O_3$. It should be noted that this method requires the use of samples cooled (quenched) to room temperature, and this may affect the correctness of the results obtained.

In the literature, there are reports of work concerned with the measurement of oxoacidic properties of ionic melts by gravimetric measurements of the solubility of acidic gases in these media [76, 77]. The solubility of sulfur(VI) oxide in molten sodium phosphates was determined by the gravimetric method [76]. A correlation was obtained between the melt basicities and the solubility of gaseous acid SO_3 in them. Iwamoto reported the estimation of the basic properties of molten salt by measurements of acidic gas solubilities (carbon dioxide and water) in them [77]. However, similar methods cannot be used widely, owing to two factors. The first of these consists in the fact that the solubility of any gas in a liquid phase obeys Henry's law. Let us consider the following system of reactions:

$$CO_2(\text{dissolved}) \rightleftharpoons CO_2(\text{gas}), \quad K_{CO_2}.$$

$$CO_2 + O^{2-} \rightarrow CO_3^{2-}, \quad K.$$

(1.2.10)

From these equations it follows that the acidic gas solubility in the melt consists of two terms: the solubility according to the Henry law (which is constant for the same partial pressure of the gas) and the concentration of the gas fixed by oxide ions which increases with the base concentration. If K_{CO_2} is large, then the amount of gas absorbed is equal to the melt basicity, or (more correctly), to excessive concentration of O^{2-} (as compared with acidic compounds) in the melt. If the melt contains any excess of acid then the gas solubility will be practically constant since there is no fixation of the gas by oxide-ions completely consumed after introducing the excess of acid into the melt. So, acidic solutions of different acidities cannot be distinguished by the method of acidic gas solubility. Therefore, one should know first whether the melt is basic before applying the method of acidic gas solubility. That is, to measure the melt basicity by this method, we have to estimate (or know) it. The second reason consists in pollution of the basic melt by oxoanions formed owing to the reaction of the acidic gas with oxide ions. These anions will be stable and their existence will change the melt properties, so the method is not applicable for controlling the basicity *in situ*, i.e. without interfering in the chemical processes in ionic melts.

To finish this topic, the paper of Kurmaev and Amirova [78] should be noted. They studied the solubility of chlorides and oxochlorides of multivalent metals, such as $VOCl_3$, $TiCl_4$ and $SiCl_4$ in pure KCl, NaCl and the KCl–NaCl eutectic melts. The results of these determinations (in mass %) are presented in Fig. 1.2.3. It can be seen that the increase of concentration of acidic cation (Na^+) leads to a decrease in the acidic gas solubility. In the case of $VOCl_3$, whose boiling point is near 127 °C, the dissolution is accompanied by the endothermic effect, whose nature was not studied in the paper, but did not result in the formation of any stoichiometric compounds. The solubility of the oxochloride was found to increase with the rise in temperature. The solubility of $TiCl_4$ (boiling point, 137 °C) in the melts was lower than for $VOCl_3$: although possessing practically the same boiling point, it forms adducts of the

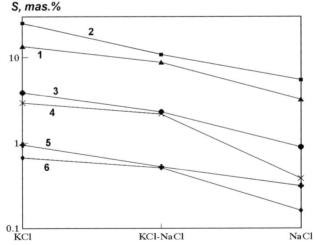

Fig. 1.2.3. The solubilities of some acidic gases in chloride melts (mass %): $VOCl_3$ (1, 850 °C, 2, 900 °C), $TiCl_4$ (3, 850 °C, 4, 900 °C), $SiCl_4$ (5, 850 °C, 6, 900 °C).

composition of Me_2TiCl_6. Naturally, in the case of $SiCl_4$ (boiling point 58 °C) the solubilities in the said melts are considerably less than those of $TiCl_4$, and the authors did not explain this by the formation of complex chlorides. In contrast to $VOCl_3$, the solubilities of both gases in the chloride melts decrease with the temperature.

The solubility values discussed above give us a qualitative insight into the comparative acidic properties of the said melts, which decrease in the sequence $NaCl-(KCl-NaCl)-KCl$, although the authors connected the solubility of the gases only with the alkali-metal cation radius.

In contrast to the preceding discussion, the so-called kinetic methods of melt acidity determination are usable for estimations of the concentration of acids in the melts based on alkali metal nitrates. Duke and Yamamoto investigated the interactions taking place in solutions of pyrosulfate ions in the molten equimolar mixture KNO_3-NaNO_3 [79, 80]. The chemical reaction

$$S_2O_7^{2-} + NO_3^- \rightarrow 2SO_4^{2-} + NO_2^+,\tag{1.2.11}$$

leads to the formation of NO_2^+ ion, which then reacts with nitrate ions according to equation (1.2.6). The latter reaction is slower than equation (1.2.11), and, therefore, its rate gives information about the melt acidity.

The equation for the rate of NO_2 evolution from acidic nitrate melts may be written as

$$-\frac{d(T_A)}{dt} = k[NO_2^+]\tag{1.2.12}$$

where k is the rate constant of reaction (1.2.6) and T_A the total acidity determined by the sum

$$T_A = \left[S_2O_7^{2-}\right] + \lfloor NO_2^+ \rfloor,\tag{1.2.13}$$

Taking into account the expression for the equilibrium constant of equation (1.2.11)

$$K = \frac{[NO_2^+][SO_4^{2-}]^2}{[S_2O_7^{2-}]},\tag{1.2.14}$$

the value of the derivative (1.2.12) may be expressed as

$$-\frac{d(T_A)}{dt} = \frac{kKT}{K + [SO_4^{2-}]^2} = k'T,$$
(1.2.15)

where k' is the pseudo-first-order rate constant

$$k' = \frac{kK}{K + [SO_4^{2-}]^2},$$
(1.2.16)

The reciprocal equation may be presented as

$$\frac{1}{k'} = \frac{1}{k} + \frac{[SO_4^{2-}]^2}{kK},$$
(1.2.17)

and the linear dependence can be constructed if $1/k'$ is plotted as a function of $[SO_4^{2-}]^2$. The slope of this plot is equal to $1/kK$ and the intercept will be $1/k$. So, such a routine gives a possibility of determining both the equilibrium constant and the rate constant if the former value is large enough.

The above description allows us to conclude that the kinetic method can be used only to determine thermodynamic parameters of strong acids, and only in strongly acidic media. Also, an available decomposition reaction, whose rate is limited by the stage of the reaction product breakdown, can be chosen to detect the oxoacidic properties of the studied ionic melt.

The most convenient method used widely for acidity (basicity) determination in various ionic media is the potentiometric method using an indicator electrode which is reversible to oxide ions. Usually, the measurements are performed in cells with a liquid junction. The most common scheme of an electrochemical cell used to measure the equilibrium concentration of oxide ions in molten salts is as follows:

Reference half-element‖Melt + O^{2-}|Electrode reversible to O^{2-}.
(1.2.18)

The reference half-element is an electrode immersed in the solution of the potential-defining ion in the "pure" melt studied. For this purpose metals immersed in a solution of the corresponding salts of a definite concentration are often used: a possible variant is the use of a similar oxygen electrode in the solution, with known and constant concentration of oxide ions (concentration cells). The quality of such electrodes is determined by their ability to maintain a constant potential with good accuracy that provides high reproducibility of the electrochemical data obtained.

Usually, the studies mentioned are performed assuming that the potential of the liquid junction is equal to zero or, in any case that its magnitude is considerably less then the accuracy of the potentiometric measurements. As a rule, these assumptions are reasonable and the explanation consists in the following. In the solutions possessing electric conductivity owing to addition of electrolytes, potentials of the liquid junction even in concentration cells are usually of the order of several decades of mV, and they bring considerable errors in potentiometric investigations. However, halide, sulfate and nitrate melts formed by salts of alkali- or alkaline-earth metals are, in most cases, typical ionic liquids. The process of formation of the potential of the liquid junction in such solutions is more affected by ionic particles of the solvent than by ions formed by the process of dissociation of the dissolved substance. If indifferent electrolytes are added to the solutions of low conductivity they drastically decrease the potentials of the liquid junction at the contacting boundaries, and in the case of completely ionized liquids they become negligible. This allows us to neglect them in thermodynamic calculations based on the measurements of e.m.f. of high-temperature cells with diaphragms.

The magnitudes of potentials of the liquid junction, estimated for cells with diaphragms of the following composition:

$$0.01 MeCl_n + 0.99(KCl-NaCl)\|KCl-NaCl, \qquad (1.2.19)$$

TABLE 1.2.2
Values of potentials of the liquid junction in concentration cells (equation (1.2.19))

$MeCl_n$	ε_d, mV at temperatures (K)			
	950	1000	1050	1100
AgCl	+0.21	+0.14	+0.06	− 0.21
$MgCl_2$	−	+0.05	− 0.15	− 0.22
$ZnCl_2$	+0.08	0.00	− 0.08	− 0.15
$CdCl_2$	+0.05	− 0.05	− 0.16	− 0.25
$PbCl_2$	− 0.13	− 0.25	− 0.38	− 0.48
UCl_3	− 0.25	− 0.43	− 0.61	− 0.76
HfF_4	− 0.75	− 1.00	− 1.24	− 1.44
AgF	+0.16	+0.10	+0.03	− 0.11
PbF_2	− 0.24	− 0.39	− 0.54	− 0.68

were calculated by Smirnov on the basis of known experimental data on diffusion coefficients, etc. [81]. These data are listed in Table 1.2.2.

As seen from this table, the potentials of the liquid junction are very small (as a rule, they are less than 1 mV, which is the usual accuracy of the measurements at ~ 1000 K). An increase of potentials of the liquid junction is observed only for the case of multi-charged cations or cations of large size, whereas substitution of chloride ion by fluoride did not result in appreciable increase of these values.

Potentiometric investigations of oxoacidic properties of molten salts are performed using various kinds of oxygen electrodes: metal-oxide, gas, and membrane ones. As a rule, these electrodes have been examined extensively, and their features are well described in the literature on molten salts electrochemistry.

In concluding this brief description of the potentiometric method, we should mention its main advantages as compared with the above-described experimental routines:

– relative simplicity of experimental design and measurements together with very good reproducibility of the results obtained;
– possibility to perform the measurements *in situ*, i.e. without any effect on the studied processes in the molten media. This makes the use of the potentiometry promising for the control of the process conditions in various technological schemes;
– the e.m.f. measured is linearly dependent not upon the concentration but upon its logarithm. This gives the possibility of studying both the most basic and the most acidic solutions using the same experimental method.

We now proceed to a consideration of the oxoacidity investigations performed in particular groups of ionic melts.

1.2.3 IONIC SOLVENTS BASED ON ALKALI METAL NITRATES

Since there are many publications devoted to different aspects of oxo-compounds and oxide ion chemistry in melts based on alkali metal nitrates, these melts are traditionally said to be the most studied. This is mainly because the relatively low melting points of both individual alkali-metal nitrates and their mixtures allows one to use very simple experimental techniques, e.g. to perform studies in air, and to use Pyrex crucibles. Moreover, since the temperatures of the said melts are lower than 500 °C (at higher temperatures any heated substance begins to radiate light) there exists the possibility of visually controlling acid–base processes in these media.

The temperature range of the stable liquid state of molten alkali metal nitrates is relatively narrow: it is limited by the melting point on the one side and by the decomposition temperature on the other. The process of nitrate melt decomposition is well known to become intensive enough at temperatures of the order of 430 °C for $LiNO_3$, whereas molten KNO_3 and $NaNO_3$ start to yield gaseous oxygen at temperatures near 450–500 °C [82] which is explained by the weaker acidities of K^+ and Na^+ cations compared with that of Li^+. The decomposition results in the formation of the corresponding nitrite: for the case of KNO_3 melt this may be represented as follows:

$$2KNO_3 \rightarrow 2KNO_2 + O_2 \uparrow . \qquad (1.2.20)$$

The known investigations of oxoacidity in nitrate melts are devoted to the determination of the constants of acid–base equilibria using the above-described kinetic and potentiometric methods.

Duke and Yamamoto determined the rate and equilibrium constants for the reaction of $K_2S_2O_7$ with the KNO_3–$NaNO_3$ eutectic mixture [79, 80] by the kinetic method whose features are described in detail above. The rate constants were found to be 0.038 at 235 °C, and 0.096 at 275 °C, the corresponding values of the equilibrium constants (1.2.11) were 2.6×10^{-2} and 3.8×10^{-2}. The activation energy of the direct reaction (1.2.11) was estimated to be 54 kJ mol^{-1}. Besides the parameters of acid–base processes, the solubility of Na_2SO_4 in the said melt was found: it was equal to (mol kg^{-1}) 2.5×10^{-1} at 235 °C, 2.82×10^{-1} at 250 °C, 3.42×10^{-1} at 275 °C, and 4.21×10^{-1} at 300 °C. The errors of solubility determination were less than 1×10^{-2} mol kg^{-1}.

The same authors studied the interaction of dichromate ions with the molten KNO_3–$NaNO_3$ eutectic [79]

$$Cr_2O_7^{2-} + NO_3^- \rightarrow 2CrO_4^{2-} + NO_2^+ \cdot NO_3^- . \qquad (1.2.21)$$

Since the chosen acids are relatively weak ones, to shift equilibrium (1.2.21) to the right, the cations precipitating chromate ions were added to the melt. Under these conditions the derivative of dichromate ion concentration with time can be expressed in the following manner:

$$-\frac{d[Cr_2O_7^{2-}]}{dt} = \frac{kK_1[Cr_2O_7^{2-}][Me^{2+}]}{K_{sp}^2},$$

(1.2.22)

where

$$K_{sp} = \left[Cr_2O_7^{2-}\right]\left[Me^{2+}\right]$$

and

$$K_1 = \frac{[Cr_2O_7^{2-}][NO_2^+]}{[CrO_4^{2-}]^2}.$$

(1.2.23)

The rate constant was found to be practically independent of the melt temperature (3.0×10^{-5} at 250 °C, and 3.4×10^{-5} at 300 °C: in both cases, the Pb^{2+} concentration was 0.044 mol kg^{-1}) and to rise with increasing the precipitant concentration. The solubility products of chromates of barium (4.0×10^{-6}) and calcium (1.7×10^{-4}) were determined.

There are some literature data devoted to the interactions of dichromate ions with oxoanions of the common formula RO_3^- which are stronger bases than NO_3^- (chlorates and bromates). The interaction of dichromate ion with chlorate ions in the nitrate melt runs in two stages [83, 84]:

$$Cr_2O_7^{2-} + ClO_3^- \rightarrow 2CrO_4^{2-} + ClO_2^+,$$

(1.2.24)

$$ClO_2^+ + Cl^- \rightarrow 2Cl_2 \uparrow + O_2 \uparrow,$$

(1.2.25)

reaction (1.2.25) is the limiting stage. The rate law for this reaction can be written as

$$-\frac{d(T_A)}{dt} = k[ClO_2^+][Cl^-],$$

(1.2.26)

where $T_A = [Cr_2O_7^{2-}] + [ClO_2^+]$. To shift reaction (1.2.24) to the right, Ba^{2+} cations were added to fix CrO_4^{2-} ions in slightly soluble $BaCrO_4$ (its solubility in mol kg^{-1} was 1.6×10^{-3} at 239 °C, 1.8×10^{-3} at 249 °C and 2.0×10^{-3} at 259 °C). The rate constant (mol^{-1} min^{-1}) of reaction (1.2.25) was found to be changed from 0.208 (250 °C) to 0.417 (270 °C).

To complete a series of investigations Duke and Schlegel studied the interaction between dichromate and bromate ions in the molten KNO_3–$NaNO_3$

eutectic [85]

$$Cr_2O_7^{2-} + BrO_3^- \rightleftharpoons 2CrO_4^{2-} + BrO_2^+, \tag{1.2.27}$$

$$BrO_2^+ \rightarrow products(Br_2 \uparrow + O_2 \uparrow). \tag{1.2.28}$$

The rate constant increases from 0.13 to 0.58 by changing the melt temperature within the 230–260 °C range. The equilibrium constant of reaction (1.2.27) increases not only with the temperature of the melt but also with the increase of the ratio Na^+/K^+ in the nitrate melt. The observed effects take place owing to the decrease of BrO_2^+ oxocation stability as the temperature rises and the melt acidity increases (Na^+ is more acidic than K^+).

The results on the acid–base constants may be summarized by the following sequence: $BrO_2^+(K = 3.5 \times 10^{-8})$–$ClO_2^+(K = 1.6 \times 10^{-10})$–$NO_2^+(K = 8.5 \times 10^{-14})$ [83, 84] where the cation acidity increases.

Considering the "total acidity" magnitude T_A introduced in Refs. [79–85], it should be emphasized that the use of this sum complicates the analysis of the result, since the acidity of molten nitrates is dependent only on the equilibrium concentration of NO_2^+ cation, and the acidity of any substance is defined by its ability to create NO_2^+ ions in the melts or—which is the same thing—to take up oxide ions from nitrate ion. The rate of reaction (1.2.6) is dependent only on NO_2^+ concentration, and rates of reactions (1.2.25) and (1.2.28) are linearly connected with the concentrations of ClO_2^+ and BrO_2^+, respectively, since the acidities of these cations are considerably weaker than those of NO_2^+.

A similar method was used by Slama to study the acidic properties of Cu^{2+} (325–375 °C) and Co^{2+} (350–400 °C) cations in molten KNO_3, KNO_3–$NaNO_3$, and $NaNO_3$ [86, 87] at concentrations of cation acids changing between 0.03 and 0.15 (molar fractions). The said cations were introduced into the melts as additions of $Co(NO_3)_2 \cdot 6H_2O$ and $Cu(NO_3)_2 \cdot 3H_2O$, respectively.

The melt acidities were determined from the rate of reaction (1.2.6); nitronium cations in the nitrate media were formed by the reaction

$$Me^{2+} + NO_3^- \rightarrow MeO(\downarrow) + NO_2^+, \tag{1.2.29}$$

where Me^{2+} denotes Co^{2+} and Cu^{2+}. The equilibrium constants calculated in the temperature ranges given above were changed from $pK = 2.4$ to 1.5 for Co^{2+} and from $pK = 2.8$ to 1.7 for Cu^{2+}. These examinations showed that the acidities of the cations studied were intermediate between those of the weak acid $Cr_2O_7^{2-}$ ($pK = 9.4$) and a strong acid $S_2O_7^{2-}$ ($pK = 1.22$) [79–85].

There is no doubt that the relatively high acidities of the cations obtained in [86, 87] can be explained not only by their "true" acidity but also by the formation of solid phase of oxide; the precipitation of the reaction product favours the shift of the reaction to the corresponding side and appreciably affects its rate.

As for the ionic melts based on alkali metal nitrates, it should be noted that most communications concern potassium nitrate melt as solvent near its melting point (336 °C). Investigations at a temperature of 350 °C are reported in [67, 88–112]. Most of these works were conducted by Shams al Din and his collaborators El Hosary and Gerges. In Refs. [88–93] they reported the results of the potentiometric determination of constants of the oxide ion addition to some oxides in the highest oxidation degree and to the corresponding anions of Group V of the Periodic Table (phosphates, arsenates and vanadates). The scheme of the potentiometric cell used with the gas platinum–oxygen electrode is presented below:

$$Ag|AgNO_3 \text{ (2 mass\%)} + KNO_3 || KNO_3 + O^{2-} | (Pt(O_2)). \qquad (1.2.30)$$

The mentioned oxides were shown to destroy the nitrate melt with their transformation to salts of the corresponding meta-acids; these processes run according to the equation

$$R_2O_5 + NO_3^- \rightarrow 2RO_3^- + NO_2^+. \qquad (1.2.31)$$

where R denotes P, As or V.

The nitronium cation arising as a product of this interaction decomposes by process (1.2.6) with the evolution of NO_2 and O_2, and the shift of equilibrium location of equation (1.2.31) to the right. The gas evolution proceeds while the e.m.f. values of the cell with the gas platinum–oxygen electrode decrease down to $\sim + 0.45$ V. This value is in good agreement with the data from other work [88–93] where the authors observed nitrate melt decomposition, and may be treated as the e.m.f. corresponding to the upper limit of the nitrate melt acidity. In addition to the above, the results of Refs. [88–93] give a possibility of estimating the intrinsic acid–base product of KNO_3 at 250 °C by a value of the order of $\sim 10^{-16}$ mol^2 kg^{-2}. It should be noted, however, that if we take into account the change of the slope of the E–pO calibration plot in the acidic region, which usually takes place with this kind of oxygen electrodes, then the value of the intrinsic acid–base product of this melt should decrease considerably.

Thus, after the termination of the intensive gas evolution from the acidic melts corresponding to the pseudo-equilibrium state of the system, the investigators performed the potentiometric titration of the products formed in the nitrate melt. Consecutive additions of the base-titrant Na_2O_2 into the formed solution give rise to two sequential acid–base processes, which we shall now discuss. The first stage is the formation of pyro-salts according to the reaction

$$2RO_3^- + O^{2-} \rightleftharpoons R_2O_7^{4-}. \tag{1.2.32}$$

This equilibrium is described by the constant of 3.1×10^{15} for the case of phosphates, and 4.7×10^{13} for acidic arsenates. For metavanadate solutions an estimation of the value of the similar constant failed owing to the levelling of the acidic properties of this Lux acid. The next stage of the titration is the formation of neutral salts as a result of acid–base neutralization of the formed pyro-anions, this stage is described by the following equation:

$$R_2O_7^{4-} + O^{2-} \rightleftharpoons 2RO_4^{3-}. \tag{1.2.33}$$

For these interactions the equilibrium constants are 6.6×10^5 (R = P), 4.3×10^4 (R = As), and 4.3×10^6 (R = V). The results presented show that the oxoacidic properties of the Group V oxides increase in the sequence: $V_2O_5 < P_2O_5 < As_2O_5$.

As for vanadium (V) oxide, whose acidic properties are considerably weaker than those of other oxides studied, the recent mass-spectrometric investigations [113] show that its interaction with KNO_3^- melt takes place in agreement with the "solid–liquid" mechanism (the melting point of vanadium (V) oxide is near $690\,°C$). This reaction results in the formation of polyvanadates of the compositions of $K_2V_6O_{16}$ and $K_3V_5O_{14}$ (at $400\,°C$, the exposure is 2 h). Keeping the mixture at $550\,°C$ during 8 h provides the formation of KVO_3 as the main interaction product. The cases considered allow us to conclude that the interaction of acidic oxides with molten salts based on alkali metal nitrates leads to the formation of polynuclear oxoacids at relatively low experimental temperatures, whereas at temperatures close to the temperature of the "pure" melt decomposition takes place with the formation of the neutral salts. This occurs since, under the conditions mentioned, potassium nitrate is a very strong base and the completeness of running the neutralization reaction is provided by the complete removal (solvolysis) of the acid of the solvent.

Besides these reactions, the papers [88–90] reported investigations of the equilibria which, according to the authors' assumption, take place in the titration of ordinary hydro-derivatives of the mentioned oxoacids formed by $H_2PO_4^-$, HPO_4^{2-}, $H_2AsO_4^-$ and $HAsO_4^{2-}$ ions. These compounds were added to the melts as weights of pure acidic salts. On the basis of the following titration procedure the equilibrium constants of the following acid–base interactions were estimated:

$$2H_2RO_4^- + O^{2-} \rightleftharpoons 2HRO_4^{2-} + H_2O(\uparrow), \tag{1.2.34}$$

$$2HRO_4^{2-} + O^{2-} \rightleftharpoons 2RO_4^{3-} + H_2O(\uparrow). \tag{1.2.35}$$

At the same time it is well known that even pure salts formed by the corresponding anions decompose at temperatures considerably lower than 350 °C. Depending on their composition, this process leads to the formation of meta- and pyro-salts [114], and, therefore, under the conditions when water vapour is absent from the atmosphere over the melt, the possibility of practical realization of equilibria (1.2.34) and (1.2.35) is doubtful. Instead of the assumed reactions (1.2.34) and (1.2.35) in KNO_3 melt, there occurred equilibria (1.2.32) and (1.2.33). The difference of the equilibrium constants obtained is caused by a discrepancy in the stoichiometric coefficients of the assumed reactions and those actually taking place in the ionic melt.

The other wide field of oxoacidity of nitrate melts concerns the equilibria occurring in oxide solutions of Subgroup VIb elements of the Periodic System in their highest degree of oxidation, + 6 [67, 94, 95]. The investigations performed by Shams el Din et al. show that oxides of chromium (VI) and molybdenum (VI) dissolved in the nitrate melt destroy it with the evolution of nitrogen oxides. This process causes destruction of molten KNO_3 for the reasons mentioned above: the melt acidity exceeds its upper value, owing to the thermal instability of NO_2^+. The destruction process terminates when dichromate, $Cr_2O_7^{2-}$, and trimolybdate, $Mo_3O_{10}^{2-}$, are formed in the melt owing to its partial destruction. The e.m.f. values of the cell (equation (1.2.30)) after the end of the melt decomposition is near $+0.45$ B, which agrees with the magnitudes obtained in the solutions of the highest oxides of Group V [88–93]. As for the behaviour of Mo(VI) and W(VI) oxides, Afanasiev and Kerridge investigated their reactions with molten KNO_3, the study being performed by the mass-spectrometric method [113]. The main product of the interaction taking place upon MoO_3 addition to the nitrate melt heated to 400 °C was shown to be trimolybdate ion, $Mo_3O_{10}^{2-}$. The increase of the melt temperature up to 500 °C resulted in the formation of dimolybdate ion existing in the melt as $K_2Mo_2O_7$ and, finally, heating of the melt up to 550 °C leads to the formation of neutral potassium molybdate K_2MoO_4. Addition of tungsten(VI) oxide to molten KNO_3 heated to temperatures within the 400–450 °C range did not cause melt destruction, whereas heating to temperatures near 550 °C resulted in interactions leading to the formation of neutral potassium tungstate, K_2WO_4. These results show that near the KNO_3 decomposition temperature located in the 500–550 °C range, the acid of the solvent (NO_2^+) or its adduct to the constituent particles of the solvent ($NO_2^+ \cdot NO_{3-} = N_2O_5$) becomes extremely unstable. This leads to neutralization of the products possessing acidic properties, by the melt, which becomes more and more basic as the temperature rise. Similarly, an increase of the melt temperature results in a rise of the solubility of the above-mentioned acidic oxides. For the case of tungsten(VI) oxide it should be noted that its low solubility in the melt at 350–400 °C may be considered as the main origin of apparent acidic properties measured instrumentally.

Titration of the products formed by the interactions of chromium(VI) and molybdenum(VI) oxide with KNO_3 runs in one stage and leads to the formation of neutral chromates and molybdates, respectively. The investigators calculated the constants of oxide ion addition to $Cr_2O_7^{2-}$ and $Mo_3O_{10}^{2-}$ according to the following reactions:

$$Cr_2O_7^{2-} + O^{2-} \rightleftharpoons 2CrO_4^{2-}, \quad K = 1.8 \times 10^{12}, \tag{1.2.36}$$

$$Mo_3O_{10}^{2-} + 2O^{2-} \rightleftharpoons 3MoO_4^{2-}, \quad K = 9.9 \times 10^{22}. \tag{1.2.37}$$

Tungsten oxide was found to be a considerably weaker acid, unable to destroy the pure nitrate melt. Therefore, its titration occurs in one stage, which is described by the equation

$$WO_3 + O^{2-} \rightleftharpoons WO_4^{2-}, \quad K = 9.5 \times 10^{13}. \tag{1.2.38}$$

The constants of the following equilibria:

$$CrO_3 + O^{2-} \rightleftharpoons CrO_4^{2-}, \quad K = 4.24 \times 10^{18}, \tag{1.2.39}$$

$$MoO_3 + O^{2-} \rightleftharpoons MoO_4^{2-}, \quad K = 4.62 \times 10^{15}, \tag{1.2.40}$$

were estimated in an indirect way. This allows us to conclude that the oxoacidic properties of the Group Vb oxides increase in the sequence $WO_3 < MoO_3 < CrO_3$.

The magnitudes of the equilibrium constants reported in the papers [67, 88–103] cannot be considered as correct, since there is a shift of the calculated magnitudes with the increase of the initial concentration of the base-titrant, also, these values change within several orders of the calculated constant magnitude. This shift is observed because of an incorrect choice of the substances participating in the chemical equilibrium, or owing to the change in slope of the E–pO calibration plot, which usually takes place for the platinum–gas oxygen electrode in acidic solutions. However, since the direct calibration by a strong Lux base weights cannot result in the obtaining of this section of the calibration plot, it often remains unaccounted for in calculations of the constants of the equilibria the strongest Lux acids participate in.

Oxoacidic properties of oxo-compounds of chromium molybdenum in molten equimolar KNO_3–$NaNO_3$ mixture were studied by Kust [115], who used the standard potentiometric cell with a liquid junction. The measurement routine consisted in certain additions of neutral chromate or molybdate to the solution of the corresponding salt of composition $Na_2R_2O_7$. Determination of the equilibrium O^{2-} concentrations in these buffer solutions allowed the authors to estimate the equilibrium constants as

$$K = \frac{[R_2O_7^{2-}]_0[O^{2-}]}{[RO_4^{2-}]^2}. \tag{1.2.41}$$

For R = Mo these constants were 2.5×10^{-8} at 533 K, 8.5×10^{-8} at 583 K and 2.1×10^{-7} at 617 K. For R = Cr they were equal to 1.2×10^{-8} at 544 K, 2.4×10^{-8} at 587 K and 3.3×10^{-8} at 615 K. The temperature dependencies of the equilibrium constants allowed the estimation of the reaction enthalpies (71.4 kJ mol^{-1} for Mo and 41.9 kJ mol^{-1} for Cr) and entropies (-16.8 J mol^{-1} K^{-1} for Mo and -72 J mol^{-1} K^{-1} for Cr). The changes of the reaction entropy in the sequence W–Mo–Cr from $+10$ J mol^{-1} K^{-1} to -72 J mol^{-1} K^{-1} were explained by the type and degree of solvation of the central ion (Cr, Mo, W) in the formed oxo-complexes.

The polymeric anions of molybdenum in the same melt at 603 K were studied by Schlegel and Bauer [116]. Similar to Shams El Din, they found molybdenum oxide to react with the ionic melt with the evolution of NO_2 and O_2, and undissolved MoO_3 to pass to the solution because of interaction of the solid oxide with the formed MoO_4^{2-} in the manner

$$MoO_3 + MoO_4^{2-} \rightleftharpoons Mo_nO_{3n+1}^{2-}. \tag{1.2.42}$$

The potentiometric curves obtained by addition of known weights of Na_2CO_3 to the formed solution of Mo(VI) oxo-compounds are characterized by the e.m.f. drop at an initial carbonate concentration twice as low as the initial concentration of solid MoO_3, whereas addition of Na_2O_2 leads to the interactions with an e.m.f. drop at the "acid–base" ratio greater than 0.5 but lower than 0.66. Such behaviour of the acidic solutions was explained by the formation of dimolybdates in them. The deviations taking place upon Na_2O_2 addition were explained by its slight solubility in the nitrate melt. Naturally, it is completely dissolved in concentrated solutions of strong acids owing to the neutralization reaction, but in diluted solutions of the titrated acid its solubility decreases.

Relative acidic properties of a number of metal cations in molten KNO_3 at 350 °C were examined by the potentiometric titration of the corresponding carbonates with weights of Lux acids $K_2Cr_2O_7$ [96] and $NaPO_3$ [97]. The assumed acid–base reaction was as follows:

$$Me_{n/2}CO_3(\downarrow) + A \rightleftharpoons \frac{n}{2}Me^{n+} + CO_2 \uparrow +B. \tag{1.2.43}$$

These examinations resulted in the sequence of the cations K^+, $Na^+ < Li^+$, $Ba^{2+} < Sr^{2+} < Ca^{2+} < Pb^{2+}$ with increasing acidity. However, although the sequence agrees with the metal electronegativity scale, the methods of estimation of cation acidity in the nitrate melt cannot be considered as successful, for the following reasons. First, the cations studied were introduced into the melt in the form of carbonate salts, and these carbonates possess different solubilities in molten potassium nitrate. For example, one can expect that alkali metal carbonates will be soluble, lead carbonate will decompose, and the oxide formed will precipitate. As for alkaline-earth metal carbonates, they will precipitate without decomposition. The products of the acid–base titration are neutral phosphates and chromates, and these salts formed by cations of polyvalent metals are shown by Schlegel [83, 84] to be slightly soluble in the nitrate melt. The position of Ba^{2+} cation in this sequence may be evidence of such distortions: its acidic properties are considerably weaker than those of Li^+, but, nevertheless, owing to insolubility of $BaCO_3$ in molten KNO_3, the titration process is impeded. In these studies this was ascribed not to the slight solubility but to acidity—the latter being incorrect. It is obvious that the positions of other studied alkaline earth metal cations are distorted.

The relative basic properties of various Lux bases in molten KNO_3 at 350 °C have been examined by the potentiometric method [98, 99]. According to these studies, all the bases were divided into two groups. The first group called the "group of oxide ion" comprises such strong bases as OH^-, O^{2-} and O_2^{2-}. Their dissociation with the formation of oxide ion in melts is practically complete. The second group called the "carbonate ion group" includes weak bases having the composition CO_3^{2-}, HCO_3^-, $(COO)_2^{2-}$, CH_3COO^- and $HCOO^-$. Their decomposition at the experimental temperature (350 °C) results in the formation of carbonate ions in the solutions. However, introducing some of these bases into the melt does not result in the creation of the equivalent concentration of the base. To illustrate this effect, we consider oxalate ion decomposition in molten KNO_3. This reaction is known to occur as follows:

$$(COO)_2^{2-} \rightarrow CO_3^{2-} + CO \uparrow. \tag{1.2.44}$$

However, the interaction does not end after this stage. Carbon monoxide produced as a product of the decomposition reacts with nitrate ions, which lead to the formation of nitrite ions possessing appreciable basic properties in molten salts:

$$CO \uparrow + NO_3^- \rightarrow NO_2^- \rightarrow CO_2. \tag{1.2.45}$$

Therefore, the addition of organic salts leads to their decomposition into carbonate ions plus the formation of uncontrolled concentrations of nitrite ions owing to the melt reduction. No direct correlation is observed between the weight of such a base and the concentration of oxide ion created, owing to the base dissociation in the melt. This fact makes most of the studied bases unavailable for investigations of oxoacidity in molten salts.

One of the most interesting problems of the chemistry of oxocompounds in nitrate melts concerns the form of the existence of oxide ions in these melts. Shams El Din and El Hosary [100] assumed that oxide ions in solution are stabilized by two nitrate ions that result in the formation of the so-called pyronitrate ion:

$$O^{2-} + 2NO_3^- \rightarrow N_2O_7^{4-}, \tag{1.2.46}$$

which is a weak base, and whose oxobasicity is comparable with that of carbonate ions.

However, the potentiometric measurements performed in the solutions of strong bases (Na_2O_2, NaOH) demonstrate that they are more basic than the solutions of alkali metal carbonates with the same concentration. In the case of the formation of pyronitrate, one cannot observe this difference owing to the levelling of basic properties of oxide-ion donors in molten alkali metal nitrates. Indeed, Shams El Din and El Hosary [96] show that $N_2O_7^{4-}$ ions are not formed, even in strongly basic solutions in nitrate melts. Oxide ions in molten nitrates are subject to oxidation followed by the formation of peroxide or superoxide ions. This especially concerns the nitrates formed by cations of weak acidity (K^+, Na^+). The addition of stronger cation acids stabilizes oxide ions, owing to their fixation into oxide complexes. The same is observed for the measurements made in Pyrex containers, where stabilization of oxide ions takes place owing to the formation of different silicate derivatives.

Refs. [102, 103] contain some information about the behaviour of gas and metal-oxide oxygen electrode in nitrate melts. The "anomalous" reversibility of the gas platinum–oxygen electrode $Pt(O_2)$, usually observed for basic solutions in molten nitrates, is attributed to a potential-defining process of discharging of peroxide ions at the inter-phase boundary "platinum/melt/oxygen". This is the so-called peroxide function of gas oxygen electrodes. As for metal-oxide electrodes, which belong to the electrodes of the second kind, they maintain their reversibility to oxide ions over a wide range of oxide ion concentrations. The slope of E–pO plots for these electrodes agrees with the generally accepted $1.15RT/F$.

Coumert and his co-workers studied acid–base equilibria in solutions of various phosphates and molybdates in molten nitrate media [104, 105]. These anion acids were found to form heteropolyacids in such solutions. The formation of the mentioned polynuclear complexes was confirmed by cryoscopic examinations reported in Ref. [106].

The authors of Refs. [107, 108] have described peculiarities of acid–base potentiometric titration in ionic melts. The former work [107] is devoted to introducing "automatic" potentiometric titration in molten nitrates. This routine consists of the following. A rod made of a solidified solution of

$K_2Cr_2O_7$ in KNO_3 is put down at a constant rate into a melt containing a Lux base. The e.m.f. values of the potentiometric cell are registered and printed using a recorder. This method may be employed to analyse ionic melts' basicity in industrial molten systems since it allows one to estimate the concentration of bases in the melts studied, to an accuracy of the order of 3.5%. However, the thermodynamic parameters cannot be estimated on the basis of the e.m.f.–time plots obtained, since the e.m.f. recorded is not referred to the equilibrium state.

Hoshino et al. investigated the effect of acidic oxides on the thermal stability of molten sodium nitrate [109]. They found that an increase of oxoacidic properties of the corresponding oxide resulted in a decrease in the decomposition temperature of the solution formed in the nitrate melt. This fact is explained by a decrease in the activation energy of the decomposition reaction (1.2.6) because of fixation of oxide ions formed as a result of the autodissociation process (equation (1.44)) by the Lux acids. Nevertheless, the phenomenon observed may be explained without kinetic speculations. The simplest explanation is as follows. The decomposition reaction is shifted towards the formation of nitrogen oxides and oxide ions according to Le Chatelier's rule: the removal of a participant of a chemical reaction favours the shift of equilibrium of this reaction to the side providing the formation of this reagent. The oxoacidic properties of the oxides studied were found to decrease in the sequence $SiO_2 > TiO_2 > ZrO_2 > Al_2O_3 > MgO$.

The interaction of PbO with molten nitrates was studied by Poppel and Khoroshevskaya [117]. The oxide is known to react with nitrate ions, forming an oxynitrate of composition $5PbO\cdot Pb(NO_3)_2$ and oxide ions, represented by the following equation:

$$6PbO \downarrow +2NO_3^- \rightleftharpoons 5PbO\cdot Pb(NO_3)_2 \downarrow +O^{2-}. \qquad (1.2.47)$$

If this reaction actually takes place then it should be sensitive to the cation composition of the nitrate melt. The cause of this lies in the different acidic properties of the melt cations, and the degree of their interaction with oxide ions formed as a product of interaction (equation (1.2.47)) defines the completeness of this reaction, according to Le Chatelier's rule. So, the extent of running reaction (1.2.47) is expected to reduce with the increase of alkali-metal cation radius, in the sequence Li–Na–K–Rb–Cs. Indeed, the most acidic $LiNO_3$ reacts with PbO starting from 300 °C, whereas the interaction of PbO with $NaNO_3$ is initiated only at 400 °C. This temperature is a threshold one, since above it PbO undergoes oxidation by the nitrate ion according to the reaction

$$3PbO \downarrow +2NO_3^- \rightleftharpoons Pb_3O_4 \downarrow +NO_2^-. \qquad (1.2.48)$$

An increase of the oxoacidic properties of the constituent cation of the nitrate melts favours a reduction of the nitrite ion stability that causes a shift of reaction (1.2.48) to the right. This reaction was found to proceed completely in molten lithium nitrate, whereas in molten KNO_3 and $NaNO_3$ the transformation is incomplete. This can be explained by the higher degree of interaction of Li^+ with nitrite ions as compared with other alkali-metal cations: NO_2^- is known to have appreciable basic properties in molten salts. It should be mentioned that even at 600 °C the concentration of alkali-metal oxide in the latter molten nitrates does not achieve the level needed for formation of plumbates, but in the case of $LiNO_3$ the formation of α-Li_2PbO_3 as brown–yellow particles is observed. Baraka *et al.* performed investigations of the reversibility of metal-oxide electrodes of the second kind (metal-insoluble acidic the metal oxide) in molten nitrates. The oxygen electrodes ($Nb|Nb_2O_5$, $Ta|Ta_2O_5$, $Zr|ZrO_2$) [110, 112] were found to be reversible to oxide ions with the slope of E–pO plot of $1.15RT/F$. On the basis of the e.m.f. measurements fulfilled in the solutions of different acids and bases, the constructed was the following acidity row in the nitrate melt at 350 °C: $NH_4VO_3 > NaPO_3 > NaH_2PO_4 > K_2Cr_2O_7 > K_2HPO_4 > Na_4P_2O_7 > Na_2HAsO_4 > K_2CO_3 > Na_2O_2$, where the oxoacidic properties decrease.

Similar to Shams El Din, Baraka *et al.* [110, 112] mention a number of acids and bases whose existence is doubtful in the melt under the experimental conditions (see equations (1.2.34) and (1.2.35)). This means that in this work the investigations of Shams El Din *et al.* have been used as a model without an analysis of the correctness of some of their assumptions from the modern point of view. It is also obvious that ammonium metavanadate should decompose with the formation of V_2O_5, which should be reduced with ammonia to VO_2, etc. and the acidic oxide formed should destroy the nitrate melt with the formation of metavanadate. As for acidic salts, they decompose to the corresponding pyro- and meta-salts yielding water. Therefore, the mentioned row of acidity cannot be accepted "as it is".

Attia reported a potentiometric investigation of the stoichiometry of acid–base reactions of phosphorus acids with oxide ions in molten KNO_3 at 350 °C using a $Nb|Nb_2O_5$ oxygen electrode [118]. There are no data which can be considered as new in comparison with the studies performed by Shams El Din *et al.* and Baraka *et al.*

In concluding our consideration of oxoacidity studies in molten nitrates it should be mentioned that, owing to their relatively low melting points and the simplicity of experimental design, they are most available media for the investigation of various acid–base processes. This simplicity was attractive for a number of investigators, and has led to numerous publications. Nevertheless, there are many errors and incorrect treatment of the experimental results that reduces the significance of some work.

The most widespread mistake is the assumption that thermally unstable substances exist in an unchanged state at elevated temperatures in nitrate melts, and processes such as dehydration and decarboxylation are not taken into account. This omission is partially reflected in the conclusion that there is no correlation between the pH of aqueous solutions of some salts (usually, phosphates) and the pO values of the corresponding solutions of these salts in molten KNO_3. The reason for this is as follows.

The neutralization of phosphoric acid in aqueous solutions occurs as follows:

$$H_3PO_4 \rightarrow H_2PO_4^- \rightarrow HPO_4^{2-} \rightarrow PO_4^{3-}, \tag{1.2.49}$$

whereas the corresponding steps in the titration in molten salt can be represented by the sequence

$$P_2O_5 \rightarrow PO_3^- \rightarrow P_2O_7^{4-} \rightarrow PO_4^{3-}. \tag{1.2.50}$$

Therefore, there should be correlations between the acidities of the aqueous solutions of an anion of the corresponding anhydroacid in molten salt, e.g. NaH_2PO_4 and $NaPO_3$, or Na_2HPO_4 and $Na_4P_2O_7$. The anhydroacids $NaPO_3$ and $Na_4P_2O_7$ are easily obtained by heating the corresponding acidic salts, whereas the reverse processes in aqueous solutions cannot be realized, so the phosphates mentioned exist in water as salts of the separate acids HPO_3 and $H_4P_2O_7$. In the case of meta- and pyro-arsenates, which are more disposed to hydrolysis, a correlation of the acidities of aqueous solutions and solutions in the ionic melt should occur. Hence, this discrepancy is only apparent, since it is caused only by kinetic limitations (the rate of the polyanion salt hydrolysis).

The analysis of the literature data, taking into account the generalized solvosystem concept introduced in Part 1, gives a possibility of revealing the levelling of acidic properties of the strongest acids in molten nitrates—in particular, KNO_3. This effect has not been observed before. The NO_3^- ion as a constituent part of this melt is capable of dissociation with the formation of O^{2-}, by the process in equation (1.44), which is the intrinsic acid–base equilibrium of the solvent [45]. Therefore, in molten KNO_3 there exists a possibility for the levelling of both acidic (equation (1.2.5)) and basic properties, similar to that described in [119]:

$$O^{2-} + 2K^+ \rightleftharpoons K_2O. \qquad (1.2.51)$$

This signifies the impossibility of the existence in the solutions formed in molten nitrates, of acids stronger than NO_2^+ ions and bases with more strong basic properties than those equivalent to the concentration solutions of K_2O. In agreement with equation (1.2.5) the acid–base titration curves for the strongest acids should coincide if the initial concentrations of the acids are close. This allows us to give another interpretation of the results obtained by Shams El Din *et al.* [97–99] for the acid–base titration performed in molten KNO_3. To do this, let us consider Fig. 1.2.4. The potentiometric curves presented here are practically coincident (it should be emphasized, however, that the initial concentrations of acids are somewhat different). This gives evidence that the levelling of oxoacidic properties in the nitrate melt takes place in all the titrations whose results are shown in Fig. 1.2.4. So, in all the

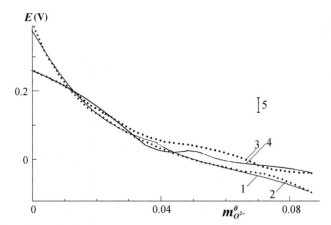

Fig. 1.2.4. Potentiometric titration curves of solutions of the oxoacids CrO_3 (1) [97–99], MoO_3 (2) [97–99], WO_3 (3) [97–99], PO_3^- (4) [96] in molten KNO_3 at 350 °C: all the concentrations are close to 0.1 mol kg^{-1} (5, the scatter of e.m.f.).

cases Shams El Din *et al.* [97], El Hosary *et al.* [98] and Shams El Din and El Hosary [99] have investigated the same acid–base reactions attributing to their different stoichiometric coefficients. This was the reason for the directed shift of the calculated values of the equilibrium constants:

$$NO_2^+ + O^{2-} \rightleftharpoons NO_3^-. \tag{1.2.52}$$

The described levelling of the acidic properties in molten nitrates makes them unavailable for the determination of the constants of Lux acid–base equilibria where the strongest acids take part. The melts based on molten alkali metal halides are the most convenient solvents for this purpose. Since the most essential physico-chemical properties of ionic melts (density, charges and ionic radii) are close,[1] application of the data obtained in molten chlorides for the description of acid–base equilibria in molten nitrates is more correct than is the use of a similar approach to room-temperature molecular solvents, since the properties determining their acid–base properties are numerous (e.g. dielectric constants, donor–acceptor properties).

1.2.4 MOLTEN ALKALI METAL SULFATES

Molten alkali metal sulfates are typical ionic liquids, which means that their dissociation to the constituent ions is practically complete. For example, Na_2SO_4 breaks down as shown:

$$Na_2SO_4 = 2Na^+ + SO_4^{2-}. \tag{1.2.53}$$

and the value of the product $[Na^+]^2[SO_4^{2-}]$ is too large. Nevertheless, the intrinsic acid–base equilibrium (equation (1.2.2)) is shifted to the left. Since oxide ion is formed as one of the solvent autodissociation products, molten sulfates are classified among solvents of the first kind.

The studies of sulfate systems are not extensive: there are only a few papers in the literature. This is explained by the relatively high melting points of single sulfates (859 °C for Li_2SO_4, 884 °C for Na_2SO_4, 1074 °C for K_2SO_4) and their low-melting mixtures. Thus, the weakly acidic K_2SO_4–Na_2SO_4 system is characterized by a series of solid solutions with a minimum melting point at 830–836 °C and 22.5–30 mol% of K_2SO_4. A system such as

---

[1] The thermochemical radii of NO_3^- and Cl^- ions are 0.165 and 0.167 nm, respectively.

$Cs_2SO_4-K_2SO_4$ possesses a minimum melting point at 918 °C and 56 mol% of Cs_2SO_4. The binary system $Cs_2SO_4-Na_2SO_4$ melts at 615 °C [120] and 30 mol% Cs_2SO_4. All the temperatures mentioned are above the decomposition temperature of alkali metal pyrosulfates, which are close to 450 °C [114, 121]. This causes acidic solutions in sulfate melts prepared by the addition of a strong Lux acid to the melt to be extremely unstable. Indeed, the autodissociation reaction (1.2.2) should result in the formation of SO_3 as the acid of the melt; under the temperatures mentioned, this substance exists as a gas. Sulfate melts keep SO_3 in the dissolved state, owing to formation of polysulfates in acidic solutions: pyrosulfate ion can be mentioned as an example of such substances, $S_2O_7^{2-} \equiv SO_3 \cdot SO_4^{2-}$. The decomposition temperature of pyrosulfates dissolved in ionic media should be lower than those of the pure compounds. If the sulfate melts used for investigations of oxoacidity are kept in a flow of an inert gas (argon or nitrogen) then the process of evaporation of SO_3 from these melts

$$SO_{3,melt} \leftrightarrow SO_{3,gas}, \tag{1.2.54}$$

according to Le Chatelier's rule should be shifted completely to the right, since in the flow of inert gases $p_{SO_3} = 0$. This means that acidic solutions in sulfate melts are pseudo-equilibrium systems and true equilibrium in them is not achieved.

The melting points and the temperature of the melts under investigation may be reduced by the addition of Li_2SO_4 to the mentioned single and binary melts, but after this addition the melts become more hygroscopic and effectively more acidic. The lowest-melting ternary mixtures of alkali metal sulfates are $Cs_2SO_4-K_2SO_4-Li_2SO_4$ with the minimum at ~ 600 °C [122], and $K_2SO_4-Li_2SO_4-Na_2SO_4$ (0.135:0.78:0.085) with the melting point of 512 °C [123]. The acid–base reactions in these melts can be studied at temperatures considerably lower than those used in the classic work of Lux [17], and, consequently, the acidic and basic solution will be more stable. Owing to the relatively high acidic properties of these melts they cannot be used as reference melts for the construction of the generalized oxoacidity scale of molten salts.

The first study of the oxoacidic properties of molten sulfates was performed by Lux [17] on an equimolar $K_2SO_4-Na_2SO_4$ mixture at 950 °C. The measurements of activity of oxide ions in the melt were made by the potentiometric method with the use of an $Au(O_2)$ gas oxygen electrode.

The measurements were performed in the following cell:

$$\text{Rh(O}_2)\left|\begin{array}{c}\text{melt with the stable}\\ \text{O}^{2-}\text{ concentration}\end{array}\right\|\begin{array}{c}\text{melt with the variable}\\ \text{O}^{2-}\text{ concentration}\end{array}\right|\text{Pt(O}_2).$$

(1.2.55)

Additions of Na_2O were equivalent to those of Na_2O_2, and Na_2CO_3 was found to decompose completely with the formation of Na_2O under the experimental conditions. The process of decomposition is efficiently retarded in concentrated solutions, since CO_2 is hardly removed from strongly basic media. The basic solutions were characterized by practically stable potentials, and only slightly shifted towards the neutral point of the ionic solvent studied. In contrast, acidic solutions obtained by the addition of $NaPO_3$ to the pure sulfate melt are extremely unstable; dissolution of the acid is accompanied with fast evolution of SO_3, and in a short time the e.m.f. value achieves that of the neutral point. The equilibrium constant of the following reaction

$$PO_3^- + O^{2-} = PO_4^{3-} \tag{1.2.56}$$

at 950 °C was estimated to be 2×10^5.

The papers of Flood et al. [18–20] are devoted to the investigations of acid–base reactions in molten pyrosulfate media. On the basis of the data obtained in Ref. [19], an obvious conclusion is that the strengthening of the constituent cation's acidity results in a reduction in the decomposition temperature of molten pyrosulfates.

Kaneko and Kojima studied the oxoacidic properties of $Na_2S_2O_7$ and Na_2O_2 in the molten eutectic mixture K_2SO_4–Li_2SO_4–Na_2SO_4 at 550 °C [124, 125]. The difference of pO values in their 0.01 mol kg^{-1} solutions is equal to 10 pO units. This value allows us to determine that the width of the acid–base range for the standard solutions in this melt is equal to 14 pO units. The acid–base range of stable existence of VO^{2+} ion is found to lie within the interval of pO = 8.5–10.6.

Rahmel investigated the acidic properties of some Lux acids (MoO_3, $K_2Cr_2O_7$, $NaPO_3$, $Na_4P_2O_7$ and V_2O_5) in molten K_2SO_4–Li_2SO_4–Na_2SO_4 at 625 °C [126]. The examination was performed by potentiometric titration of the acids with Na_2CO_3 to detect the equivalence points. The e.m.f. drops have been observed at an "acid/base" ratio of 1:1, the titration curve of MoO_3 includes an e.m.f. drop of 200–250 mV, and the titration of $Cr_2O_7^{2-}$ ion is characterized by a drop of 450 mV. As for pyrophosphate ion, its titration is described by an e.m.f. drop at the equivalence point of the order of 350 mV. The titration of vanadium(V) oxide leads to very blurred drops at the assumed equivalence points, and complete neutralization takes place at a ratio of 1:3. The formation of basic vanadates is not observed, owing to the insufficient basicity of carbonate ion as a Lux base. The determination of the stoichiometry of the interactions is the main result of the study in Ref. [126], but there is no information on the corresponding equilibrium constants.

1.2.5 SILICATE MELTS

Investigations of molten ionic silicates are mainly performed for a variety of applied purposes, since these systems are widely used in branches of industry such as slags and glasses.

The silicate slag systems are involved in most pyro-metallurgical processes which result in the formation of a number of ferrous and non-ferrous metals (Cu, Ag, Zn, Cd Sn, Pb, Sb, Bi). These systems are oxide- and sulfide–oxide melts, and, therefore, the processes of their interaction with raw materials are dependent on the melt properties (oxidation ability, solubility of metals), and on their oxoacidic properties. Any metallurgical slag contains MgO, CaO, FeO and SiO_2 as one of main components. The oxidation ability of the slags increases with a rise of their basicity (increase of equilibrium O^{2-} concentration), caused by the following electrochemical process:

$$Me + O_{slag}^{2-} - 2\bar{e} = MeO, \tag{1.2.57}$$

and the oxide formed may either remain in the slag or dissolve in the metal. Therefore, an increase in the concentration of acidic oxides (SiO_2) reduces the oxidation action of a slag upon the commercial metal obtained. As for sulfur, its dissolution in molten metals decreases with the basicity of the slag, but this increases the oxide ion concentration in the obtained metal. The simplest explanation of this fact consists in the following interfacial equilibrium:

$$MeS_{metal} + O_{slag}^{2-} = MeO_{metal} + S_{slag}^{2-}, \tag{1.2.58}$$

which shows that the change of the slag basicity affects the distribution of the simple halogenide anions between liquid ionic and metallic phases. Therefore, the development of pyrometallurgical processes requires variation of the slag's basicity, to obtain the optimal technology of metal production.

The formal basicity of a slag is expressed as the difference in the equivalent concentrations of basic and acidic oxides in it, e.g.:

$$R = N_{CaO} + N_{MgO} + N_{MnO} - 2N_{SiO_2} - 4N_{P_2O_5}, \tag{1.2.59}$$

which means that ortho-silicates (SiO_4^{4-}) and pyrophosphates $(P_2O_7^{4-})$ are considered as neutral forms of Si and P oxo-compounds in melts of such a kind, i.e. one molecule of SiO_2 can fix two oxide ions, and that of P_2O_5 reacts with four O^{2-} ions. A detailed description of slag chemistry can be found in the monograph [127].

Among the investigations concerned with the oxoacidic properties of the melts used for glass production we consider the works of Flood *et al.* who studied the influence of the oxoacidic properties of silicate melts of the PbO–SiO_2 system, where $(N_{PbO} = 0.4–1)$, on their phase and chemical composition at 1100–1200 °C [128–130]. These studies were performed in a potentiometric cell with the use of gas oxygen electrodes $Ag(O_2)$ in the concentration cells. It was found that in the system studied the anions SiO_4^{4-} co-existed with those of the composition $(SiO_3)_3^{6-}$ and $(SiO_{2.5})_6^{6-}$ [129]. A reduction of the Na_2O/SiO_2 concentration ratio in the molten glasses was shown to result in an appreciable oxobasicity decrease, whereas magnesium oxide had no appreciable effect on the acid–base properties of the studied melts [130].

Minenko *et al.* [131] reported an investigation of the reversibility of the platinum–gas oxygen electrode $Pt(O_2)$ in molten silicates of the compositions PbO–SiO_2, Na_2O–CaO–SiO_2, MeO–PbO–SiO_2, with Me being an alkaline-earth metal cation. The studied electrode was well reversible to oxide ions in the silicate melts and its use for the oxobasicity measurements was, therefore, substantiated.

The use of a membrane oxygen electrode for monitoring the oxoacidic properties of molten silicates is described by Perkins [132]. Itoh and Yokokawa [133] examined the basicity of silicate melts of the Na_2O–Al_2O_3–SiO_2 system by the e.m.f. method in cells with the liquid junction, the oxoacidic properties of Al_2O_3 were shown to be weaker than those of SiO_2.

Bobrova *et al.* [134] performed investigations of the corrosion processes for aluminosilicate glasses of the composition $Na_2O \cdot xAl_2O_3 \cdot 2SiO_2$, where $x = 0–0.4$ in molten alkaline-earth metal nitrates. The degree of the interaction of the nitrate melts with the glasses increases with the radius of the alkali metal cation, which is accompanied by the strengthening of its acidity

$$Na_2O_{glass} + Me^{2+} \rightarrow MeO + Na^+. \tag{1.2.60}$$

Didtchenko and Rochow investigated the effect of the charge and radius of some metal cations [135] on the basic properties of molten lead meta-silicate, $PbSiO_3$, at 800, 850 and 900 °C. The said cations were added to the melts in the form of the oxides; Tl_2O, PbO, CdO, ZnO, Bi_2O_3. The melt basicity was stated to increase with the increase of the cation radius for doubly charged cations.

1.2.6 KCl–NaCl EQUIMOLAR MIXTURE

This ionic chloride melt, with a minimum melting temperature of 658 °C [136] serves as a solvent of the second kind for the Lux acid–base equilibria. Its effect on these processes is determined, first, by the acid–base interactions of the constituent cation of the melt (Na^+, K^+) with different Lux bases. Since these cations possess relatively weak acidic properties, this melt is suitable for the determination of the constants for the addition reactions of oxide ion, both to strong and weak Lux acids. It can also be used for the construction of the general acidity scale as a standard (reference) ionic solvent. Preparation of this mixture is extremely simple; it can be obtained immediately in the crucible by mixing and melting of non-hygroscopic KCl and NaCl. The melt is not one of the "capricious" ones; it is hardly polluted by admixtures and holds no water traces. Owing to these features the equimolar KCl–NaCl mixture became the most investigated molten chloride medium in relation to oxoacidity. For the stated reasons there are many reports devoted to the studies of the reactions of oxoanions and acidic oxides with Lux bases in this melt, and these interactions are not complicated by the formation of insoluble oxide phases.

1.2.6.1 Oxocompounds of chromium(VI)

A priori, chromium(VI) oxide, CrO_3, should be the strongest acid of all the Cr oxo-compounds. However, it is unstable at temperatures of the order of 1000 K. Its melting point is near 196 °C, and decomposition begins after a small temperature increase. The splitting out of oxygen runs in several stages:

$$CrO_3 \rightarrow Cr_3O_8 \rightarrow Cr_2O_5 \rightarrow CrO_2 \rightarrow Cr_2O_3. \qquad (1.2.61)$$

The decomposition process terminates in the formation of Cr_2O_3 at temperatures of the order of 500 °C [137]. Therefore, it cannot exist at temperatures close to the minimal melting point of the KCl–NaCl system (658 °C). On being introduced into the chloride melt, CrO_3 oxidizes it

$$CrO_3 + 2Cl^- \rightarrow CrO_2 + O^{2-} + Cl_2. \tag{1.2.62}$$

Simultaneously, there takes place the fixation of the formed oxide ions by the excess of CrO_3, resulting in the appearance of dichromate ions in the melt. This process leads to a decrease of the melt's acidity and stabilization of Cr(VI) oxo-compounds. Partial neutralization makes some difficulties for investigations of its oxoacidic properties, because of uncontrolled changes of the initial acid concentration. Therefore, potassium dichromate is considered to be a more convenient Lux acid of chromium, although there is no explicit possibility of determining directly the equilibrium constant of the following reaction:

$$2CrO_3 + O^{2-} \rightleftharpoons Cr_2O_7^{2-}. \tag{1.2.63}$$

The $K_2Cr_2O_7$ is known to melt at 397.5 °C [114], and hence the chloride melt–$K_2Cr_2O_7$ system may be considered as a binary liquid one. Subsequent heating of molten potassium dichromate causes its decomposition with the formation of Cr compounds in lower oxidation states; according to the reference data [138], this begins at ~600 °C. Nevertheless, diluted solutions of $K_2Cr_2O_7$ are stable enough even at 700 °C.

Shapoval *et al.* investigated the oxoacidic properties of Cr(VI) oxocompounds [139–141] and used CrO_3 as an initial oxoacid. To avoid the above-mentioned decomposition, chromic anhydride was added to the melt having dissolved the Lux base. However, in this case only partial neutralization took place: a considerable quantity of CrO_3 existed in the melt at high pO values, and obviously underwent decomposition according to reaction (1.2.62). Assuming that the neutralization is realized in two stages, the first of which is described by equations (1.2.63) and (1.2.36), respectively, the authors established the corresponding respective equilibrium constants (in molar fractions) as $K = 2.88 \times 10^3$ and 1.4×10^{-1}. This led to the conclusion that, in contrast with the molten nitrates, CrO_3 was a weak acid in the molten KCl–NaCl eutectic, and the $Cr_2O_7^{2-}$ ion showed appreciable basic properties. Some years ago it was stated that the reason for such a conclusion lay in incorrect estimations of the pO of the acidic solutions. The pO values used for determination of the equilibrium constant were calculated from the calibration E–pO plots obtained in the base excess region, whereas the parameters of such a plot in the acidic region were considered in the above said papers to remain unchanged. However, the slope of $E-pO$ plots for the platinum–oxygen electrode in the acidic solutions is characterized by lower values than in the basic ones, and, therefore, estimations of pO using only the calibration data in the basic region leads to an underestimated pO and equilibrium oxide-ion concentrations. Similarly, the use of $Pt(O_2)$ as an indicator oxygen electrode requires saturation of the studied melt by gaseous oxygen, which is performed in a manner similar to that in the use of the platinum–hydrogen electrode in room-temperature protolytic solvents. However, passing gaseous oxygen through halide melts causes their oxidation, followed by the formation of oxide species and evolution of free halogen. In the ionic halides containing Lux acids the oxidation process becomes especially favourable from the thermodynamic point of view, since oxide ions formed as a result of the process of oxidation are combined with the acid. Actually, at sufficiently high values of pO of acidic solutions of Cr(VI) oxo-compounds, there takes place the process of titration of the acids with gaseous oxygen. Schematically, this can be represented by the equation:

$$Cr_2O_7^{2-} + \tfrac{1}{2}O_2 \uparrow +2Cl^- \rightleftharpoons 2CrO_4^{2-} + Cl_2 \uparrow . \tag{1.2.64}$$

For these reasons, the results of the investigations performed without passing oxygen through the halide melts, are not distorted by the interference of any side processes, and the constants obtained are more correct. Such investigations of properties of $K_2Cr_2O_7$ in molten KCl–NaCl at 700 °C were reported in Refs. [142, 143], and the titration of the said acid was performed with quantities of NaOH, Na_2O_2 (strong Lux bases) and Na_2CO_3 (a weak Lux base). The potentiometric cell with the membrane oxygen electrode $Pt(O2)|YSZ$ was as follows:

$$Ag|Ag^+(0.02 \text{ mol kg}^{-1}), KCl-NaCl\|KCl-NaCl, O^{2-} |YSZ|Pt(O_2). \tag{1.2.65}$$

The initial weight of $K_2Cr_2O_7$ introduced into the melt increases the values of pO up to 8; then subsequent addition of a base results in a decrease in pO, and there is a pO drop of the size of four pO units in all the potentiometric curves. These drops are observed at a $[Cr_2O_7^{2-}]_0/[O^{2-}]_0$ ratio of 1:1 and the ligand number value equal to 1 (Fig. 1.2.5a). This may serve as an argument for the realization of both equilibrium (1.2.36) and

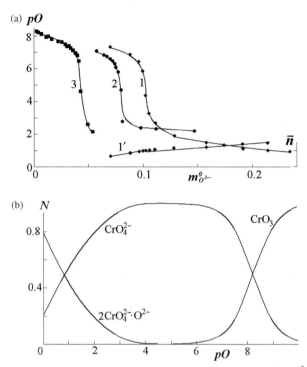

Fig. 1.2.5. Results of potentiometric investigations of acidic properties of $Cr_2O_7^{2-}$ in the molten KCl–NaCl equimolar mixture at 700 °C: (a) potentiometric curves 1, 0.101 mol kg^{-1} of $Cr_2O_7^{2-} + O^{2-}$ (1′, the dependence of the ligand number $\bar{n}$ against the initial concentration of the Lux base); 2, 0.080 mol kg^{-1} $Cr_2O_7^{2-} + CO_3^{2-}$; 3, 0.043 mol kg^{-1} $Cr_2O_7^{2-} + O_2^{2-}$; (b) fractions of oxocompounds of Cr^{VI} in diluted solutions (<0.05 mol kg^{-1}) at different pO.

reaction (1.2.39) in the chloride melt. The $Cr_2O_7^{2-}/CrO_4^{2-}$ buffer solutions formed at a ligand-number value of 0.5–0.7 (the half-titration section) are characterized by the buffer number $\beta \approx -0.05$.

The solutions of potassium dichromate examined in Ref. [142] are sufficiently concentrated (of about 0.1 mol kg^{-1}). Treatment of the titration results shows that $Cr_2O_7^{2-}$ is the main acidic form of oxocompounds of Cr(VI) in such solutions, and the acid–base process taking place in the titration procedure agrees with equation (1.2.36). Dilution of these solutions to concentrations of the order of 0.04 mol kg^{-1} of $Cr_2O_7^{2-}$ and lower favours the formation of CrO_3 as the main acidic form of Cr(VI) oxoacids [143]. This gives evidence of dichromate dissociation—it is broken down into the equimolar mixture of CrO_3 and CrO_4^{2-}

$$Cr_2O_7^{2-} \rightleftharpoons CrO_3 + CrO_4^{2-}. \tag{1.2.66}$$

This result is not so unexpected, since aqueous solutions of dichromate salts or dichromate chromic acid at room temperature are known to be subjected to hydrolysis [144] with the formation of mononuclear acidic salts of Cr(VI), which cannot be obtained in the pure form. Both processes occur on dilution and increase of pH, owing to fixation of the acid formed.

$$H_2Cr_2O_7 + H_2O \rightleftharpoons 2H_2CrO_4, \tag{1.2.67}$$

$$Cr_2O_7^{2-} + H_2O \rightleftharpoons 2HCrO_4^-. \tag{1.2.68}$$

The addition of strong Lux bases into the melt studied after the equivalence point results in the following interaction of the chromate formed with the excess of O^{2-}. This is confirmed by an increase of the ligand number $\bar{n}$ by 1.5 times as compared with its value at the equivalence point, although this interaction causes no additional pO drops in the potentiometric curves.

Since the result above is in good qualitative agreement with the behaviour of neutral chromate ions in basic solutions in nitrate melts [67], it was assumed that basic chromates of composition of $2CrO_4^{2-} \cdot O^{2-}$ were formed as a result of the said interaction in the chloride melt, the corresponding pK value was estimated as 1.60 ± 0.20 (on the molality scale). Taking into account the fact that in strongly basic solutions the errors of potentiometric measurements, and consequently, the determined constants increase considerably, the obtained potentiometric data give no exhaustive evidence of formation just of basic chromate of composition $2\,CrO_4^{2-} \cdot O^{2-}$. The calculated value of pK only shows that CrO_4^{2-} in ionic melts possesses appreciable acidic properties owing to coordinational non-saturation of Cr(VI) with respect to oxide ions. While using Na_2CO_3 as a base-titrant there is no formation of basic chromate and this means that the said basic chromates are stronger Lux bases than carbonate ions in inert atmospheres. We should add that neutral chromate demonstrates an obvious ability to form basic products in alkaline aqueous solutions: according to the data of [144] the composition of this basic compound crystallized from the aqueous solution is $Na_4CrO_5 \cdot 13H_2O$.

The dependence of the fraction of different acids of Cr(VI) against pO is presented in Fig. 1.2.5b. These data show that not more than two forms of Cr(VI) oxocompounds may coexist in appreciable quantities in the melts based on alkali metal chlorides. Therefore, a simple scheme of calculations provides the estimation of the thermodynamic parameters of the said equilibrium to a sufficiently high accuracy. Moreover, as seen from Fig. 1.2.5b, the formation of the basic chromate is initiated at pO<<3 and these values are not achieved when Na_2CO_3 is used as a base-titrant.

1.2.6.2 Oxoacids of molybdenum(VI)

Similar to chromium(VI) oxide, MoO_3 shows extremely strong acidic properties in the molten chlorides studied and reported in some papers. Shapoval *et al.* [145] performed investigations of the interactions of MoO_3 with O^{2-} in a KCl–NaCl equimolar mixture by the method of potentiometric titration using the platinum–oxygen electrode as an indicator. The titration process was performed in two manners: from the acid to the base, and from the base to the acid. As for solid molybdenum(VI) oxide, it should be noted that it is stable enough at conventional temperatures for the experiment (700–727 °C), its melting point is located at 795 °C and the boiling point is near 1155 °C [121]. This allows us to assume that difficulties similar to those described in the previous section would be absent in the acid–base potentiometric titration. Nevertheless, in MoO_3 solutions in molten ionic chlorides there is observed the intrinsic acid–base interaction between particles of the dissolved substance, where one MoO_3 particle reacts as a Lux acid and another serves as a base

$$2MoO_3 \rightleftharpoons MoO_2^{2+} + MoO_4^{2-}. \tag{1.2.69}$$

This process seems to take place because of the extremely strong acidic properties of MoO_3. The cation particle MoO_2^{2+} adds chloride ions forming the oxochloride MoO_2Cl_2 which is volatile under the experimental conditions, and evaporates from the chloride melt as white smoke

$$MoO_2^{2+} + 2Cl^- \rightarrow MoO_2Cl_2 \uparrow. \tag{1.2.70}$$

Shapoval asserted that they observed a blue colouration of the melt containing molybdenum oxide, which seemingly arose owing to partial reduction of MoO_3 to lower oxidation degrees of molybdenum with the simultaneous oxidation of chloride ions to chlorine, which evaporated from the melt. The molybdenum oxides formed were of the general composition Mo_nO_{3n-1}.

The reaction (1.2.70) leads to a reduction of the equilibrium concentration of the acid (MoO_2^{2+}) in the KCl–NaCl melt: that, in its turn, causes a shift of equilibrium (1.2.69) to the right. As a result, the $[MoO_3]/[MoO_4^{2-}]$ ratio decreases, which results in a reduction in pO, as follows from the equation

$$pO = pK_{MoO_3/MoO_4^{2-}} + \log([MoO]_3/[MoO]_4^{2-}). \tag{1.2.71}$$

Reaction (1.2.70) runs while MoO_2^{2+} ions are present in the chloride in appreciable concentrations. The use of $Pt(O_2)$ gas electrode for the investigations requires the passing of oxygen through the melt that results in its oxidation owing to the reactions similar to equation (1.2.64), and favours the shift of pO to the region of higher oxide-ion concentrations. Consequently, there is a possibility of realization of three potential-defining electrode processes on metallic platinum immersed in alkali-metal chloride melt containing MoO_3 and saturated with gaseous oxygen

$$nMoO_3 + 2\bar{e} \rightarrow Mo_nO_{3n-1} + O^{2-} \, (n = 4\text{–}9), \tag{1.2.72}$$

$$Cl_2 \uparrow + 2\bar{e} \rightarrow 2Cl^-, \tag{1.2.73}$$

$$\tfrac{1}{2}O_2 \uparrow + 2\bar{e} \rightarrow O^{2-}. \tag{1.2.74}$$

Despite the possibility of the simultaneous realization of these processes, the potentiometric curves of acid–base titration of MoO_3 have no distinctive features as compared with the conventional ones [145]. However, the reverse titration curve differs appreciably from the direct titration results. Using the latter data the equilibrium constant of the formation of dimolybdate in the molten KCl–NaCl eutectic was calculated in Ref. [145] as

$$2MoO_3 + O^{2-} \rightleftharpoons Mo_2O_7^{2-}, \quad K = 1460 \pm 390. \tag{1.2.75}$$

The curve of the reverse titration contains an additional drop of e.m.f. (pO) that corresponds to a "MoO_3/MoO_4^{2-}" buffer system and allowed estimation of the equilibrium constant of the following reaction as $K = 880 \pm 880$.

$$Mo_2O_7^{2-} + O^{2-} \rightleftharpoons 2MoO_4^{2-}, \tag{1.2.76}$$

This value is characterized by the error equal to the value determined, i.e. such a constant is statistical zero. Such a high error may be explained by the action of at least two factors. At first, it is obvious enough that the statistical treatment was performed on the assumption of the normal distribution of the calculated constant value, but not its logarithm (i.e. $-pK$). The second factor follows from the above-mentioned evaporation of an excessive quantity of MoO_3 from the acidic melt, and it consists in the addition of MoO_3 in strongly acidic melt after the first equivalence point detected by the potentiometric method, which causes uncontrolled loss of the titrant and considerable distortions of the experimental results. Therefore, the reverse titration method should be recognized as unsuitable for acid–base systems with an unstable acidic form.

Investigation of the acidic properties of MoO_3 in an equimolar molten KCl–NaCl mixture at 700 °C by the potentiometric titration method with the use of a membrane oxygen electrode $Pt(O_2)|YSZ$ was reported in Ref. [142]: the potentiometric cell scheme used in this study was similar to that given in equation (1.2.65). The titration was performed from the acid to the base using NaOH and Na_2CO_3 as basic titrants. After the addition of a definite weight of MoO_3 the melt was kept in an argon flow until intensive evaporation of MoO_2Cl_2 ended and a pseudo-equilibrium state in the solution was achieved. The latter was detected after stabilization of the e.m.f. value of the cell (equation (1.2.65)) corresponding to pO $\approx$ 9.5: after the e.m.f. stabilization the titration was performed. The most typical titration curves are presented in Fig. 1.2.6a. Treatment of the results of potentiometric investigations was performed beginning with determination of the abscissa ($m_{O^{2-}}^0$) of the drop at the equivalence point. Owing to the loss of MoO_2Cl_2, the $\bar{n}'$ value calculated using the initial concentration of MoO_3 was less than 1. A fraction of acid equal to $1 - \bar{n}'/2$ is removed from the melt as the oxochloride, MoO_2Cl_2, and an equivalent quantity of MoO_4^{2-} was created in the solution owing to the processes (1.2.69) and (1.2.70).

The curves of the potentiometric titration of MoO_3 with additions of Lux bases include a pO drop of 5 pO units at the equivalence point, the position of this drop corresponds to a $[MoO_3]_0/[O^{2-}]_0$ ratio close to 1.4. Taking into account the removal of MoO_2Cl_2 with simultaneous neutralization of the remaining MoO_3, this fact testifies to the running reaction (1.2.40). As for the buffer number, it should be noted that the values of β do not exceed -0.04 for the section of the potentiometric titration curves corresponding to the ligand number of the order of 0.5.

Subsequent addition of both NaOH, and Na_2CO_3 does not cause pO drops, or a bend in the potentiometric titration curves. However, analysis of the dependence of the ligand number against the initial concentration of NaOH (see Fig. 1.2.6a, dependence $1'$) allows us to state that the addition of strong

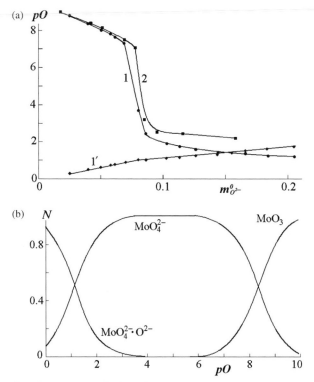

Fig. 1.2.6. Results of potentiometric investigation of the acidic properties of MoO_3 in the molten KCl–NaCl eutectic at 700 °C: (a) potentiometric curves 1, 0.080 mol kg^{-1} $MoO_3 + O^{2-}$ (1′, the dependence of $\bar{n}$ against the initial molality of the titrant); 2, 0.084 mol kg^{-1} $MoO_3 + CO_3^{2-}$; (b) fractions of Mo^{VI} oxoacids in the solutions at different pO values.

acids after the detected equivalence point gives rise to the formation of the adduct with the composition $MoO_4^{2-} \cdot O^{2-}$ (pK = -1.39 ± 0.3).

It should be noted that the basic molybdates of composition $Me^{II}MoO_5$ can be obtained by the conventional solid-phase synthesis [146]. As is the case for Cr(VI) oxo-compounds, in the solutions of molybdenum oxoacids there exist two compounds, which are the conjugate acid and base (Fig. 1.2.6b).

1.2.6.3 Oxocompounds of tungsten(VI)

All that has been said in the previous section allows us to assume that, similar to MoO_3, strongly acidic solutions of W(VI) oxocompounds should be unstable in molten chloride melts—although pure tungsten(VI) oxide is stable enough at 700 °C (the melting point is 1470 °C and the boiling point 1827 °C [136]). Indeed, tungsten(VI) is prone to formation of volatile oxochlorides of compositions $WOCl_4$ and WO_2Cl_2 [147], which exist in gaseous state at temperatures of the order of 1000 K. The former oxochloride evaporates at 232 °C, whereas the latter one passes to the gaseous state at temperatures near 500 °C.

Investigation of the oxoacidic properties of WO_3 in the molten KCl–NaCl eutectic [148, 149] by means of the gas platinum–oxygen electrode as an indicator was performed by Shapoval et $al.$ The potentiometric curve of the direct titration was found to occur in one stage, described by equation (1.2.38); the equilibrium constant value was estimated as $K = (5–6) \times 10^4$. The initial addition of WO_3 in the chloride melt caused evaporation of the oxochloride WO_2Cl_2 and chlorine, and the oxide solution became blue owing to partial reduction of W(VI) to lower degrees of oxidation of tungsten. The reverse titration resulted in a two-step potentiometric curve, the pO drops corresponding to the following reactions of oxide ion addition:

$$2WO_3 + O^{2-} \rightleftharpoons W_2O_7^{2-}, \tag{1.2.77}$$

$$W_2O_7^{2-} + O^{2-} \rightleftharpoons 2WO_4^{2-}, \tag{1.2.78}$$

and these processes were characterized by the constant values 2×10^6 and 1×10^3, respectively. The corresponding confidence ranges are not presented, probably because of high errors in the determination of the equilibrium constant owing to the evaporation of the tungsten oxochlorides. In this case, we see a complete analogy with the results for investigation of the properties of Mo oxocompounds, which were discussed above. In the opinion of the authors, it follows from the obtained data that the platinum–oxygen electrode demonstrates sufficient reversibility to oxide ions in the chloride melt containing tungsten oxo-species [148], and in solutions of moderate basicities the slope of the E–pO plot is close to $1.15RT/F$. However, this conclusion is doubtful since the next investigations performed by the same authors showed that the slope, of $1.15RT/F$, was not characteristic of the $Pt(O_2)$ oxygen electrode even in solutions of weak bases, e.g. carbonate [139]. Actually, the said calibration can be performed in a narrow range of Na_2WO_4 concentrations, since there arises a plateau of dependence owing to either slight dissociation or saturation of the melt by the tungstate compound. The change of the slope in strongly acidic solutions was not taken into account in the calculation of the constants, which caused the underestimation of their values. Similarly, the oxygen function of the platinum electrode is not maintained in the solutions containing compounds of tungsten in intermediate degrees of oxidation.

Although the results obtained can be considered as approximate rather than accurate, they give some information about the relative strength of oxocompounds of the Group VIb elements. Nevertheless, these results were used as initial data for the development of the processes of reduction of these oxocompounds in chloride melts [150–152]. Theoretical bases and principles for monitoring the electrochemical processes of deposition of the free refractory metals and their compounds with some non-metals (carbides, borides, silicides, etc.) from molten ionic media [153] were developed.

Combes and Tremillon [154] studied the oxoacidic properties of tungsten(VI) oxide and the solubility of calcium tungstate in a molten equimolar KCl–NaCl mixture at 1000 K. A potentiometric cell with the membrane indicator electrode Ni,NiO|YSZ was used for the detection of the equilibrium oxide ion concentration. Investigation of the equilibria taking place in $CaWO_4$ solutions in KCl–NaCl allowed them to determine the solubility of CaO in the said melt at 1000 K as 0.084 mol%. The solubility of Scheelite ($CaWO_4$) was determined to be $10^{-3.5}$ mol kg^{-1}, and the equilibrium constant of reaction (1.2.38) was estimated as $\sim 10^{10}$.

Reverse titration results in the formation of an adduct, whose composition was assumed by Combes and Tremillon [154] to be $W_3O_{10}^{2-}$. The equilibrium constant of the reaction

$$W_3O_{10}^{2-} + O^{2-} \rightleftharpoons 3WO_4^{2-} \tag{1.2.79}$$

was equal to 5×10^{12}. From the data in Ref. [154], it may be concluded that the potentiometric titration curve of WO_3 is not reversible because of a considerable difference of the solution concentration with respect to W(VI). The direct titration is performed at a constant total concentration of tungsten, and the solutions are dilute enough to prevent the formation of the polynuclear oxo-complexes of tungsten ($W_2O_7^{2-}$, $W_3O_{10}^{2-}$). On the contrary, in the reverse titration, the concentration of WO_3 increases permanently and the titration finishes with moderately concentrated solutions of tungsten oxocompounds that favours the formation of the polynuclear complexes. Naturally, the values of the constants calculated on the basis of data chosen from the acidic region, are distorted because of uncontrolled loss of the titrant (evaporation of the tungsten oxochlorides from the acidic solutions).

Cherginets and Banik reported a study of the acidic properties of WO_3 by the potentiometric titration method using a gas membrane oxygen electrode $Pt(O_2)|YSZ$ [142]. The addition of WO_3 to the molten KCl–NaCl equimolar mixture was accompanied by evaporation of WO_2Cl_2 from the melt

$$2WO_3 + 2Cl^- \rightleftharpoons WO_4^{2-} + WO_2Cl_2 \uparrow. \tag{1.2.80}$$

According to the titration data, the loss of the Lux acid because of the sublimation of WO_2Cl_2 was within 15–20% of the total quantity of introduced WO_3. After termination of the evaporation, the solution was titrated with weights of both NaOH and Na_2CO_3. All the potentiometric curves were characterized by a pO drop at the equivalence point, the magnitude of the drop was determined to be near 6 pO units (see Fig. 1.2.7a). The pO drop value allows us to conclude that WO_3 is the strongest acid of the oxides of elements belonging to the Group VIb.

On the basis of the obtained titration data, the constant of oxide-ion addition to WO_3 was calculated according to equation (1.2.38) to be $pK = -9.31 \pm 0.2$, which is in good agreement with the data of Combes and Tremillon [154]. The potentiometric curves were characterized by the buffer number of the order of -0.035 at ligand-number values near $\bar{n} \approx 0.5$. The addition of excessive weights of a strong base, NaOH, after the equivalence point resulted in an increase of the ligand number above 1 (see Fig. 1.2.7a, dependence $1'$) that testified to the subsequent interaction with the formation of the basic tungstate having the composition $WO_4^{2-} \cdot O^{2-}$ ($pK = -1.36 \pm 0.5$). The latter was referred to strong bases, whose basic properties were nearly equal to those of the basic molybdate (see the previous subsection). Since there is no formation of the basic tungstate in the potentiometric curves obtained with the use of Na_2CO_3 as a titrant, one can conclude that the basic tungstate is a more basic compound than is the carbonate. As known from the literature, the basic tungstates are formed in solid-phase reactions of the most basic oxides with WO_3 [155]. It is seen from the obtained data, that the basicity of the melt where the formation of the basic tungstate is observed, is at least not less than that of pure BaO.

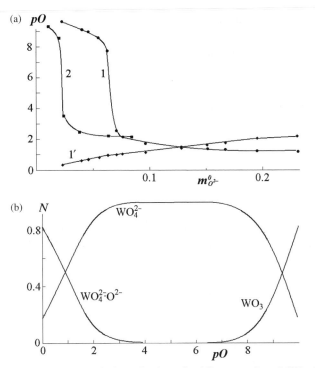

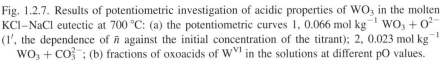

Fig. 1.2.7. Results of potentiometric investigation of acidic properties of WO_3 in the molten KCl–NaCl eutectic at 700 °C: (a) the potentiometric curves 1, 0.066 mol kg^{-1} $WO_3 + O^{2-}$ (1′, the dependence of $\bar{n}$ against the initial concentration of the titrant); 2, 0.023 mol kg^{-1} $WO_3 + CO_3^{2-}$; (b) fractions of oxoacids of W^{VI} in the solutions at different pO values.

1.2.6.4 Oxoacidic properties of phosphates

The strongest anhydroacid of phosphorus(V) is P_2O_5, which exists in the gaseous state under the experimental conditions (its sublimation temperature is near 360 °C [121]), so this compound cannot be used immediately as an initial oxoacid in high-temperature chloride. Therefore, it may be assumed that the strongest—and, at the same time, stable—oxoacid of P(V) is sodium metaphosphate $NaPO_3$, which has the melting point 627 °C [121]. This compound may be considered as the first stage of neutralization of P_2O_5 in molten salts [88]:

$$P_2O_5 + O^{2-} \rightleftharpoons 2PO_3^-. \tag{1.2.81}$$

Shapoval *et al.* reported an examination of the acidic properties of metaphosphate ions in the molten KCl–NaCl eutectic at 700 °C [141]. This investigation was performed by the potentiometric method with the use of the gas platinum–oxygen electrode as an indicator. The potentiometric titration with weights of a strong base resulted in the neutralization process occurring in two stages, listed below:

$$2PO_3^- + O^{2-} \rightleftharpoons P_2O_7^{4-}, \tag{1.2.82}$$

$$P_2O_7^{4-} + O^{2-} \rightleftharpoons 2PO_4^{3-}. \tag{1.2.83}$$

The equilibrium constants of these equilibria (in molar fraction scale) were $K = 2.88(\pm 1.2) \times 10^3$ and $2.5(\pm 1.4) \times 10^1$ for the first and the second steps, respectively.

The formation of the basic phosphates, to which Delimarskii *et al.* [141] ascribed the composition of $PO_3^- \cdot O^{2-}$, took place with excess of the titrant (O^{2-}), whereas the use of a weak base, Na_2CO_3, did not allow this basic compound to be obtained. Ba^{2+} and Li^+ were found to show appreciable acidic properties on the background of the molten KCl–NaCl eutectic [139]: the corresponding equilibrium constants are 8.1×10^1 and 3.53×10^2, respectively. Under the experimental conditions there was no precipitation of solid oxide from the said chloride melt.

Tazika *et al.* investigated the acid–base equilibria occurring in solutions of phosphorus oxo-compounds in molten KCl–NaCl at 700 °C [156]. The potentiometric cell with the gas membrane oxygen electrode $Pt(O_2)|YSZ$ was used for the determination of the equilibrium oxide ion concentrations. The formation of polynuclear anion complexes (polyphosphates) was found in Ref. [156] to take place at a Na_2O/P_2O_5 ratio within $1.67-3$ that corresponded to the formation of polyphosphates such as $Na_6P_4O_{13}$, $Na_5P_3O_{10}$ and $Na_4P_2O_7$. The acid–base interactions in this system could be expressed by the following common scheme:

$$P_nO_{3n+1}^{-(n+2)} + PO_4^{3-} = P_{n+1}O_{3n+5}^{-(n+5)}. \tag{1.2.84}$$

The constants of the corresponding equilibria were determined, and the solubilities of the polyphosphates were found.

The results of potentiometric titration of sodium metaphosphate with weights of Na_2CO_3 and $NaOH$ were reported in Ref. [157]. The most indicative potentiometric curves are collected in Fig. 1.2.8a, together with the plot of the ligand number $\bar{n}$ against the initial concentration of the base-titrant.

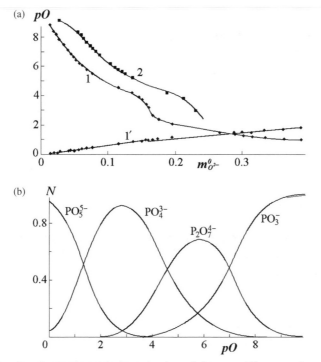

Fig. 1.2.8. Results of potentiometric investigation of the oxoacidic properties of $NaPO_3$ in the molten $KCl-NaCl$ eutectic at 700 °C: (a) titration curves 1, 0.160 mol kg^{-1} $PO_3^- + O^{2-}$ (1′, the dependence for $\bar{n}$); 2, 0.245 mol kg^{-1} $PO_3^- + CO_3^{2-}$; (b) fractions of oxoacids of P^V at different pO values.

The titration with sodium hydroxide showed that the neutralization of PO_3^- in the chloride melt proceeds in two consecutive stages, which were described by equations (1.2.82) and (1.2.83), which were in good agreement with the data of Ref. [141]. The equilibrium constants of oxide-ion addition (pK) were estimated as -8.01 ± 0.1 and -3.75 ± 0.1, respectively. The maximum buffer number value, β, did not exceed -0.03 in the buffer section $PO_3^- - P_2O_7^{4-}$ whereas in the section $P_2O_7^{4-} - PO_4^{3-}$ its values were sufficiently higher, at -0.06 to 0.07. This allowed them to consider the solutions of phosphorus oxocompounds as most suitable for the preparation of standard buffer systems in molten salts.

It should be added that the potentiometric curves of metaphosphate titration with weights of Na_2CO_3 showed several small and hardly reproducible peaks, which were seemingly caused by the presence of traces of phosphorus(V) polyanions in the studied solutions, although the corresponding constants of the acid–base equilibria contained very high errors [157], and, therefore, the values obtained were statistically insignificant. Hence, the formation of appreciable quantities of $P_4O_{13}^{6-}$ and $P_3O_{10}^{5-}$ should not be anticipated; at least, this concerns diluted solutions in the molten equimolar KCl–NaCl mixture with concentrations of the order of 0.1 mol kg^{-1}.

The initial addition of $NaPO_3$ to the melt increases the pO value up to 10, and after the achievement of the first equivalence point, which corresponds to the formation of pyrophosphate ion, the pO value reduces sharply to 5.5–6. The next section of the potentiometric curve is characterized by a relatively flat pO decrease. Here, $P_2O_7^{4-} + PO_4^{3-}$ solutions possessing considerable buffer numbers are formed. This section ends with the second, relatively small pO drop of 2–2.5 pO units at the equivalence point.

The addition of excess of the strong Lux base after the equivalence point leads to an increase in the ligand number, which approaches 2. This is caused by the formation of the basic phosphate, as below, and the corresponding pK value is equal to -1.31 ± 0.1.

$$PO_4^{3-} + O^{2-} = PO_4^{3-} \cdot O^{2-}, \tag{1.2.85}$$

Consequently, in strongly basic solutions, orthophosphate ion shows appreciable basic properties. The dependence of the fraction of different phosphorus oxoacids on pO is depicted in Fig. 1.2.8b. As can be seen, in the case of the phosphorus oxocompounds three acid–base forms can coexist in the region of moderate acidity (pO $= 4$–6) which corresponds to the formation of $P_2O_7^{4-}/PO_4^{3-}$ buffer solutions. Besides the mentioned anions, in these solutions PO_3^- exists in concentrations of 10% of the concentration of the main oxoacids: pyro- and ortho-phosphate. This causes some increase in the error of determining the equilibrium constant (1.2.83) far from the corresponding equivalence point. For this reason, it is necessary to calculate the average values of the acid–base constant of phosphorus oxocompounds near the equivalence points, although this leads to an increase in the estimation of errors owing to the reduced accuracy of determination of the titrant and the oxoacid concentrations.

1.2.6.5 Oxoacids of vanadium(V)

The reference data show that vanadium(V) oxide, V_2O_5, decomposes at temperatures above 700 °C [121], i.e. its decomposition starts near its melting point of 680 °C. Therefore, at temperatures not higher than 700 °C this substance can be used as an initial vanadium oxoacid for investigations in chloride melts [157], which are known to possess relatively weak reduction properties. However, this oxoacid should oxidize bromide and iodide melts, being reduced to lower oxidation states of vanadium, e.g. VO_2. Because of this, $NaVO_3$ should be used instead of V_2O_5 for oxoacidity investigations in these media, since it is stable enough and its oxidation properties are noticeably weaker.

To determine the acidic properties of vanadium oxocompounds, we performed potentiometric titration of V_2O_5 in an equimolar mixture of molten KCl–NaCl at 700 °C with known weights of Na_2CO_3 and NaOH (see Fig. 1.2.9a) [157]. The addition of initial weights of V_2O_5 to the molten

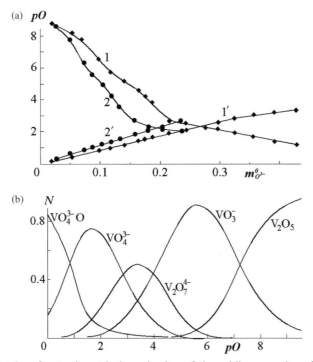

Fig. 1.2.9. Results of potentiometric investigation of the acidic properties of V_2O_5 in the molten KCl–NaCl equimolar mixture at 700 °C: (a) titration curves 1, 0.160 mol kg^{-1} $V_2O_5 + O^{2-}$ (1′, dependence for $\bar{n}$); 2, 0.245 mol kg^{-1} $V_2O_5 + CO_3^{2-}$ (2′, dependence for $\bar{n}$); (b) fractions of oxoacids of V^V at different pO values.

chloride mixture causes no appreciable changes of the melt colour, which confirms the absence of any redox processes. These additions lead to the increase of pO values up to 8.7–8.8, in comparison with pO = 4 for the pure chloride melt, that testifies to very strong acidic properties of V_2O_5. The potentiometric curves and the dependence of the ligand number on the initial molality of the base-titrant are presented in Fig. 1.2.9a. They contain two sharp drops in pO, which correspond to the ligand number values of 1 and 2, respectively. This gives the possibility of assuming that the process of VO_3^- neutralization has two stages

$$V_2O_5 + O^{2-} \rightleftharpoons 2VO_3^-,\tag{1.2.86}$$

$$2VO_3^- + O^{2-} \rightleftharpoons V_2O_7^{4-}.\tag{1.2.87}$$

These stages are characterized by the equilibrium constant values (pK) equal to -6.5 ± 0.2 and -5.23 ± 0.3, respectively. The buffer solutions, whose formation corresponds to the section of the titration curves with $\bar{n} \sim 1$ and the neutralization stage (1.2.86), are characterized by the buffer number of -0.03, whereas for the section conforming to the equation (1.2.87) this value is markedly higher (up to -0.05).

In the case when NaOH is used as the base-titrant, its addition after the second equivalence point (meta-vanadate/pyro-vanadate) gives rise to the third slightly pronounced pO drop, which corresponds to the formation of ortho-vanadate from pyrovanadate by the following equation:

$$V_2O_7^{4-} + O^{2-} \rightleftharpoons 2VO_4^{3-}.\tag{1.2.88}$$

This reaction is characterized by a pK value of -2.04 ± 0.4 and the buffer number equal to $\beta \approx -0.12$. The equivalence point of this neutralization step can be detected more sharply using the differential potentiometric titration curve.

It should be noted that a small pO drop characteristic of this reaction can be observed for the titration curves obtained with the use of carbonate as a titrant. This demonstrates that the oxoacidic properties of $V_2O_7^{4-}$ are stronger than those of carbon dioxide at low CO_2 pressures ($p_{CO_2} \sim 0$)—that is, the equilibrium of the reaction

$$V_2O_7^{4-} + CO_3^{2-} \rightleftharpoons 2VO_4^{3-} + CO_2 \uparrow\tag{1.2.89}$$

is shifted to the right in the inert atmosphere.

We now return to the equivalence point corresponding to equation (1.2.88). Addition of NaOH after the corresponding equivalence point is accompanied by a continuing increase of the ligand number. This points to the formation of basic vanadates in the melt—Shams El Din and El Hosary [88] ascribed the formula $VO_4^{3-} \cdot O^{2-}$ to this oxoanion. Hence, the interaction can be presented as:

$$VO_4^{3-} + O^{2-} \rightleftharpoons VO_4^{3-} \cdot O^{2-}, \qquad (1.2.90)$$

the corresponding pK value is estimated as -0.79 ± 0.4 and the buffer number $\beta \approx -0.14$ in the middle of the buffer section.

From the known values of the acid–base constants for vanadium oxocompounds one can establish the relationship between different acid–base forms in the studied chloride melt over a wide range of pO. The corresponding dependence is presented in Fig. 1.2.9b. It is seen that in the case of V^V oxo-compounds more than two acid–base forms can coexist in a relatively wide pO range, which gives rise to additional errors in the equilibrium constant calculations. Nevertheless, in the given case this does not lead to an appreciable directed shift of the calculated values of the equilibrium constants, or to an increase of the confidence ranges for reactions (1.2.86) and (1.2.87).

Moreover, investigation of the oxoacidic properties of the V_2O_5 derivatives by the potentiometric titration method with the use of Ni,NiO|YSZ oxygen electrode is reported in Ref. [158]. The data obtained show that V_2O_5 is a dibasic acid. The potentiometric determination was performed from the base to the acid. The equilibrium constants (pK) corresponding to reactions (1.2.86) and (1.2.87) were equal to -9.3 ± 0.3 and -8.3 ± 0.3, respectively (in molar fractions).

1.2.6.6 Oxoacids of boron(III)

Boron(III) oxide is known to be the strongest acid among the oxo-compounds of B^{III}; moreover, it is sufficiently up to temperatures of about 1000 °C (the melting point is 450 °C and the boiling point is 1700 °C [121]). However, because B_2O_3 is practically completely polymerized even in the molten state, sodium tetraborate (borax) $Na_2B_4O_7$ is classified among the most convenient boron compounds for investigation of the Lux acid–base interactions. The melting point of $Na_2B_4O_7$ is close to 742 °C and, on being heated to $\sim 1545\,^{\circ}\text{C}$ it is subjected to decomposition [121]. In contrast to the oxoanions considered above, borax is appreciably polymerized even in the liquid state, although its degree of polymerization is lower than that of B_2O_3. In our work [157], examination of the acidic properties of $Na_2B_4O_7$ in the molten equimolar KCl–NaCl mixture was performed with weights of NaOH and Na_2CO_3, from the acid to the base. The corresponding potentiometric and derived dependences are presented in Fig. 1.2.10a. Initial addition of the acid results in an increase of pO values up to ~ 6–6.5. Subsequent addition of the Lux bases up to the equivalence point, which corresponds to $\bar{n} = 1$, gives the possibility of calculating the equilibrium constant of oxide ion addition to tetraborate ions,

$$B_4O_7^{2-} + O^{2-} \rightleftharpoons 4BO_2^-$$
(1.2.91)

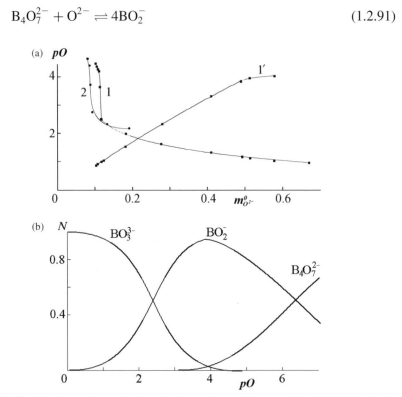

Fig. 1.2.10. Results of potentiometric investigation of the acidic properties of $Na_2B_4O_7$ in the molten KCl–NaCl equimolar mixture at $700\,^{\circ}\text{C}$: (a) titration curves 1, $0.116\ \text{mol kg}^{-1}$ $B_4O_7^{2-} + O^{2-}$ (1′, dependence for $\bar{n}$); 2, $0.090\ \text{mol kg}^{-1}$ $B_4O_7^{2-} + CO_3^{2-}$; (b) fraction of oxocompounds of B^{III} at different pO values.

as $pK = -4.82 \pm 02$, which points to the relatively strong oxoacidic properties of $B_4O_7^{2-}$. The buffer-number values in the acidic section of the potentiometric curves vary from -0.02 to -0.04, and the pO drop at the equivalence point is close to 3 pO units.

However, as in the cases of the other oxoanions considered above, the formed metaborate, BO_2^-, shows appreciable acidic properties. After the pO drop at the equivalence point the oxo-anions continue to fix oxide ions, and the value of the ligand number rises to 5. This means that the second stage of the neutralization process takes place. Neutral orthoborate ion is the product resulting from such an interaction, and it is a strong Lux base. The buffer solutions "metaborate/orthoborate" exist, mainly, within a $1-2$ pO range, and possess buffer numbers of the order of $\beta \approx -0.4$ to -0.6. The pK value describing the following equilibrium:

$$BO_2^- + O^{2-} \rightleftharpoons BO_3^{3-}, \tag{1.2.92}$$

is equal to -2.37 ± 0.2. By analogy with most cases of Lux acids and bases in molten chlorides, not more than two anionic forms co-exist in the solution at the same pO. Therefore, the calculated thermodynamic values are not overburdened by additional errors owing to neglect of certain acid–base forms which actually exist under the equilibrium conditions. The formation of BO_4^{5-} anions with an excess of the strong base is not observed, since the basicities of the studied solution do not permit the study of such an interaction—the data of Claes *et al.* show that this anion of boron may be obtained only in molten alkali-metal hydroxides [159].

1.2.6.7 Acidic properties of Ge(IV) and Nb(V) oxocompounds

Oxides of germanium(IV) and niobium(V) are stable at temperatures of the order of 700 °C, and under these conditions they are solids: depending on its crystalline modification GeO_2 has its melting point in the temperature range 1086–1116 °C, whereas niobium oxide exists as a single crystalline phase with the melting point at 1512 °C [121].

Potentiometric investigation of the oxoacidic properties of the oxides mentioned above was reported in Refs. [160, 161]. *A priori*, before performing this study it was assumed that niobium(V) oxide would be the complete analogue of V_2O_5 and that its acidic properties would be even stronger. The basis for such an assumption may be found in Group VIb, where there is an appreciable increase of the acidic properties in the sequence Cr–Mo–W.

As for germanium(IV) oxide, this was considered to be slightly soluble in the molten chloride—similar to SiO_2—and its acidic properties were expected to be too weak. However, one feature was not mentioned. While considering aqueous solutions we can see that GeO_2 possesses appreciable solubility in water. For example, at 100 °C approximately 1.0 g of GeO_2 is dissolved in 100 g of water, and this allows us to assume that under the experimental conditions conventional for molten chlorides this oxide will be sufficiently soluble. Indeed, this is the case for the molten equimolar KCl–NaCl mixture.

The oxoacidic properties of both germanium and niobium oxides were examined by the potentiometric titration method with the use of a $Pt(O_2)YSZ$ membrane oxygen electrode in a cell similar to that in equation (1.2.65). The titration curves and the dependence of the ligand number on the initial molality of the titrant (NaOH) for GeO_2 and Nb_2O_5 are presented in Fig. 1.2.11. In contrast to V_2O_5, niobium(V) oxide was found to be practically insoluble in the molten equimolar KCl–NaCl mixture at 700 °C. Although the addition of extra Nb_2O_5 resulted in a pO increase to 8, the first weight of the Lux base shifted the oxygen index down to pO $\approx$ 3. However, additions of the base to the melt containing Nb_2O_5 were fixed in part by the oxide precipitate. This was accompanied with an increase in the ligand number to 4, which gives evidence of the consecutive formation in the precipitate of all the forms of acidic (NbO_3^-, $Nb_2O_7^{4-}$), neutral (NbO_4^{3-}) and basic ($NbO_4^{3-} \cdot O^{2-}$) niobates as the products of the acid–base reactions:

$$Nb_2O_5 + O^{2-} = 2NbO_3^-, \qquad (1.2.93)$$

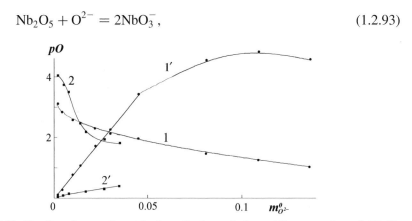

Fig. 1.2.11. Results of potentiometric investigation of the acidic properties of Nb_2O_5 (1, 0.01 mol kg^{-1}) and GeO_2 (2, 0.051 mol kg^{-1}) and the corresponding dependences of the ligand number $\bar{n}$ (1′ and 2′, respectively) in the molten KCl–NaCl eutectic at 700 °C.

$$2NbO_3^- + O^{2-} \rightleftharpoons Nb_2O_7^{4-}, \tag{1.2.94}$$

$$Nb_2O_7^{4-} + O^{2-} \rightleftharpoons 2NbO_4^{3-}, \tag{1.2.95}$$

and

$$NbO_4^{3-} + O^{2-} \rightleftharpoons NbO_4^{3-} \cdot O^{2-}. \tag{1.2.96}$$

In contrast, germanium oxide was found to be appreciably soluble in the molten equimolar KCl–NaCl mixture and its solutions show weak acidic properties. Its acid–base titration using NaOH as a titrant results in the appearance of a small pO drop at the equivalence point (~ 1 pO unit), which corresponds to the ligand number value ≈ 0.5. This allows us to conclude that the acid–base titration runs in one step, and the final product is di-germanate ion, $Ge_2O_5^{2-}$

$$2GeO_2 + O^{2-} \rightleftharpoons Ge_2O_5^{2-}, \tag{1.2.97}$$

the pK value for this equilibrium is equal to -4.18 ± 0.4. As for the buffer number of "germanium oxide/di-germanate" solutions in the molten KCl–NaCl, it should be noted that it is not great.

1.2.7 MOLTEN KCl–LiCl (0.41:0.59) EUTECTIC

This salt mixture, possessing the melting point at 361 °C [136], is practically uninvestigated as a medium for running homogeneous acid–base processes. However, such studies are of essential interest when determining the regularities of the effect of ionic melt acidity on the acid–base equilibria taking place in its background. We have reported examinations of the acidic properties of $K_2Cr_2O_7$ in this melt at 700 °C by the method of potentiometric titration using a membrane oxygen electrode Pt(O_2)|YSZ [162, 163]. For this purpose, the solution was chosen with a concentration of the acid of approximately 0.1 mol kg^{-1}, since from what we have said above (Section 2.6.1) it follows that dichromate ion, but not CrO_3, would be the main acidic form under these conditions.

The curves of the potentiometric titration of $Cr_2O_7^{2-}$ with NaOH in molten KCl–LiCl and KCl–NaCl eutectics are shown in Fig. 1.2.12. The data presented show that the potentiometric titration curve in the molten KCl–LiCl

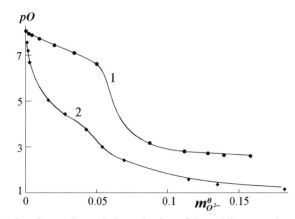

Fig. 1.2.12. Results of potentiometric investigation of the acidic properties of $K_2Cr_2O_7$ with NaOH in KCl–NaCl (1, 0.085 mol kg^{-1}) and KCl–LiCl (2, 0.09 mol kg^{-1}) eutectic mixtures at 700 °C.

eutectic lies under the corresponding curve for the KCl–NaCl eutectic: the shift is approximately 3 pO units. This means that the oxoacidic properties of dichromate ion in the Li$^+$-containing melt are considerably weaker than those in the molten KCl–NaCl. Indeed, on the basis of the data obtained, the equilibrium constant of equation (1.2.36) can be estimated as pK = -3.09 ± 0.3.

The cause of the decrease of the calculated pK value is the competitive action of the cation acid Li$^+$, whose acidic properties are sufficiently strong, even on the background of the equimolar KCl–NaCl mixture [141]. In other words, the said shift of the constant value may be considered as the difference of constants of the following equilibria:

$$K_2Cr_2O_7 + Na_2O \rightleftharpoons K_2CrO_4 + Na_2CrO_4, \qquad (1.2.98)$$

and

$$K_2Cr_2O_7 + Li_2O \rightleftharpoons K_2CrO_4 + Li_2CrO_4. \qquad (1.2.99)$$

where the equilibrium of reaction (1.2.99) is shifted to the left as compared with reaction (1.2.98).

1.2.8 MOLTEN NaI

Sodium iodide with the melting point at 661 °C is referred to as one of the substances having wide practical use. First of all, it is the initial material for obtaining NaI(TlI) scintillation single-crystals. Molten NaI was studied as a medium for running Lux acid–base interactions in the 1970s by Rybkin and Banik. Their investigations were devoted to the construction of the acidity scale (the row of acidity) in the said melt, and to semi-quantitative analysis of certain acid–base interactions at 700 °C. All these studies were performed with the use of a membrane oxygen electrode $Pt(O_2)|ZrO_2(CaO)$ as the indicator to detect oxide-ion concentration.

The acidity row in NaI at 700 °C was called an "acidity scale" by the investigators, and was constructed [64] for solutions of molality 0.01 mol kg^{-1}. It is represented by the following sequence (the differences of pO indices in the pure solvent and in the studied solution are presented in brackets): HPO_4^{2-} (3.0) → $P_2O_7^{4-}$ (2.52) → PO_3^- (1.60) → $H_2PO_4^-$ (1.50) → $B_4O_7^{2-}$ (1.05) → VO_3^- (0.98) → SO_3^{2-} (0.15) → NaI (0) → SO_4^{2-} (−0.16) → WO_4^{2-} (−0.20) → PO_4^{3-} (−0.88) → IO_3^- (−1.54) → OH^- (−2.28) → CO_3^{2-} (−2.39). From this sequence, the authors concluded that carbonate and hydroxide ions are Lux bases of the same strength. Nevertheless, this conclusion cannot be recognized as correct since solutions of the same concentration of 0.01 mol kg^{-1} have been used for the construction of the acidity row. However, it follows from equations (1.2.3) and (1.2.4) that with respect to the formation of O^{2-} these additions are non-equivalent because dissociation of 1 mol of CO_3^{2-} leads to the formation of 1 mol of O^{2-}, whereas decomposition of 1 mol of OH^- results in the formation of 0.5 mol of oxide ion. Indeed, when performing experiments on the determination of the comparative strength of bases (SO_4^{2-}, CO_3^{2-} and OH^-) by their potentiometric titration with sodium pyrophosphate, Rybkin and Banik showed hydroxide ion to be a stronger base than carbonate [164].

As for the correctness of the above experiments, it should be mentioned that the positions of some Lux acids in this scale are incorrect: this concerns V_2O_5, acidic phosphates, sodium meta- and pyro-phosphate, whose dissolution in the melt is accompanied by the evolution of gaseous iodine. In the case of V_2O_5 the reason consists in its strong oxidation properties in acidic solutions. As regards the iodine evolution upon dissolution of the various phosphates, in the case of acidic phosphates it is explained by the pyrohydrolysis of the melt by water formed by dehydration of the acidic salts—for example, potassium dihydrogenphosphate reacts with iodide melts as follows:

$$H_2PO_4^- + 2I^- \rightleftharpoons PO_4^{3-} + 2HI \uparrow, \qquad (1.2.100)$$

and thermal decomposition of the evaporated HI leads to the formation of free iodine.

    An interesting feature of strongly acidic solutions in molten iodides is the evolution of iodine, which leads to partial or complete neutralization of the Lux acid. This seems to be a consequence of the transport of gaseous oxygen through the solid-electrolyte membrane, although this phenomenon has not been investigated yet. In any case, the iodine evolution points to the contact of oxidant (e.g. gaseous oxygen) with iodide melts. The oxide ions formed as a result of this oxidation are fixed by the dissolved acid that decreases the pO down to pO = 5–6. This leads to a discrepancy between the chemical composition of the substance introduced into the melt and the one really existing in it under equilibrium conditions. Therefore, to study the effect of a melt anion on Lux acid–base equilibria there should be chosen substances leading to a shift of the pO of the melt to the moderately acidic region (pO within 5–6). This condition is correct for the use of $NaVO_3$, $Na_4B_4O_7$ and $Na_4P_2O_7$.

    The relative basicity of some Lux bases (OH^-, CO_3^{2-}, SO_4^{2-}) in molten NaI at 700 °C was investigated by the potentiometric titration method with weights of sodium pyrophosphate from base to acid [165]. The first of these bases reacted with the acid-titrant in the proportion 2:1, the others were neutralized in the proportion 1:1. Depending on the value of the pO drop at the equivalence points the studied bases were arranged in the sequence OH^- – CO_3^{2-}–SO_4^{2-} with decreasing basicities. The constants of the acid–base equilibria were not determined. Among the results presented, the most doubtful is the validity of the process of sulfate ion decomposition by the Lux acid. The reason consists in the fact that SO_3 formed as a result of the acid–base reaction

$$P_2O_7^{4-} + SO_4^{2-} \rightleftharpoons 2PO_4^{3-} + SO_3 \uparrow, \qquad (1.2.101)$$

under the experimental conditions is not only a volatile substance, but is also a very strong acid and a very strong oxidant. Owing to the above, the acid–base interaction should not be limited by process (1.2.101), but it also includes the additional redox reaction

$$SO_3 + 2I^- \rightleftharpoons SO_2 \uparrow + I_2 + O^{2-}, \tag{1.2.102}$$

that leads to the appearance of additional quantities of oxide ion in the melt and appreciable deviations from the 1:1 stoichiometry.

Hence, the studies described above did not result in the obtaining of any quantitative characteristics which could give the possibility of discussing the strength of acids in molten sodium iodide, or the acidic properties of pure NaI. Construction of the general acidity scale for high-temperature ionic media calls for the choice of indicator acid–base pairs whose interaction with acids and bases in different ionic solvents will run without effects specific for each solvent, apart from the difference in oxoacidity. This problem is close to the similar one involved in the construction of the common proton acidity scale in room-temperature solvents using the indicator method introduced by Hammett [48]. There are three kinds of acids in molten ionic media: cations, uncharged particles and anions. It is obvious, that the first class of acids reacts with the melt anions, forming complexes of different stability, whose composition can be described by the formula MeX_4^{2-} in the case of halide melts. The strength of such acids in molten media having various anion compositions (e.g. $SrCl_4^{2-}$ in chloride melts and $SrBr_4^{2-}$ in bromide ones) should differ considerably.

Therefore, for the case of ionic melts of different anion composition, anionic and uncharged acids should be considered to be more promising indicator acid/base pairs. However, a limitation of the use of uncharged acids consists in the fact that they are foreign particles with respect to ionic melts, to the same degree as ions in room-temperature molecular solvents. This leads to the conclusion that anionic acids are most suitable for the construction of general acidity scales in ionic melts of different anion compositions. Nevertheless, not all the anionic acids considered above will be applicable to the construction of oxoacidity scales in the molten ionic media based on alkali-metal halides of different anion composition. For example, the oxoacids of Group VIb possess high oxidation properties at elevated temperatures, which means that they should oxidize molten chlorides and bromides. In the general case, more substantiated is the use of oxo-derivatives of boron(III) and vanadium(V) as indicator Lux acids.

Investigations of the oxoacidic properties of boron(III) compounds in molten NaI and KCl–NaCl at 700 °C were reported by Cherginets and

Banik [166, 167]. The acidic properties of $Na_2B_4O_7$ were studied by the potentiometric titration method using NaOH and Na_2CO_3 as base-titrants, and the equilibrium oxide ion molality was detected by the potentiometric cells with the membrane oxygen electrode $Pt(O_2)|YSZ$. The process of titration was performed both from the acid to the base and from the base to the acid. The most characteristic potentiometric curves and the dependencies of the ligand number on the titrant molality are presented in Fig. 1.2.13.

The pO drop at the equivalence point corresponds to the ligand number 1, which confirms the running of the interaction (1.2.91) in molten NaI. This interaction is characterized by a pK value of -5.02 ± 0.6. It should be emphasized that with the excess of the Lux acid the equilibrium is achieved slowly enough. This seems to be caused by the fact that sodium tetraborate is practically completely polymerized, although the polymerizations' effects on the titration process are less pronounced than in the case of B_2O_3. After the equivalence point, with the pO drop of 3–3.5 pO units, the equilibrium conditions are achieved in a shorter period, since under these conditions all the polymerized particles are destroyed. The excess of the base over that necessary for the formation of BO_2^- results in the fixation of O^{2-} by the formed metaborate, which demonstrates the oxoacidic properties. The dependence of the ligand number on the initial titrant molality allows us to assume that the final product is sodium orthoborate, BO_3^{3-} (equation (1.2.92)), whose pK value is close to 2. All that we have said above shows that the oxoacidic properties of tetraborate ions in molten sodium iodide are stronger

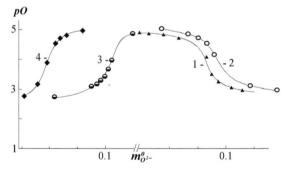

Fig. 1.2.13. Potentiometric curves of direct and reverse titration for $B_4O_7^{2-} + O^{2-}$ in molten NaI at 700 °C: 1, 0.096 mol kg^{-1} $B_4O_7^{2-} + O^{2-}$; 2, 0.099 mol kg^{-1} $B_4O_7^{2-} + CO_3^{2-}$; 3, 0.096 mol kg^{-1} $O^{2-} + B_4O_7^{2-}$; 4, 0.076 mol kg^{-1} $CO_3^{2-} + B_4O_7^{2-}$.

than those in the molten equimolar KCl–NaCl mixture, owing to lower acidic
properties of the iodide melt.

The curves from the direct and reverse potentiometric titration of metavanadate ions with the Lux bases NaOH and Na_2CO_3 are presented in
Fig. 1.2.14a. The neutralization process runs in two stages according to the
equations (1.2.87) and (1.2.88), to which two small diffuse pO-drops (bends)
in the potentiometric curves correspond. Nevertheless, these bends become
apparent on the differential potentiometric titration curve, shown in
Fig. 1.2.14b. The reverse titration results in two steps of neutralization,
also, that confirms the reversibility of the process of the acid–base
neutralization of vanadium(V) oxocompounds in molten NaI. Comparison
with the corresponding dependencies for the molten equimolar KCl–NaCl

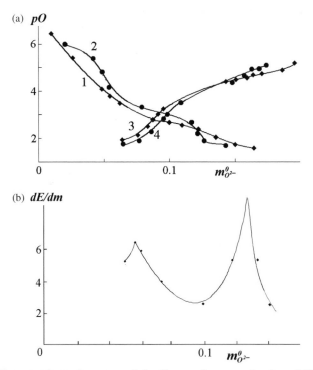

Fig. 1.2.14. The potentiometric curves of the direct and reverse titration of $VO_3^- + O^{2-}$ in
molten NaI at 700 °C: (a) 1, 0.105 mol kg^{-1} $VO_3^- + O^{2-}$; 2, 0.120 mol kg^{-1} $VO_3^- + CO_3^{2-}$;
3, 0.085 mol kg^{-1} $O^{2-} + VO_3^-$; 4, 0.089 mol kg^{-1} $CO_3^{2-} + VO_3^-$. (b) The differential
potentiometric curve, corresponding to potentiometric curve 2 in Fig. 1.2.14a.

mixture (the second and the third stages of the neutralization) shows that the behaviour of the potentiometric titration curves is practically the same.

It should be emphasized that, as with the chloride melt, pyrovanadate ion in NaI is a stronger Lux acid than is CO_2, and in this case equilibrium (1.2.89) is shifted to the right with an inert atmosphere over the melt. The estimations show that the equilibrium constants in molten KCl–NaCl and NaI are practically the same, but in the iodide melt the pK values are lower by 0.2 logarithmic units, thus confirming the weaker acidity of the NaI melt as compared with the equimolar KCl–NaCl mixture. This means that homogeneous acid–base equilibria with the participation of anion acids can be proposed as indicator reactions for constructing the general oxoacidity scale in ionic melts of different anion compositions, including halide and oxygen-less media. However, to investigate ionic solvents whose acidic properties may be varied within wide limits, the indicator acids should be strong enough. In the opposite case, it is possible to level the acidic properties by the solvent cations, owing to a considerable increase of pK (which means a decrease in the absolute magnitude) corresponding to the shift of the acidity scale (in pI_L units) in the acidic solvent.

1.2.9 OTHER ALKALINE-METAL HALIDES

Rybkin and Seredenko reported the construction of empirical scales of oxoacidity (acidity rows) in molten KCl at 800 °C and CsI at 650 °C [62, 63]. Estimation of the oxoacidic properties was performed for buffer solutions obtained by the addition of equimolar quantities of conjugated acid and base in the melt. E.m.f. (pO) measurements were performed in the potentiometric cell with the use of a membrane oxygen electrode Pt(O_2)|YSZ.

The "acidity scale" in molten KCl was constructed on the basis of pO values of the buffer solutions of concentration equal to 0.01 mol kg^{-1}. The pK value of the equimolar mixture of $NaPO_3$ and Na_3PO_4 was chosen as the reference point (and the conditional pO equal to zero was attributed to this mixture). As is easily seen, this mixture is nothing other than a relatively strong Lux acid, $Na_4P_2O_7$. The order of acidity can be presented by the following sequence of the buffer solutions: KPO_3 (0.54) → Na_3PO_4 + $Na_4P_2O_7$ (0.12) → Na_3PO_4 + $NaPO_3$ (0) → $Na_2W_2O_7$ + Na_2WO_4 (−0.10) → Na_2WO_4 (−0.32) → $Na_2B_4O_7$ (−0.72) → K_2SO_4 + $K_2S_2O_7$ (−0.84) → Na_2SO_4 + $Na_2S_2O_8$ (−1.56) → KNO_3 + KNO_2 (−3.16) → K_2CO_3 + Na_2O (−3.74).

In this work, the difference in acidity between the most acidic and the most basic solution was found to be 4.28 pO units.

The relatively high temperature of the experiment (800 °C) leads to decomposition of certain unstable conjugate acids and to a shift of the pO of the buffer solutions towards higher basicities. This is seen clearly for the cases of $Na_2W_2O_7 + Na_2WO_4$ ($pO' = -0.32$), $K_2S_2O_7 + K_2SO_4$ ($pO' = -0.72$) which are classified among the strongest acids. The results of the investigation contain some inaccuracies, e.g. for the $Na_4P_2O_7 + Na_3PO_4$ solution, whose acidity is sufficiently higher than that of pure $Na_4P_2O_7$ ($\equiv NaPO_3 + Na_3PO_4$).

The practical significance of the discussed study in molten KCl consists not in the length of the scale, but in the introduction of some buffer solutions ($SO_4^{2-}/S_2O_8^{2-}$, $WO_4^{2-}/W_2O_7^{2-}$, $PO_4^{3-}/P_2O_7^{4-}$) which, according to the authors' data, are characterized by well-reproducible magnitudes of pO and can be used as standard buffer solutions to calibrate oxygen electrodes of different kinds in molten alkali-metal halides. Comparison of the data obtained using two kinds of oxygen electrodes such as $Pt(O_2)$ and $Pt(O_2)|ZrO_2$ (CaO) in the study [62] leads to the surprising conclusion that the use of the gas oxygen electrode $Pt(O_2)$ with passing O_2 through the melt, allows one to obtain an order (scale) of acidity, which reflects more precisely the determined relative acidic properties of various acids and bases.

Another investigation by the same authors concerns the construction of the acidity scale in molten CsI at 650 °C [63] on the basis of a similar experimental technique. The acidity series obtained is KPO_3 (4.12) $\rightarrow$ CsI (0.63) $\rightarrow$ Na_3PO_4 $+NaPO_3$ (0) $\rightarrow$ Na_2SO_4 (-0.07) $\rightarrow$ K_2CO_3 (-0.32) $\rightarrow$ KOH (-0.71) $\rightarrow$ Na_2WO_4 (-1.51) $\rightarrow$ Cs_2CO_3 (-2.12). The acid–base range in molten CsI estimated from these data is equal to 6.24 pO units. It is noticeable that the solution of potassium metaphosphate possesses a pO' exceeding the corresponding value in the buffer mixture $NaPO_3 + Na_3PO_4$ (with $pO' = 0$) by 4.12 pO units. Such a drastic increase of the oxoacidic properties of KPO_3 with respect to $NaPO_3 + Na_3PO_4$ stresses the above-mentioned drawbacks of the method of construction of "experimental" acidity scales. Similarly, there are considerable distinctions in basicities for solutions of Cs_2CO_3 in CsI, and the buffer solution $K_2CO_3-Na_2O$ in KCl: -2.12 against -3.74. The most probable reason for such a discrepancy is the low reproducibility of the e.m.f. of the buffer solutions which, in their turn, is caused by the existence of oxygen-containing admixtures in the melts studied. The concentration of the latter was not measured, and they were not removed from the melt after melting.

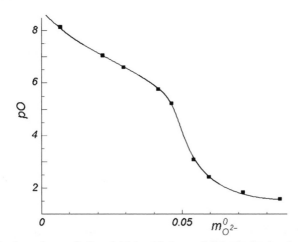

Fig. 1.2.15. The dependence of pO vs initial oxide ion molalities in the titration of $P_2O_7^{4-}$ with NaOH in molten NaCl at 830 °C.

An investigation of the oxoacidic properties of $Na_4P_2O_7$ in molten NaCl was reported in Ref. [168]. An example of a potentiometric curve is presented in Fig. 1.2.15. The behaviour of this curve shows that homogeneous $P_2O_7^{2-}$ / PO_4^3 solutions possess sufficiently higher buffer numbers in the middle of the acidic section. The pO drop at the equivalence point corresponds to a $\bar{n}$ value equal to 1, which confirms the suitability of equation (1.2.83) for description of the neutralization reaction. The points belonging to the acidic section can be used for averaging the pK estimated in the said work as -5.74 ± 0.15, i.e. by 2 units lower than the constant value obtained in the KCl–NaCl equimolar mixture at 700 °C (-3.75 ± 0.15 [157]). So, the acidity of pyrophosphate ion strengthens considerably together with a rise of the chloride melt temperature.

1.2.10 CONCLUSION

Thus, the whole of this part of the book shows that the current state of investigations of homogeneous acid–base equilibria in high-temperature ionic solvents cannot be considered satisfactory.

The correctness of the constants of the acid–base equilibria determined in molten alkali metal nitrates are doubtful, because the mentioned melts are referred to the solvents of the first kind for the oxoacidic reactions. This requires combining the titration of solutions of the solvent acid NO_2^+ with the titrations of the particular acids. For this reason, the results obtained can hardly be attributed to the acidic properties of the strong acids studied. Moreover, they only concern the properties of the solvent acid because of the above-described phenomenon of the levelling of acidic properties by the ionic solvents.

The correct quantitative data can be obtained only for molten alkali metal halides and, mainly for the molten KCl–NaCl eutectic as the reference melt for oxoacidity scales. These data are collected in Table 1.2.3.

Comparison of the data obtained with the use of the potentiometric titration technique shows that, as a rule, the equilibrium constants obtained using membrane oxygen electrodes exceed the corresponding values obtained by means of the gas oxygen electrode $Pt(O_2)$ by 3–4 orders of magnitude. This takes place owing to the change of the potential-defining process at the electrode when passing from the acidic solutions to the basic ones.

There is a trend towards the formation of polynuclear complexes in the solutions of oxocompounds of Group V of the Periodic Table in molten salts, whereas the oxocompounds of Group VI are not prone to the formation of

TABLE 1.2.3
Constants of the Lux acid–base equilibria in molten alkali metal halides at 700 °C

Equilibrium	$-pK$, molalities		
	KCl–NaCl	KCl–LiCl	NaI
$2\,PO_3^- + O^{2-} \rightleftharpoons P_2O_7^{4-}$	8.01 ± 0.1		
$P_2O_7^{4-} + O^{2-} \rightleftharpoons 2\,PO_4^{3-}$	5.93 ± 0.1		
$PO_4^{3-} + O^{2-} \rightleftharpoons PO_4^{3-}{\cdot}O^{2-}$	7.24 ± 0.1		
$CrO_3 + O^{2-} \rightleftharpoons CrO_4^{2-}$	8.41 ± 0.1		
$Cr_2O_7^{2-} + O^{2-} \rightleftharpoons 2\,CrO_4^{2-}$	7.18 ± 0.1	3.09 ± 0.3	
$2CrO_4^{2-} + O^{2-} \rightleftharpoons 2CrO_4^{2-}{\cdot}O^{2-}$	1.60 ± 0.2		
$MoO_3 + O^{2-} \rightleftharpoons MoO_4^{2-}$	8.32 ± 0.2		
$MoO_4^{2-} + O^{2-} \rightleftharpoons MoO_4^{2-}{\cdot}O^{2-}$	1.39 ± 0.3		
$WO_3 + O^{2-} \rightleftharpoons WO_4^{2-}$	9.31 ± 0.2		
$WO_4^{2-} + O^{2-} \rightleftharpoons WO_4^{2-}{\cdot}O^{2-}$	1.36 ± 0.5		
$B_4O_7^{2-} + O^{2-} \rightleftharpoons 4\,BO_2^-$	4.82 ± 0.1		5.02 ± 0.6
$BO_2^- + O^{2-} \rightleftharpoons BO_3^{3-}$	2.37 ± 0.2		~ 2
$V_2O_5 + O^{2-} \rightleftharpoons 2\,VO_3^-$	6.50 ± 0.4		
$2\,VO_3^- + O^{2-} \rightleftharpoons V_2O_7^{4-}$	5.23 ± 0.2		5.40 ± 0.3
$V_2O_7^{4-} + O^{2-} \rightleftharpoons 2\,VO_4^{3-}$	2.04 ± 0.5		1.68 ± 0.3
$VO_4^{3-} + O^{2-} \rightleftharpoons VO_4^{3-}{\cdot}O^{2-}$	0.79 ± 0.5		
$2\,GeO_2 + O^{2-} \rightleftharpoons Ge_2O_5^{2-}$	4.18 ± 0.4		

anions of the $Cr_2O_7^{2-}$ kind—at least, in diluted solutions. Complexes such as $R_2O_7^{2-}$ can be formed, mainly, in the sufficiently concentrated solutions, that occur in the titration from the base to the acid. Quantitative description of the Lux acid–base equilibria in molten salts is more complicated, just because of the formation of such polynuclear complexes—since, in this case, the equation of the material balance is not a linear function of the equilibrium molality of the acid studied.

As seen from the content of this Part of the book, the formation of basic products of addition of oxide ions to the "normal" neutral anions is the rule rather than the exception, and is connected with the higher basic properties of O^{2-} as compared with OH^-, which is the strongest of the possible bases in aqueous medium. Hydrolysis is the most obvious reason: it is well-known that the constant of the following equilibrium

$$O^{2-} + H_2O \rightarrow 2OH^-, \qquad\qquad (1.2.103)$$

exceeds a value of 10^{22} [21].

Certain acid–base equilibria considered above can serve as a basis for the determination of the relative acidity of high-temperature ionic solvents, both from the potentiometric titration data and the acidities of the corresponding buffer solutions. In any case, there are many data which can be used for the estimation of the pO and the buffer numbers of such solutions.

Part 3. Acid–Base Ranges in Ionic Melts. Estimation of Relative Acidic Properties of Ionic Melts

The known investigations of the acidic properties of high-temperature ionic melts as solvents for performing the Lux [17] acid–base interactions were carried out by two methods. The first approach implied the construction of "experimental" acid–base scales, and the second approach consisted in determination of the constants of acid–base equilibria in molten ionic media by various methods (kinetic, potentiometric, etc.). The scientific meaning of the first approach is narrower than that of the second one. The latter gives a possibility of estimating the effect of ionic-melt composition on acid–base reactions, and of obtaining the parameters connected with the melt acidity, whereas the former approach only allows one to estimate the relative positions of acids and bases and to assign them as weak or strong ones.

The first attempt to construct an "experimental" acid–base scale was made by Lux [17] who was the founder of oxoacidity as a branch of molten salts chemistry. Lux constructed the acid–base scale in the molten equimolar K_2SO_4–Na_2SO_4 mixture by adding equivalent quantities of acids (PO_3^-) and bases (Na_2O, Na_2O_2) to the pure melt, with subsequent registration of the e.m.f. Further, the e.m.f. values were recalculated as pO ones. The neutral point of this melt was found to be located at the oxide ion molality equal to 2×10^{-4} mol kg^{-1}, which corresponded to pO ≈ 3.7. Addition of acids led to a reduction in this concentration, whereas addition of bases exerted the reverse effect, i.e. $m_{O^{2-}}$ increased. Although this study gave rise to a number of similar works, the results obtained could not be considered as definite ones, since no equilibrium conditions were achieved in the studied acidic (basic) solutions. This was explained by evaporation of the acid of solvent (SO_3) from acidic solutions, and the base of solvent ($O^{2-} \equiv Na_2O$) from alkaline ones. Therefore, the addition of an acid or a base to a K_2SO_4–Na_2SO_4 melt caused, at first, an appreciable shift of e.m.f. from the corresponding value for the neutral point. Then, the e.m.f. was shifted back to the neutral point, owing to evaporation of the excess of SO_3 (acidic solutions) or Na_2O (basic solutions). This reason forced Lux to make the connection of e.m.f. with the moment of acid (base) addition that, of course, affected the accuracy of the values obtained and the validity of the conclusions.

Investigations of ionic-melt acidities by constructing the acid–base scales have been reported in some papers. Thus, Shams El Din *et al.* [90] built the acidity scale in molten KNO_3 at 350 °C. The acidity studies were performed by adding weights of substances corresponding to a molality of 0.01 mol kg^{-1} and measuring the equilibrium e.m.f. [90]. Then the studied acids and bases were arranged in the sequence according to the e.m.f. increase that corresponded to the rise of the solution acidity. After being arranged, the e.m.f. data were used to calculate the acid (base) number as compared with the pure nitrate melt. The bases possessed positive numbers, whereas for acids these values were negative. The acidity scale was constructed to be as follows: PO_3^- $(-11.2) \rightarrow H_2PO_4^-$ $(-11) \rightarrow Cr_2O_7^{2-}$ $(-10.75) \rightarrow CrO_4^{2-}$ $(-3.05) \rightarrow$ $HPO_4^{2-} (-1.7) \rightarrow P_2O_7^{4-}$ $(-0.85) \rightarrow NO_2^-$ $(-0.17) \rightarrow NO_3^-$ $(0) \rightarrow PO_4^{3-}$ $(1) \rightarrow HAsO_4^{2-}$ $(1.25) \rightarrow CO_3^{2-} (3.3) \rightarrow O^{2-} (4.65)$.

Addition of alkali metal halides to KNO_3 was shown to cause an increase of the equilibrium oxide ion concentration in the sequence $Cl^- \rightarrow Br^- \rightarrow I^-$, although these substances themselves could not generate oxide ions. The basicity strengthened with the increase of atomic number in Group VII in the manner: Cl^- $(0) \rightarrow Br^-$ $(2.3) \rightarrow I^-$ (3.3). This result may be explained by an increase of the reductive properties in this Group, that causes accumulation of nitrite ions in the said solutions, as NO_2^- ions are known to be sufficiently strong bases in molten salts. The main drawback of the method considered consists in errors which arise from a number of random factors. Thus, the positions of metaphosphate, PO_3^-, and dihydrogenphosphate, $H_2PO_4^-$, ions in the above sequence are easily explained since decomposition of dihydrogen-phosphate in the melt results in the formation of water which, in turn, gives rise to hydrolysis of the nitrate melt and to increase of the melt basicity. The determined relative positions of HPO_4^{2-} and $P_2O_7^{4-}$ are incorrect, since the decomposition of 0.01 mol of the former ion gives 0.005 mol of the latter. Therefore, the solution of the latter compound should be more acidic, since it contains twice as much acid than the former compound. Similar random errors led to displacement of neutral or slightly basic substances such as CrO_4^{2-} and NO_2^- into the acidic region. On the contrary, an acid $HAsO_4^{2-}$, equivalent to $\frac{1}{2}$ $As_2O_7^{4-}$ according to the data of Ref. [90], was classified among basic substances. The results of Ref. [90] characterize the length of the acid–base range in the nitrate melts discussed in detail in Part 1. The length of this range may be estimated to be within 18–20 of pO units, which hardly agrees with the data obtained in other ways [46].

Rybkin *et al.* constructed "experimental" acidity scales in molten KCl at 800 °C [62], CsI at 650 °C [63], NaI [64], and the equimolar KCl–NaCl mixture at 700 °C [65]. These studies are characterized by analogous disadvantages; nevertheless, they give rough knowledge about the strength of the Lux acids and bases in molten ionic media. Naturally, all that has been said in Part 1 allows us to understand that a solvent of the second kind possesses no acid–base product, and the term "acid–base range" is meaningless for these media and it cannot be connected with the data obtained in Refs. [62–65].

The width of the acid–base range in a particular medium has been shown in Part 1 to be understood as the difference in acidity (basicity) indices of the standard solutions of a strong acid and a strong base. If a solvent possesses its intrinsic acid–base equilibrium (solvent of the first kind) then its auto-dissociation products should be considered as the strongest acid and base. If we attempt to add a stronger acid or base (in pure form) to this solvent, then these substances will be transformed into the equivalent concentrations of the auto-dissociation products owing to solvolysis. This phenomenon has been mentioned above as "the levelling of acidic and basic properties"; it is non-specific, and is applicable to all acids and bases which are stronger than the corresponding constituent parts (particles) of the solvent. The specific levelling/differentiation of acidic (basic properties) is dependent on the chemical composition of the solvent and the nature of dissolved substance, and is observed upon the displacement of acid (base) from one solvent to another owing to the change of the interaction strength in the system "acid (base) residue–solvent".

Thus, owing to the levelling of acidic and basic properties, additions of the strongest acids and bases cannot result in a widening of the acid–base range in the solvents of the first kind. However, in oxygen-less ionic melts, which are solvents of the second kind, there is no intrinsic acid–base equilibrium resulting in the formation of oxide ions. Therefore, the basicity index (pO) in the "pure" melt varies within a rather wide range. As mentioned above, the solvents of the first kind are characterized by a neutrality point located at $pA = pB = \frac{1}{2}pK_L$, where $pA^* = -\log[A^*]$, $pB^* = -\log[B^*]$.

To elucidate the validity of the acidity scales obtained in molten alkali metal halides which belong to the solvents of the second kind, we shall consider the "principal" acid–base equilibrium, which is achieved by the addition of a strong acid to an ionic solvent of the second kind (an oxygen-less melt)

$$A+ \quad O^{2-} \quad = B$$

Before addition	–	$m'_{O^{2-}}$	–
Initial moment	m^0_A	$m'_{O^{2-}}$	–
Equilibrium	$m^0_A - x$	$m'_{O^{2-}} - x$	x

$$(1.3.1)$$

The equilibrium concentrations of the participants of this reaction are reciprocally dependent on one another, and the connection between them can be expressed using the Law of Mass Action

$$K_A = \frac{x}{(m^0_A - x)(m'_{O^{2-}} - x)}.$$

$$(1.3.2)$$

For the case of strong acid solutions whose concentrations considerably exceed the oxide-ion molality in the pure melt, the following assumptions are correct: $x \ll m^0_A$, $x \sim m'_{O^{2-}}$. Taking this fact into account one can rewrite equation (1.3.2) as

$$K_A = \frac{m'_{O^{2-}}}{m^0_A(m'_{O^{2-}} - x)}.$$

$$(1.3.3)$$

Transformation of this equation to logarithms and indices (negative logarithms) gives the relationship

$$pO \equiv -\log(m'_{O^{2-}} - x) = pm'_{O^{2-}} - pm^0_A - pK_A.$$

$$(1.3.4)$$

Hence, the pO value in a moderately concentrated solution of a strong acid is dependent on three parameters. They are the initial molality of oxide ions in "pure" melts (which, in their turn, are defined by the concentration and chemical composition of the oxygen-containing admixtures dissolved in the melt), the initial molality of the acid, and the equilibrium constant of equation (1.3.2). It is obvious that in the standard solution of a strong base, pO is equal to 0 and, in this case, the length of the acid–base range can be estimated according to equation (1.3.4). Since there are no parameters dependent on the solvent properties, and the initial oxide-ion molality in equation (1.3.4) and the equilibrium constant varies in wide ranges, the acidity scale is characterized by only one point for the standard solution of a strong base (namely, pO = 0). From the other side, this scale is practically unlimited and any experimentally found "upper" limit of acidity is not the real "boundary" of the acidity scale in such a kind of solvent.

Another disadvantage of the "experimental" acidity scales is the absence of serious quantitative data because, as a rule, a number of equilibria are reciprocally affected in the studied solutions. In this case, the estimation of some equilibrium constants based on one point is impossible. Therefore, more precise data on the acidity scales may be obtained by establishing equilibrium constants of acid–base reactions and the effect exerted by the cation and anion composition of an ionic melt on them. The regularities obtained on the basis of these parameters will help us to treat some aspects of the problem in question more correctly.

Thus, the method of constructing the acidity scales based on determination of the position of acid–base pairs using one experimental point (weight) cannot be considered correct. Also, similar scales for solvents of the first kind are distorted appreciably because of the levelling of acidic properties of strong acids which, according to the said method, should possess the same acidic properties.

Another approach to this problem was proposed by Tremillon *et al.* [169–171]. Their idea consists in the construction of the general acidity scale for chloride melts and determination of the positions of standard solutions of strong Lux bases on this scale. As media for acid–base reactions, chloride melts should be classified among solvents of the second kind, and, therefore, the positions of such solvents on the acidity scale should be described by only one point—the acidity function for the standard solution of a strong base.

Tremillon introduced the common oxoacidity function, Ω, as a measure of the relative acidity (basicity) of solutions of various substances in molten media. This parameter is similar to the well-known protic acidity functions, H_0 (Hammett function) and pA (Izmailov's acidity function) considered in Part 1. In terms of the primary medium effects, the function Ω may be represented as the sum of the primary medium effect of O^{2-} for the given medium ($-\log \gamma_{0,O^{2-}}$) and the instrumentally measured pO (pO_L) in this medium:

$$\Omega = -\log \gamma_{0,O^{2-}} + pO_L. \tag{1.3.5}$$

To construct the general acidity scale in molten media Combes *et al.* [169, 170] proposed to estimate the acidity functions for ionic melts by determining the constant of the following equilibrium:

$$H_2O + 2Cl^- \rightleftharpoons O^{2-} + 2HCl \qquad (1.3.6)$$

in this melt and to compare it with the similar constant obtained in the reference melt chosen previously.

According to this proposition, the common oxoacidity function, Ω, may be represented by the formula

$$\Omega = 14 + \log \frac{p^2_{HCl}}{p_{H_2O}}, \qquad (1.3.7)$$

and hence this oxoacidity function [169, 170] can be expressed using the equilibrium constant of reaction (1.3.6) in the following way:

$$\Omega = [14 - pK_{HCl/H_2O}] + pO. \qquad (1.3.8)$$

It is obvious that the expression enclosed in the brackets by the author of the present book is nothing but the primary medium effect of O^{2-} expressed *via* the difference in the values of the equilibrium constants of equation (1.3.6) for the media compared: the molten equimolar KCl–NaCl mixture, which was chosen as a reference melt, and for which pK_{HCl/H_2O} was found to be 14 at $\sim 700\,°C$, and the melt studied. As to the physical sense of the common acidity function Ω, this is equal to the pO of the solution in the molten equimolar KCl–NaCl mixture, whose acidic properties (oxide ion activity) are similar to those of the solution studied. Moreover, from equation (1.3.7) it follows that solutions in different melts possess the same acidic properties (Ω) if they are in equilibrium with the atmosphere containing HCl and H_2O and $p^2_{HCl}/p_{H_2O} = $ constant. This explanation confirms that the Ω function is similar to the Hammett function. Therefore, Ω values measured for standard solutions of strong bases in molten salts allow the prediction of the equilibrium constants on the background of other ionic solvents from the known shift of the acidity scales or the Ω value for the standard solution of a strong Lux base in the solvent in question. According to the assumption made in Refs. [169, 170] this value may be obtained if we know the equilibrium constant of the acid–base reaction (1.3.6) in the solvent studied.

Combes *et al.* [169, 170] substantiate the expediency of using reaction (1.3.6) by the fact that in molten ionic chlorides the molar fraction of chloride ion is equal to 1, and the activities of HCl and H_2O in a chloride melt are proportional to their partial pressures in the atmosphere over this melt.

The creation of constant partial pressures of both gases over different ionic solvents is easily achieved by passing an inert gas (N_2, Ar) through aqueous solutions of HCl of a certain concentration kept at a constant temperature. Under these conditions, passing this constant-composition gaseous mixture through different melts allows one to obtain solutions of the same Ω irrespective of the melt composition.

The above-mentioned papers [169, 170] report the determination of the upper limit of basicity which can be achieved in alkali metal chlorides and the melts containing polyvalent metal chlorides as constituent parts of the melt. Owing to their small ionic radii and high charge, these cations demonstrate elevated acidic properties. Increase of the initial molality of oxide ions finally results in the precipitation of the solid oxide, formed by such acidic cations, from the saturated solution. The oxoacidity function corresponding to this limit may be described by the following equation:

$$\Omega_{min} = 14 - pK_{HCl/H_2O} - \log s_{O^{2-}}. \qquad (1.3.9)$$

Investigations of $KCl-NaCl + xCeCl_3$ melts where the x values are within the range 0.05–0.30 (molar fraction) [169] show that an increase of the acidic cation concentration in the melt leads to a rise of the corresponding Ω_{min} value. Nevertheless, it remains in the range 13–14, whereas the acidity scale shift (Ω value of the standard solution of a strong base) is within 10–12. Strengthening of the melt acidity caused by the increase of the molar fraction of $CeCl_3$ in the melt from 0.1 to 0.3 reduces the solubility of Ce_2O_3 in the melt from $10^{-2.3}$ to $10^{-1.65}$. There is no linear correlation between the oxide solubility and acidity. Similar investigations of $KCl-NaCl-MgCl_2$ and $KCl-NaCl-CaCl_2$ melts with the addition of 20 mol% of the alkaline earth metal chlorides show that for the former melt the shift of the acidity scale is approximately equal to 14; in the latter case it is only 9. There are upper limits of basicity in both melts; in the former melt, Ω_{min} is approximately equal to 14.7, and in the Ca^{2+}-doped melt, Ω_{min} is close to 11.

There is no the upper limit to the basicity in the molten KCl–LiCl (0.41:0.59) eutectic mixture, and the shift of the acidity scale against the equimolar KCl–NaCl mixture is approximately equal to 8 [169].

Watson and Perry [172], and Picard et al. [173] studied the processes of pyrohydrolysis of alkali metal halide melts doped with $ZnCl_2$ at $\sim 600\,°C$ and established that these melts were more acidic than the KCl–LiCl eutectic.

The pO scales in the melts mentioned were shifted by three units in the acidic region, compared with the KCl–LiCl eutectic.

In concluding the discussion of Tremillon's approach, it should be mentioned that most ionic melts were investigated by indirect methods, which excluded the addition of solid base into the melt. The upper limits of basicity were achieved at low O^{2-} concentrations, which complicated investigations of other acids and bases by the weight-titration method.

All known investigations of the relative acidic properties of ionic melts are based just on the determination of the equilibrium constants of reaction (1.3.6) in these media. Since the applicability of these data for estimating the equilibrium parameters and conditions was not checked experimentally, there was no reason to doubt the correctness of the use of equilibrium (1.3.6) as an indicator one for Ω estimations. Indeed, we shall now consider the data presented in Ref. [169], where the acidic properties of the KCl–LiCl eutectic at 600 °C were investigated using equilibrium (1.3.6). The shift of the oxoacidity scale as compared with the KCl–NaCl equimolar mixture was found to be close to 8. Such a considerable difference in acidities for KCl–NaCl and KCl–LiCl melts is unexpected, since the index of the Li_2O dissociation constant (pK):

$$Li_2O = 2Li^+ + O^{2-} \tag{1.3.10}$$

on the background of the equimolar KCl–NaCl mixture was found to be p$K = 3.95$ [141]. The high acidity of the KCl–NaCl–$CaCl_2$ melt (0.4:0.4:0.2) was surprising too, and the acidity–scale shift was estimated to be approximately 9 [169].

If the Ω values obtained in Refs. [169, 170] may be attributed to the primary medium effect of O^{2-}, they allow us to consider the above-mentioned melts as extremely aggressive acidic media, analogous to anhydrous sulfuric acid in comparison with water ($H_0 \approx -\log \gamma_{0,H^+} = -11.1$) [174]. In particular, such melts should dissolve appreciable quantities of solid oxides, e.g. MgO or NiO, up to concentrations of the order of $1 \ mol \ kg^{-1}$. For example, MgO is practically insoluble in the equimolar KCl–NaCl mixture (p$P = 9.33 \pm 0.06$) [175], but should be characterized by values of the order of p$P \approx 1$ (since $9.33 - 8 \approx 1$) in the KCl–LiCl eutectic that corresponds to the value of the solubility product of the order of 10^{-1}. For the KCl–NaCl–$CaCl_2$ melt, the solubility-product value should be close to 1. On the contrary, most metal oxides which are slightly soluble in the KCl–NaCl melt were shown by

Shapoval, Makogon and Pertchik [176, 177] to remain slightly soluble in the KCl–LiCl eutectic. Besides, these oxides (NiO, CoO) are incompletely dissociated under these conditions [176].

The equilibria occurring in the hydrolysis of high-temperature chloride melts, carbonate ion dissociation and magnesium oxide solubility in the equimolar $CaCl_2$–KCl mixture at 575 °C were studied in Ref. [178]. The pK value of equilibrium (1.2.3) was found to be equal to 1.4, and the solubility product index of MgO:

$$Mg^{2+} + O^{2-} \rightleftharpoons MgO, \qquad (1.3.11)$$

was p$P_{MgO} = 5.3 \pm 0.1$. The shift of the values of these constants against those obtained in melts of weak acidity (e.g. the CsCl–KCl–NaCl eutectic, or the equimolar KCl–NaCl mixture) did not agree with the values predicted on the basis of the Ω values from the papers mentioned [169,170].

The above considerations led us to the conclusion that the use of equation (1.3.6), as a rule, yields appreciably distorted Ω values which, in their turn, complicate the estimations of molten salt acidity. One of the *pro* reasons is the fact that the use of this equilibrium implies $[Cl^-] =$ constant. This is true for the case of the melts based on alkali chloride melts—among them chloride–sulfate and chloride–nitrate melts—should be mentioned. However, reaction (1.3.6) cannot be used for estimating the acidic properties of melts without chloride ions. Besides, neutral (uncharged) HCl and H_2O molecules are "foreign" admixtures for ionic melts as ionic substances in aqueous medium: in both cases, there are two kinds of principally different interactions; electrostatic (between ions) and Van der Waals or donor–acceptor interactions (between uncharged particles).

Another argument against the use of reaction (1.3.6) as an indicative one is the sufficiently strong dependence of the partial pressures of HCl and H_2O on the concentration of HCl in the aqueous solution which is used as a source for obtaining of the mixture. Errors arising in the concentrated solutions are sizeable, especially if they concern solutions with HCl concentrations exceeding 30 mass%. The last, most important reason arises from the amphoteric character of H_2O. The assumption justifying the use of reaction (1.3.6) is that HCl is an acid and H_2O is the conjugate base. It should be mentioned, however, that water shows appreciable oxoacidic properties in molten salts—it fixes oxide ions with the formation of hydroxide. Moreover, it is obvious that when water vapour is passed through the melts they interact appreciably with the vapour. As is known, in the melts based on lithium salts the dependence of HCl and H_2O solubilities on their partial pressures deviates considerably from Henry's Law [179]. Under these conditions we can attribute acidic properties not only to Li cations but also to complexes of the composition $Li(H_2O)_n^+$. The concentration of the latter is connected with the hygroscopic properties of the cations forming the melt, and, therefore, one can expect that the melts formed by hygroscopic cations such as Ca^{2+} and Li^+ should show elevated acidity when their oxoacidic properties are estimated on the basis of the reaction (1.3.6). This fact is stated in Ref. [178].

In the general case, the problem of choosing the indicator equilibrium is still far from being solved. Similarly, experimental studies suggest that in the choice of the reference melt, the KCl–NaCl equimolar mixture is an appropriate reference medium, but this system does not exist in the liquid state at temperatures below 658 °C.

1.3.1 THE OXOBASICITY INDEX AS A MEASURE OF RELATIVE OXOACIDIC PROPERTIES OF HIGH-TEMPERATURE IONIC SOLVENTS

As follows from equation (0.1), the interaction between an ionic solvent and dissolved oxide ions (mainly) to account for cation particles of the solvent is one of the most important factors, which define the effect of the ionic-melt composition on the acid–base reactions which occur in it. Such an interaction may be represented by the following basic scheme:

$$L + O^{2-} \rightleftharpoons L \cdot O^{2-}, \tag{1.3.12}$$

where "L" designates a solvent particle.

The higher the degree of the mentioned interaction, the more complete the oxide-ion fixation in complexes, which shifts acid–base reactions (1.3.6) to the left. This can be found experimentally by changing the calculated values of the equilibrium constants of the acid–base reactions in the given ionic solvent as compared with the corresponding value in the reference melt.

The results, considered below, provide information from which the relative acidic properties of ionic solvents may be derived.

Ionic solvents, among which we mention molten alkali-metal halides, and their mixtures with alkaline metal halides, are characterized by the absence of constitutional oxygen (i.e. oxide ions). As stated in Part 1, they are referred to as the "solvents of the second kind". This means that there is no intrinsic acid–base equilibrium in these media, or limitations of the strength of Lux acids. Nevertheless, the strength of bases is levelled to the basicity of the complex formed by oxide ions and those cation particles of the solvent which show the strongest acidic properties in the given medium. For example, Tremillon et al. [180], on the basis of general considerations, concluded that the properties of the Lux bases in the equimolar KCl–NaCl mixture are levelled to the basic properties of Na_2O, but not of K_2O, since the following equilibrium:

$$K_2O + 2Na^+ \rightleftharpoons Na_2O + 2K^+ \qquad (1.3.13)$$

in the said melt is shifted to the right. Because of this, bases stronger than Na_2O do not exist in the melts of the KCl–NaCl series subjected to solvolysis. This conclusion from the data of Ref. [180] agrees with the present-day knowledge about the relative acidic properties of K^+ and Na^+ as Lewis acids.

In the general case, the basicity of the standard solution ($a_{O^{2-}} = 1$) of a strong base in a solvent of the second kind is dependent on the absolute concentration of "free" oxide ions in this solution, and it is an important fundamental characteristic of the oxoacidic properties of the said ionic solvent.

Considerations similar to those lying at the basis of the determination of the acidic properties of ordinary room-temperature molecular solvents [174] can be used to determine the relative acidic properties of high-temperature ionic melts-solvents. To estimate the positions of melts within the general oxoacidity scale, it is necessary to choose the indicator equilibrium and the reference melt. The basicity of the standard solution of a strong base in the reference solvent according to the definition should be equal to zero (0).

From the above-said it follows that the difference in the experimentally determined constants (pK) of the same equilibrium in different ionic solvents should reflect the difference in their acidic properties. In order to deduce a parameter which describes the relative oxoacidic properties of ionic melts, let us consider the system of acid–base reactions of oxide-ions with the melt cations, assuming that the "absolute" concentrations of the latter determine the melt acidity:

$$i\mathrm{Me}^{m+} + \mathrm{O}^{2-} \rightleftharpoons \mathrm{Me}_i\mathrm{O}^{i \cdot m - 2}, \mathrm{K}_{\mathrm{L},i}. \tag{1.3.14}$$

where Me^{m+} is the designation of the most acidic cation of the melt. Naturally, increase of $K_{\mathrm{L},i}$ values causes a shift of the equilibrium (1.3.14) to the right. The distribution of oxide ions dissolved in the ionic melt, between different complexes formed by melt cations, can be represented by the following sum (where N_x is the molar fraction of component "x"):

$$N^0_{\mathrm{O}^{2-}} = N_{\mathrm{O}^{2-}}\left(1 + \sum_{i=1}^{n} K_{\mathrm{L},i} \cdot N^i_{\mathrm{Me}^+}\right). \tag{1.3.15}$$

The relationship between the absolute equilibrium concentration of "free" oxide ion and the initial concentration of oxide ion may be found from equation (1.3.15). We shall designate it as, I_L

$$I_\mathrm{L} = N_{\mathrm{O}^{2-}}/N^0_{\mathrm{O}^{2-}} (= m_{\mathrm{O}^{2-}}/m^0_{\mathrm{O}^{2-}}) = 1 \bigg/ \left(1 + \sum_{i=1}^{n} K_{\mathrm{L},i} \cdot N^i_{\mathrm{Me}^+}\right). \tag{1.3.16}$$

High-temperature ionic solvents are known to contain relatively high total concentrations of cations (e.g. in the KCl–LiCl eutectic, the concentration of Li^+ is approximately equal to 8.5 mol kg^{-1} of the melt). Usually, cation–anion complexes in molten salts are characterized by co-ordination numbers of the order of 4–6. This means that the maximal consumption of acidic cations does not exceed 0.4–0.6 mol kg^{-1} in diluted solutions with concentrations close to 0.1 mol kg^{-1}. This estimate is considerably lesser than the initial concentration of acidic cations in the "pure" melt. In the case of the KCl–LiCl eutectic melt, this consumption is only of the order of 5–7%, and the value of N_{Me^+} in equation (1.3.16) may be assumed to be constant. Therefore, for each ionic solvent of the second kind (kind II) the denominator in equation (1.3.16) is a constant which characterizes its acidic properties. We shall define $pI_\mathrm{L} = -\log I_\mathrm{L}$ to be the relative measure of acidic properties of a solvent and call it the "oxobasicity index of ionic melt" [37, 162, 181]. Since the direct determination of the absolute concentration of "free" oxide ions in molten salts is practically impossible, the reference melt should be chosen— for this melt, I_L is assumed to be 1 and $pI_\mathrm{L} = 0$. The equimolar KCl–NaCl mixture, of weak acidity, is one of the most studied ionic melts. Its status is similar to that of water among molecular liquids that allows us to choose the equimolar KCl–NaCl mixture to be the reference melt. The second possible reference melt is the CsCl–KCl–NaCl eutectic mixture, whose acidic properties are close to those of the KCl–NaCl equimolar mixture, but the former exists in the liquid state over a wider temperature range; its melting point is near 480 °C [182]. This feature of the ternary eutectic mixture allows us to perform experimental investigations of the relative acidities of ionic solvents at temperatures lower than 650 °C; under these conditions the equimolar KCl–NaCl mixture solidifies. Further, by analogy with the Hammett approach [174], the indicator equilibrium should be chosen. The difference of the experimentally determined constants of this equilibrium should be attributed only to the melt acidity changes and, hence, to the changes in the oxobasicity index.

For developing a more general basis for the oxobasicity index (irrespective of the anion and cation compositions of ionic melts), the following scheme [37, 183, 184] should be analysed.

The reference melt, $\Delta G^* = -RT \ln K^*$

$$A^* + O^{2-*} = B^*$$

$$-\log \gamma_{0,A} - \log \gamma_{0,O^{2-}} - \log \gamma_{0,B}. \qquad (1.3.17)$$

$$A_L + O_L^{2-} = B_L$$

The melt studied, $\Delta G_L = -RT \ln K_L$

According to Scheme (1.3.17), the Lux acid–base equilibrium in the reference melt is described by the Gibbs energy, ΔG^*, to which the value of K^* corresponds. Investigation of another melt results in the following thermodynamic parameters, the Gibbs energy is ΔG_L and the equilibrium constant is equal to K_L. The change of ΔG ($= RT \ln K$) is equal to the sum of the primary medium effects for the acid, the base and the oxide ions; these values are $-\log \gamma_{0,A}$, $-\log \gamma_{0,B}$ and $-\log \gamma_{0,O^{2-}}$, respectively. If the following approximation

$$\log \gamma_{0,A} - \log \gamma_{0,B} \approx 0, \qquad (1.3.18)$$

is correct, then all the $\Delta G(K)$ changes can be attributed only to the primary medium effect of O^{2-} – $\log \gamma_{0,O^{2-}}$, whose value defines the difference in acidity for the reference melt and the melt studied. In this case, the difference of the equilibrium-constant indices (pK) in the said melts

$$\Delta pK \equiv -\log(K_L/K^*) = (\Delta G_L - \Delta G^*)/2.3RT, \qquad (1.3.19)$$

yields the oxobasicity index value of the solvent studied, L:

$$pI_L = pK_L - pK^* \sim -\log \gamma_{0,O^{2-}}. \qquad (1.3.20)$$

The considerations stated above show that the oxobasicity index of ionic solvents is the concentration analogue of the primary medium effect for H^+ in molecular solvents [174], which allows us to compare the acidic properties of different protolytic and aprotic solvents with respect to H^+ as an acidic particle.

The necessary condition for the use of any acid–base equilibrium as an indicator one requires closely similar medium effects for the acid A and the conjugate base B upon changes of solvent, i.e. the correctness of the condition (1.3.18) should be maintained.

Homogeneous Lux acid–base equilibria similar to equation (0.1) have not been considered earlier as convenient indicator reactions allowing the determination of the oxobasicity indices of ionic melts. Nevertheless, the equilibrium constant of such an equilibrium, pK_L, in the melt studied should be expressed as a function of the concentrations and the oxobasicity index in the following manner (the 1:1:1 stoichiometry is assumed)

$$pK_L = -\log \frac{N_B}{N_A(N_{O^{2-}} \cdot I_L^{-1})} = pK_{KCl-NaCl} + pI_L, \qquad (1.3.21)$$

on the premise that the condition (1.3.18) is correct. Homogeneous equilibria, which obey this rule (A = $Cr_2O_7^{2-}$ or VO_3^-) have been considered in Part 2. The $Cr_2O_7^{2-}/CrO_4^{2-}$ acid–base pair is especially convenient for the acidity estimations of molten alkali metal halides since both anions are doubly charged particles and have larger radii in comparison with halide ions.

Finally, the suitability of the oxide solubility data for estimations of the oxoacidic properties of molten media should be considered. The solubility product of a metal oxide, MeO, P_{MeO}, can be presented as

$$P_{MeO,L} = N_{Me^{2+}}(N_{O^{2-}} \cdot I_L^{-1}) = P_{MeO,KCl-NaCl} \cdot I_L^{-1}, \qquad (1.3.22)$$

or

$$pP_{\text{MeO,L}} = pP_{\text{MeO,KCl}-NaCl} - pI_{\text{L}}.\tag{1.3.23}$$

The use of the solubility method for estimating the acidic properties of melts with the same anionic composition is sufficiently substantiated, since the Gibbs energy of MeO precipitated as a solid phase is independent of the ionic-melt composition. For Me^{2+} cations, their most probable forms of existence in halide melts are complex anions of the composition MeX_4^{2-}. Therefore, for anions of the same composition it may be assumed that $-\log \gamma_{0,\text{Me}^{2+}} \approx 0$. Taking into account the fact that $-\log \gamma_{0,\text{MeO}} = 0$ (see the first sentence of this paragraph), the condition (1.3.18) is adequately justified.

Limitations in the use of the solubility method are caused, in the first place, by the different stabilities of the metal-halide complexes formed by acidic cations with various halide ions. Therefore, the solubility data can only be used to estimate the oxobasicity indices of ionic solvents having the same anion composition. When the latter is changed, the magnitudes obtained contain considerable errors, because the Me^{2+} designation means essentially differing anion complexes of different stability, e.g. MeCl_4^{2-} and MeBr_4^{2-} complexes formed in chloride and bromide melts, respectively. So, aside from the cation composition, the solubility-product changes in high-temperature ionic solvents are also affected considerably by the anion composition of these solvents.

In order to know what kind of indicator equilibrium has been used to determine the value of the oxobasicity index of a melt, there should be introduced an additional subscript for the designation of the oxobasicity index, by analogy with the Hammett function. If this value is obtained using the solubility data, then the designation $pI_{\text{L,s}}$ is to be used ("s" means "solubility"). If an indicator homogeneous acid–base reaction is used, this designation should be $pI_{\text{L,h}}$, and for reactions with the participation of gaseous acids such as H_2O and CO_2, it is $pI_{\text{L,g}}$. This additional information will be very useful for practical applications of the oxobasicity indices when estimating equilibrium parameters of the Lux acid–base reactions, the melts having different anion and cation compositions.

The oxobasicity index allows us to connect the instrumental values of pO obtained by performing a variety of experiments, with the values of the common oxoacidity function, Ω, by a rather simple relationship

$$\Omega = pI_{\text{L}} + pO_{\text{L}},\tag{1.3.24}$$

which permits the immediate recalculation of the values of pO in the common oxoacidity functions if one has information about the oxobasicity index values.

1.3.2 OXOACIDITY SCALES FOR MELTS BASED ON ALKALI- AND ALKALINE-EARTH METAL HALIDES

As heterogeneous equilibria in "solid oxide–ionic solvent" systems are the most investigated in molten chlorides, the general acid–base scale for such melts may be constructed using the solubility method.

Investigations of the solubilities of oxides in molten alkali metal chlorides, and in the melts based on these salts at 600 °C, are not numerous. The only available data are on MgO solubility in two Ca^{2+}-containing chloride melts: $CaCl_2$–$NaCl$ [178] and $BaCl_2$–$CaCl_2$–$NaCl$ [185], which possess considerable acidic properties. We have studied oxide solubilities in the $CsCl$–KCl–$NaCl$ eutectic melt [186] as a melt having weak acidity, and in the KCl–$LiCl$ eutectic melt [187]. These investigations gave the possibility of determining the oxobasicity indices of the said melts, considering the $CsCl$–KCl–$NaCl$ melt as the reference one. The oxobasicity indices were shown above to give exhaustive information about the acidic properties of these molten media.

As follows from Part 1, the ionic melts based on molten alkali metal halides are referred to the solvents of the Second Kind (Kind II), and, therefore, the acid–base ranges for these media are half-open (see Fig. 1.1.1, scheme N3). Therefore, to form an idea of the relative oxoacidic properties of the studied chloride melts it is enough to know their oxobasicity indices. The necessary experimental parameters obtained at 600 °C are presented in Table 1.3.1. The data in this Table show that the KCl–$LiCl$ eutectic melt possesses appreciable acidic properties, the corresponding oxobasicity index being equal to 3.2.

TABLE 1.3.1
Solubility products of MgO in some chloride melts at ~ 600 °C and their oxobasicity indices (with respect to the $CsCl$–KCl–$NaCl$ eutectic)

Melt/composition	pP_m	pP_N	$pI_{L,s}$
$CsCl$–KCl–$NaCl$ (0.455:0.245:0.30)	10.78	12.68	0
KCl–$LiCl$ (0.41:0.59)	6.99	9.45	3.2
$BaCl_2$–$CaCl_2$–$NaCl$ (0.235:0.245:0.52)	6.2	8.15	4.5
$CaCl_2$–$NaCl$ (0.5:0.5), 575 °C	5.3	7.44	5.2

The oxoacidic properties of $CaCl_2$-containing melts are considerably stronger. The oxobasicity index of the $BaCl_2-CaCl_2-NaCl$ eutectic melt is higher by 1 (one unit) compared with the molten KCl–LiCl eutectic. The oxoacidic properties of the $CaCl_2-NaCl$ equimolar mixture (~ 5.2) are stronger compared with the $BaCl_2-CaCl_2-NaCl$ eutectic, owing to the higher Ca^{2+} concentration in this melt: the molar fraction of $CaCl_2$ in the $CaCl_2-NaCl$ melt is 0.5. It should be noted that the two-fold growth of Ca^{2+} concentration as compared with the $BaCl_2-CaCl_2-NaCl$ eutectic increases the oxobasicity index by 0.7.

Hence, the oxoacidic properties of the ionic melts based on alkali metal chlorides at 600 °C strengthen in the following sequence: CsCl–KCl–NaCl $\rightarrow$ KCl–LiCl $\rightarrow$ $CaCl_2-BaCl_2-NaCl$ $\rightarrow$ $CaCl_2-NaCl$.

A number of our works are devoted to the investigations of different kinds of acid–base equilibria in the ionic melts based on alkali metal halides in order to determine their oxobasicity indices at 700 and 800 °C. Unfortunately, none of the necessary experimental data have been published by other investigators. The equimolar KCl–NaCl mixture has been chosen to be the reference melt at these temperatures, although its oxoacidic properties differ by less than 0.1 from the CsCl–KCl–NaCl eutectic (see below), i.e. they are practically coincident. The solubilities of 11 metal oxides in the equimolar KCl–NaCl mixture are reported in Ref. [175]. Similar investigations in the molten CsCl–KCl–NaCl eutectic [188] allow us to conclude that the solubility products of the same oxide (in molar fraction scale) in both melts are close. This leads to the conclusion that both melts are suitable as reference ones, not only at 700 °C but also at other temperatures at which these media exist in the liquid state.

Investigations of oxide solubilities in the molten KCl–LiCl eutectic at 700 °C have been reported in Refs. [162, 189, 190]. Oxide solubilities in this melt were found to be considerably larger than those in the equimolar KCl–NaCl mixture at the same temperature (see p. 296). The value of the $pP_{N,MeO}$ shift as compared with the KCl–NaCl was calculated to be close to 3.4 logarithmic units, this value being the oxobasicity index, $pI_{KCl-LiCl,s} = 3.4$. It is remarkable that the shifts are so close to each other, which demonstrates the applicability of the parameter introduced for the melt acidity estimations—at least, for calculations of the solubility parameters. Aside from the solubility investigations, the equilibrium constant of the addition of O^{2-} to dichromate ion, $Cr_2O_7^{2-}$ in the KCl–LiCl eutectic at 700 °C was performed in Ref. [163]. The shift of pK value in this melt as compared with the KCl–NaCl equimolar mixture was approximately equal to 3.6 pK units that made it possible to estimate p$I_{KCl-LiCl,h}$ as 3.6. Therefore, a good agreement of the value obtained from solubility data and the shift of pK of the homogeneous reaction showed the oxobasicity index to be a good mean for estimation of constants of different acid–base equilibria in various ionic media.

Besides the studies discussed above, oxide solubilities have been determined in ionic melts consisting of alkali-metal chlorides and alkaline-earth metal chlorides; the concentrations of latter were close to 0.25 (molar fractions). The compositions of these melts are as follows: $CaCl_2$–KCl (0.235:0.765) [191], $SrCl_2$–KCl–NaCl (0.22:0.36:0.42) [192], $BaCl_2$–KCl–NaCl (0.43:0.29:0.28) [193],and $BaCl_2$–KCl (0.26:0.74) [194]. Practically, the same shifts of the solubility product indices with changes of the cation composition of the melts were seen in the experimental results for all the oxides studied. The addition of alkaline-earth metal chlorides to the melts having weak acidity strengthens their acidic properties, which results in the increase of oxide-solubility. The rise in oxoacidity is observed in the sequence "melt + Ba^{2+} → melt + Sr^{2+} → melt + Ca^{2+}, which agrees with the well-known tendency of strengthening the acidic properties in the same Group of the Periodic Table together with the decrease of the nuclear charge. The average shifts of p$P_{N,MeO}$ substantiated above are the same as the oxobasicity indices of the studied chloride melts.

As shown above, the solubility data are appropriate for estimations of acidity of ionic melts unless these differ in anionic composition. As for iodide melts, their acidities cannot be compared with that of KCl–NaCl in such a manner. To solve this problem, the reactions of oxide ion addition to $B_4O_7^{2-}$ and to VO_3^- anion acids were studied in the equimolar KCl–NaCl mixture and NaI melt at 700 °C [166].

These reactions were considered as indicator ones. The corresponding pK values are collected in Table 1.3.2. Some increase of acid strength in the iodide melt seen from the data in the Table causes practically

TABLE 1.3.2
pK values of some acid–base equilibria in the equimolar KCl–NaCl mixture and NaI at 700 °C

Equilibrium	KCl–NaCl	NaI
(2.91)	4.82 ± 0.2	5.02 ± 0.3
(2.87)	5.23 ± 0.2	5.40 ± 0.2

coincident pK shifts for both reactions. This magnitude is the oxobasicity index of molten sodium iodide, p$I_{NaI,h}$, equal to -0.2. The negative value means that the iodide melt possesses weaker acidic properties than the KCl–NaCl equimolar mixture, although the latter consists not only of Na$^+$ cations but also of K$^+$ cations, which are weaker as acids compared to Na$^+$. So, molten iodide melts show weaker acidity than do molten chlorides of similar cation compositions.

All above-mentioned experiments have made a basis for the construction of the oxoacidity scale of melts based on alkali metal halides at 700 °C [195]. This scale is presented in Fig. 1.3.1. The studied ionic melts may be divided into three groups. The first group includes the melts of weak acidity, such as the KCl–NaCl equimolar mixture, the ternary CsCl–KCl–NaCl eutectic possessing the p$I_{CsCl-KCl-NaCl,s}$ which is equal to -0.1, and the NaI melt.

The second group consists of the melts based on alkali- and alkaline-earth metal chlorides such as BaCl$_2$ and SrCl$_2$. This group is characterized by oxobasicity indices of the order of 2. With closely equal contents of alkaline-earth metal chloride in the melts, the acidity of Ba^{2+}-containing melts is found to be close to that of the Sr^{2+}-containing ones (see Fig. 1.3.1), i.e. the oxobasicity indices of BaCl$_2$–KCl with $N(Ba^{2+}) = 0.26$ and SrCl$_2$–KCl–NaCl with $N(Sr^{2+}) = 0.22$ differ by 0.1.

The effect of increasing the content of acidic cation in ionic melts was determined for the Ba^{2+}-containing melts, and shown to raise their oxobasicity indices—and, hence, to strengthen their acidic properties. The experimental values of the oxobasicity indices, namely: p$I_{BaCl_2-KCl-NaCl,s} = 2.01$ and p$I_{BaCl_2-KCl,s} = 1.83$ lead to the conclusion that when the Ba^{2+} concentration in chloride melts increases by two times, which

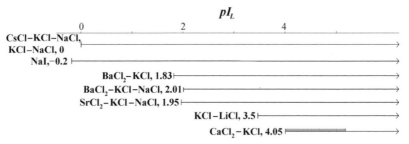
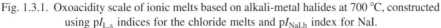

Fig. 1.3.1. Oxoacidity scale of ionic melts based on alkali-metal halides at 700 °C, constructed using p$I_{L,s}$ indices for the chloride melts and p$I_{NaI,h}$ index for NaI.

corresponds to the following expression for the melts studied

$$-(\log N_{Ba^{2+},Ba,K,Na\|Cl} - \log N_{Ba^{2+},Ba,K\|Cl}) = 0.21, \tag{1.3.25}$$

then $pI_{L,s}$ increases by 0.18. This makes it possible to estimate the oxobasicity index of melts which differ somewhat in the content of the same acidic cation, according to the following relationship:

$$pI_{L_1,s} = pI_{L,s} + \log \frac{N_{Me^{n+},L_1}}{N_{Me^{n+},L}}, \tag{1.3.26}$$

where N_{Me^{n+},L_1} and $N_{Me^{n+},L}$ are the molar fractions of the most acidic cation in the melts with the known oxobasicity index (denoted as "L") and the melt in question (with subscript, L_1), respectively.

The melts of the highest acidity belonging to the third group are the KCl–LiCl eutectic mixture and the equimolar $CaCl_2$–KCl melt. The upper limit of basicity in the latter melt is caused by the limited solubility of CaO in it (the addition of O^{2-} above such a limit results in the precipitation of CaO from the saturated melt). Similar behaviour of the $CaCl_2$–KCl equimolar mixture is observed when other oxide ion donors (e.g. NaOH, KOH) are introduced into it.

We shall now consider Fig. 1.3.2. The initial additions of the Lux base which result in $pO^* > 1.3$ are accompanied by an e.m.f. reduction, according to the Nernst equation. At pO^* values lower than 1.3 there arises a "plateau" in

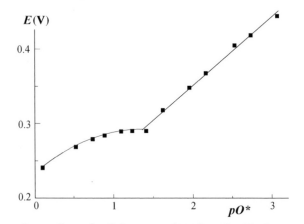

Fig. 1.3.2. Dependence of e.m.f. of the potentoimetric cell with the membrane oxygen electrode against pO^* in the molten $CaCl_2$–KCl equmilar mixture at 700 °C. The onset of the "plateau" is connected with CaO precipitation.

the calibration plot. This plateau occurs because of CaO deposition from its saturated solution in the $CaCl_2$–KCl melt. A subsequent small reduction in e.m.f. is caused by the fact that the addition of a strong Lux base (in the case considered it is KOH) results in its exchange reaction with $CaCl_2$ and precipitation of the formed CaO

$$2\,KOH + CaCl_2 \rightarrow 2\,KCl + CaO\downarrow + H_2O. \tag{1.3.27}$$

The running of this reaction reduces the $CaCl_2$ concentration in the melt that gives rise to the corresponding changes of its composition. This means that in the section of the calibration dependence (Fig. 1.3.2) at $pO^* < 1$ the $CaCl_2$–KCl melt of composition 0.235:0.765 does not exist at all. On the contrary, there is no precipitation of alkaline-earth metal oxide when the melts belong to the above-considered second group (Ba^{2+}- and Sr^{2+}-based chloride melts). The reason for this consists, first, in the higher solubilities of BaO and SrO as compared with CaO. The stronger acidities of the melts doped with alkaline earth metal cations favour a subsequent solubility increase. The second reason lies in the stabilization of oxide ions in the said melts owing to the formation of polynuclear complexes, e.g. Me_2O^{2+}. Any similar effect should be neglected in conventional solubility investigations where the contents of the acidic cations studied do not exceed $0.1\ mol\ kg^{-1}$. However, it becomes very important for highly concentrated solutions in ionic melts, and it is observed in the considered chloride melts.

For the KCl–LiCl eutectic, some of our investigations give an excellent illustration of the levelling of acidic properties by high-acidic melt cations. For this purpose, we now consider Fig. 1.3.3, where the results of Pb^{2+} and Cd^{2+} titration with KOH are presented. The behaviour of the KCl–LiCl eutectic melt doped with a $0.05\ mol\ kg^{-1}$ addition of these cations does not differ from that of the "pure" melt. That is, the changes of the oxoacidic properties of the solutions, as compared with the pure eutectic, cannot be detected. This is because the acidic properties of Pb^{2+} and Cd^{2+} cations are not stronger than those of Li^+, and the following equilibrium:

$$PbO(CdO) + 2Li^+ \rightarrow Pb^{2+}(Cd^{2+}) + Li_2O, \tag{1.3.28}$$

is shifted to the right—at least in dilute solutions of the corresponding chlorides. The dissociation constant of PbO in the KCl–NaCl equimolar mixture at $700\,°C$ was determined by potentiometric titration to be 5.13×10^{-4} ($pK = 3.29 \pm 0.04$) [175], whereas the pK value of Li_2O

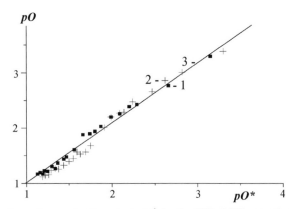

Fig. 1.3.3. The dependence of pO against pO* in the KCl–LiCl eutectic melt containing: 0.050 mol kg^{-1} of PbCl$_2$ (1) and 0.045 mol kg^{-1} of CdCl$_2$ (2). The plot obtained in the pure KCl–LiCl is presented for a comparison with KCl–LiCl.

dissociation in this melt was equal to p$K = 3.95$ [141]. This observation proves that the acidity of Li$^+$ cation exceeds that of Pb^{2+} and Cd^{2+}, and the dissociation constants of PbO and CdO were too large to be determined even by the potentiometric method [175]. The levelling of the acidic properties of the said cations may be predicted from the known oxobasicity index of the KCl–LiCl eutectic (Fig. 1.3.1): this value (p$I_{KCl–LiCl,s} = 3.4–3.5$) exceeds pK_{PbO}, which is equal only to 3.29.

Temperatures of the order of 800 °C provide an essentially broader scope for studying acid–base interactions in molten salts, since practically all the single alkali-metal halides liquefy under these conditions (excluding NaCl, whose melting point is given as close to 801 °C [121] and 808 °C in Ref. [136]). This permits us to estimate the oxobasicity indices of a large set of ionic media and to trace the effect of the anion composition of alkali metal halide melts on their acidic properties. To achieve this, investigations of some acid–base reactions in molten alkali halide melts were performed. The results of the studies of oxide solubilities (mainly, MgO and CaO) and carbonate ion dissociation are reported in some of our papers. Since the oxide solubility data are shown to be unsuitable for the estimation of oxobasicity indices in ionic media of different anion compositions, the reaction (1.2.3) of CO$_3^{2-}$ dissociation is considered a promising indicator equilibrium. It is believed to be applicable to the case of melts having the same cation-, but different anion composition [196]. The substantiation is given below.

We shall consider a single alkali-metal-halide melt, in which carbonate ion is dissolved in the form of an alkali metal carbonate. The carbonate-ion dissociation reaction occurs in such a solution as

$$Me_2^I CO_3(l) = Me_2^I O(l) + CO_2(gas). \qquad (1.3.29)$$

Assuming that the activities of CO_3^{2-} and CO_2 are independent of the anion composition of the ionic melt, there may be formulated a non-thermodynamic hypothesis that the changes of the equilibrium constant of equation (1.3.29) in various melts are caused only by the difference in their oxoacidic properties.

To estimate the relative oxoacidic properties of molten alkali-metal halides, the carbonate-ion dissociation reaction was studied for the following sets of melts at 800 °C: KCl–KBr–KI [183], CsCl–CsBr–CsI [197], NaCl–NaBr–NaI (830 °C) [198]. Also, in order to arrange all the melts in the oxoacidity scale at 800 °C, the solubility of MgO was determined for the following melts: CsCl, KCl, NaCl, KCl–NaCl, KCl–LiCl, CsCl–KCl–NaCl, which allowed the estimation of the oxobasicity indices by the solubility method.

The dissociation constants of equation (1.2.3) in KCl and CsCl melts were found to differ considerably. If these melts were of different acidities, this would be quite natural. However, KCl and CsCl were stated by the oxide-solubility investigations to show the same acidity, owing to practically the same acidities of K^+ and Cs^+ as cations of the same charge, and the close radii. Such a difference of the equilibrium constants is seemingly caused by different stabilities of single alkali-metal carbonates. Therefore, it is necessary to offset the difference in CO_2 pressures over single carbonates to estimate the relative oxoacidic properties of molten alkali metal halides. For example, the oxobasicity indices of caesium halides may be obtained from those of potassium halides using the following relationship:

$$pI_{CsX,g} - pI_{KX,g} = \log \frac{K_{(2.02),CsX}}{K_{(2.02),KX}} - \log \frac{P_{CO_2,Cs_2CO_3}}{P_{CO_2,K_2CO_3}} \qquad (1.3.30)$$

where $pI_{CsX,g}$ and $pI_{KX,g}$ are the oxobasicity indices of caesium and potassium halides, $K_{(2.02),CsX}$ and $K_{(2.02),KX}$ are the dissociation constants of carbonate ions in the said melts, P_{CO_2,Cs_2CO_3}, and P_{CO_2,K_2CO_3} are the decomposition pressures over the corresponding single carbonates. Similar considerations may be used to calculate the relative acidities of molten sodium halides. The above-described approach was applied to the treatment of a set of experimental data that resulted in the building of the oxoacidic scale of ionic melts. Its construction is based on the following considerations (Fig. 1.3.4).

According to Fig. 1.3.4, in order to determine the oxobasicity index of NaBr with respect to the CsCl–KCl–NaCl eutectic it is necessary to know, at least, two experimental magnitudes. The first set is the oxobasicity index of NaCl, and the difference of pK of CO_3^{2-} dissociation in molten NaBr and NaCl. The second set is the difference of pK of CO_3^{2-} dissociation in molten CsBr–KBr–NaBr and CsCl–KCl–NaCl and the difference of pP_{MeO} in molten CsBr–KBr–NaBr and NaBr. The former case is obviously more convenient since the investigation of the oxide solubility products in molten bromides and iodides is more complicated than the determination of the pK of CO_3^{2-} in the said melts.

Thus, the first step of the construction is to determine the oxobasicity indices of the alkali-metal chloride melts. Then the similar magnitudes for the melts having the same cation but different anion compositions were estimated as the differences of pK of carbonate ion dissociation. For example, in molten NaBr the estimation was performed according to the following formula:

$$p I_{NaBr,g} = p I_{NaCl,s} + (pK_{CO_3^{2-},NaCl} - pK_{CO_3^{2-},NaBr}). \tag{1.3.31}$$

The oxoacidity scale of molten salts at 800 °C is presented in Fig. 1.3.5. The oxoacidic properties of the chloride melts are stated to be close, whereas those of the bromide ones differ appreciably. As a rule, the bromides are more acidic than the corresponding chlorides. Similar to the single chloride melts, the molten iodides are of the same acidities, which are weaker that those of the chloride melts. Molten bromides are found to possess higher acidities in the "chloride–bromide–iodide" sequences for potassium and

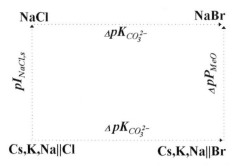

Fig. 1.3.4. Probable variants of the oxobasicity index determination in the melts of different cation and anion compositions.

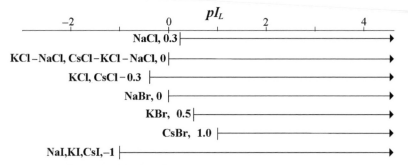

Fig. 1.3.5. Oxoacidity scale of ionic melts based on alkali-metal halides at 800 °C, constructed using $pI_{L,g}$ indices (for the chloride melts the $pI_{L,s}$ indices were used).

caesium halides. The chlorides occupy the intermediate position, and the iodide melts have the weakest acidity. This is not the case when molten sodium halides are studied. In these Na^+-based melts, a monotonic decrease of oxoacidic properties is observed in the sequence of halides given above.

A qualitative explanation of the tendencies observed may be obtained in the framework of the complexation ability of halide anions. The alkali metal cations, whose "absolute" concentration in ionic melts defines their acidities, are surrounded by anions in such media, and near the anions there exists a neighbouring cation. The higher the "cation–anion" complex stability the lower the "absolute" cation concentration in the melt, which weakens its oxoacidic properties.

The complexation ability of anions is shown in Ref. [81], to rise together with the increase of their ionic moments and polarization. In the halide-ion sequence their formal ionic moments (r_X^{-1}, nm^{-1}) decrease from F^- to I^-, whereas the polarization (α, $Å^3$) rises in the same sequence, as shown in Fig. 1.3.6. The concurrence of the actions of these factors results in the non-monotonic character of the complexation ability changes in the "F–Cl–Br–I" sequence.

It is a remarkable feature that the experimental data obtained in Na^+-based halides had led Smirnov to the conclusion that the minimum complexation ability takes place in chloride melts [81], but it is not so for potassium- and caesium-based halides. Therefore, the stability of "alkali metal cation–halide ion" is dependent not only on the anion properties but also on those of the cation. In the potassium and caesium halides this reason shifts the minimum of the complexation ability to the bromide melts, which makes these media more acidic than the corresponding chlorides and iodides.

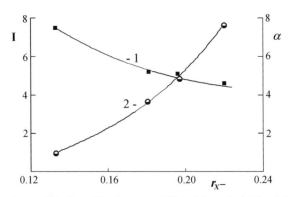

Fig. 1.3.6. Dependence of the formal ionic moment (I) and the polarization (α) of halide ions vs their radii (nm).

1.3.3 CONCLUSION

The effect of ionic melts on the acid–base processes occurring in their background is shown in this Part to be described successfully by the common oxoacidity function, Ω.

The oxoacidic properties of an ionic melt can be characterized by the oxobasicity index defined as the difference of the indices (negative logarithms) of "absolute" concentrations of oxide ions in standard solutions of a strong Lux base in the reference melt and in the melt studied.

The equimolar KCl–NaCl mixture and the ternary CsCl–KCl–NaCl (0.455:0.245:0.30) eutectic are shown to be convenient reference melts. The obvious advantage of the latter melt is its low melting point ($\sim 480\,°C$), which allows its use over a wider temperature range.

Further development of the problem of general oxoacidity scales in molten salts may be connected with finding more suitable indicator equilibria, and obtaining new data on buffer solution acidities in various melts that will allow the calibration of different oxide ion sensors against Ω values.

Oxygen Electrodes in Ionic Melts. Oxide Ion Donors

Part 4. Oxygen Electrode Reversibility in Ionic Melts

We have repeatedly emphasized that the most reliable data on Lux acid–base interactions taking place in ionic melts can be obtained using the potentiometric method, which provides detection of the equilibrium oxide-ion concentrations over wide pO ranges. These investigations imply the use of indicator electrodes reversible to O^{2-} in cells with the liquid junction. Literature sources contain much information devoted to studies of the features of reversibility of the following kinds of oxygen electrodes:

- gas oxygen electrodes that belong to electrodes of the first kind, e.g. $Pt(O_2)$ [62, 67, 90], $Au(O_2)$ [17, 199]; and
- metal-oxide oxygen electrodes are electrodes of the second kind, e.g. Cu|CuO [90], Ni|NiO and Co|CoO [200].

As mentioned earlier, investigations of the oxygen electrode operation in different ionic melts are numerous. The reversibility of the oxygen electrode is reported practically in every paper concerned with the study of Lux acid–base interactions in ionic melts.

Gas oxygen electrodes [$Pt(O_2)$ being the most widely used] are often exploited for the investigations of oxide-ion chemistry in most nitrate, chloride and sulphate melts. The construction of this kind of electrode is simple enough [62]: they are made of a noble (or in a more general case, inconsumable) metal plate (Au, Pt, Ag, Pd), placed in an atmosphere with a definite, known constant partial pressure of O_2 and partially immersed in the studied melt containing oxide ions.

It is generally accepted that at the gas oxygen electrode there takes place a process of reduction of gaseous oxygen with the transfer of two electrons per oxide ion formed. The "theoretical" process is described by the following equation:

$$\frac{1}{2}O_2\uparrow + 2\bar{e} \rightleftharpoons O^{2-}. \tag{2.4.1}$$

The Nernst equation corresponding to this electrochemical process may be written as:

$$\varphi = \varphi^0 + (1.15RT/F)\log p_{O_2}^{1/2} - (1.15RT/F)\log a_{O^{2-}}, \tag{2.4.2}$$

or

$$\varphi = \varphi^0 + (1.15RT/F)\log p_{O_2}^{1/2} + (1.15RT/F)pO. \tag{2.4.3}$$

The inertness of the metal serving as the material of the oxygen electrode is the main condition providing the reversible work of gas oxygen electrodes; however, it seldom takes place. The second condition is the agreement between the slope of E–pO calibration plot and the theoretical one predicted by equation (2.4.2).

The use of a gas oxygen electrode implies a constant partial pressure of gaseous oxygen in the atmosphere over the melt that, in some cases, results in the formation of peroxide ions according to the scheme:

$$\frac{1}{2}O_2\uparrow + O^{2-} \rightleftharpoons O_2^{2-}. \tag{2.4.4}$$

Peroxide ions are unstable in acidic solutions, and this process has no essential effect on the electrode potential. However, if the gas electrode is used for measurements in basic solutions, where the stability of O_2^{2-} ions increases, they become the predominant form of oxygen-containing anions. Under these conditions the slope of the E–pO plot is twice as large as that predicted by equation (2.4.2). From equation (2.4.4) it follows that there is a linear correlation between the concentrations of peroxide and oxide ions in these melts, and it can be assumed that the gas oxygen electrode remains reversible to oxide ions. However, their equilibrium concentration will be appreciably lower than the initial one.

We now consider Fig. 2.4.1. As may be seen, plots 1 and 2 should intersect since they possess different slopes. As a rule, under a certain oxide-ion concentration, one potential-defining process should occur at the electrode surface, and this process is characterized by the lowest Gibbs energy. Since ΔG

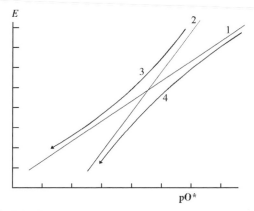

Fig. 2.4.1. E–pO* plots for gas oxygen electrode: 1, according to equation (2.4.1); 2, peroxide function; 3, thermodynamically favourable dependence; 4, experimental dependence.

is equal to $-zFE$, the most thermodynamically favourable electrode process should possess the maximum e.m.f. value. For the plots presented in Fig. 2.4.1, the E–pO plot denoted by arrow 3 is the most preferred resulting plot in the case of two competitive potential-defining processes of different slopes. However, at the gas oxygen electrode there are some electrode processes where the electron-active particles are the participants of redox reaction (2.4.4). This means that their concentrations are dependent on each other.

As is known from basic courses on electrochemistry, under the equilibrium conditions (the compensation measurements provide that the total current through the electrode is equal to zero), the magnitude of the electrode potential is defined by the ratios of partial exchange currents of all the electrode reactions at the electrode [201]. Naturally, the value of the exchange current of an electrode reaction increases together with the concentrations of the corresponding electro-active particles [201]. Therefore, in acidic solutions, the peroxide ion concentration is very low and the exchange current for a reaction (2.4.4) is negligible. The E–pO plot is described by the slope equal to $1.15RT/F$, which agrees with reaction (2.4.2) although the reduction potential of the following reduction process:

$$\frac{1}{2}O_2^{2-} + \bar{e} \rightleftharpoons O^{2-},$$
(2.4.5)

is higher (see Fig. 2.4.1). Owing to the increased oxide-ion concentration, the partial pressure of O_2 remains constant, whereas the peroxide ion concentration grows in proportion to the initial concentration of O^{2-}. This results in the increase of the exchange current of reaction (2.4.5) at the electrode surface that finally causes "switching on" this reaction as the potential-defining one.

All the above-mentioned facts allows us to conclude that for the gas oxygen electrodes the $E-pO$ plots should possess slopes close to $1.15RT/F$ ($z = 2$) at low equilibrium oxide-ion concentrations, and close to $2.3RT/F$ ($z = 1$) at high ones (see Fig. 2.4.1, arrow 4). This does actually take place.

The necessity of passing gaseous oxygen immediately through ionic melts imposes some limitations on the utilization of oxygen electrodes of the first kind for the investigations of the Lux acid–base equilibria in these media. In particular, these electrodes cannot be recommended for the following cases:

- in melts possessing reducing properties (bromides, iodides, etc.) or in the studies of acid–base reactions with the participation of substances prone to oxidation by gaseous oxygen (Cu^+, Fe^{2+}, etc.);
- in strongly basic solutions, oxide ions are oxidized by O_2 with the formation of peroxide ions, which causes change of the character of the potential-defining process at the electrode. Also, appreciable corrosion of the electrode material occurs in the basic medium; this especially concerns oxygen electrodes made of platinum metals;
- after addition of strong acids to the melts based on ionic halides, halide ions are subject to oxidation. This causes titration of the added acid with gaseous oxygen and distortion of the acid–base interaction stoichiometry;
- electrodes are broken down in the melts containing cations, which form alloys with the electrode material (e.g. Pb^{2+} ions destroy metallic platinum).

Metal-oxide electrodes are usually manufactured as metallic plates or wires with a current tap immersed in a saturated solution of the corresponding oxide in the melt studied. The potential of such an electrode can be calculated from the standard electrode potential in the melts. The activity of the metal cations in the saturated solution is determined in the following manner:

$$\varphi_{Me^{2+}} \rightleftharpoons \varphi^0_{Me^{2+}} + (1.15RT/F)\log a_{Me^{2+}} \qquad (2.4.6)$$

The equilibrium activity (concentration) is dependent on the solubility product of the oxide and the equilibrium concentration of oxide ion in the melt is given as follows:

$$a_{Me^{2+}} = P_{MeO}a_{O^{2-}}^{-1}.$$ (2.4.7)

Substitution of equation (2.4.7) into (2.4.6) results in the following Nernst equation:

$$\varphi_{Me^{2+}} = (\varphi_{Me^{2+}}^0 + (1.15RT/F)\log P_{MeO}) + (1.15RT/F)pO,$$ (2.4.8)

whose form is similar to that of equation (2.4.3).

Hence, the slopes of the $E–pO$ dependences for gas oxygen and metal-oxide electrodes should be close to $1.15RT/F$ if they are reversible to oxide ion, according to equations (2.4.3) and (2.4.8).

The use of metal-oxide electrodes for potentiometric investigations of oxoacidity is restricted, however, for the following reasons.

It follows from equation (2.4.7) that in acidic melts the metal-oxide solubility should increase owing to fixation of oxide ions by the Lux acids, and a rise in the metal cation concentration should result in essential changes to the melt's properties. A loss of reversibility of metal-oxide electrodes to oxide ions in strongly acidic melts may take place because of complete dissolution of the solid oxide in the melt; the corresponding pO value is estimated from the following equation:

$$pO_{upper} = pP_{MeO} + \log\frac{w_{MeO}1000}{w_{melt}M_{MeO}},$$ (2.4.9)

where w_{MeO} and w_{melt} are, respectively, the masses of the solid oxide and the melt in the crucible; the second term on the right-hand side is a recalculation of the oxide weight to its concentration in 1000 g of the melt. After complete dissolution of the oxide, the metal-oxide electrode is transformed into a metallic electrode of the first kind, and its potential should be constant without running additional complexation processes between the metal cations and anions introduced in the melt. From equation (2.4.9) it can be concluded that the lower the solubility of the metal oxide in the ionic melt, the wider the pO range of the reversible work of the metal-oxide electrode.

Metal-oxide electrodes cannot be used for measurements when the participants of the Lux acid–base reaction form complexes with the potential-determining cation, since there arises a systematic error of pO determination. The decrease of the electrode potential in the melt in this case is actually caused not by changes in the melt acidity, but by fixation of the metal cation in complexes whose concentration is not dependent on the melt acidity.

Metal-oxide electrodes cannot be used for investigations of the acid–base properties of strong oxidants and reducing agents because of the chemical reactions of the melt with the constructional electrode material in the former case, and with the metal-oxide powder in the latter case.

To conclude our consideration of the main kinds of oxygen electrodes, we should mention membrane oxygen electrodes with outer shells, usually prepared in the form of test-tubes made completely or partially of a solid electrolyte possessing oxide-ion conductivity. Zirconium(IV) oxide is the most widely used material for manufacturing the solid electrolyte membranes; nevertheless, some rare-earth metals [202] and thorium(IV) oxides [203] are also studied in this context. To increase the oxide-ion conductivity, the solid electrolytes are doped with oxides of metals of different valence. The above-mentioned dioxides are usually doped with alkaline-earth [204, 205] and rare-earth [206, 207] oxides, which are introduced into ceramics in concentrations not more than 10 mol%. These admixtures substitute Zr or Th ions in the crystalline lattice and such a substitution results in the appearance of vacancies of O^{2-} and the solid solution formed becomes an electrolyte possessing oxide-ion conductivity. This also causes considerable strengthening of ZrO_2 ceramics since the admixtures favour the formation of the cubic phase of zirconia that prevents the phase transition from tetragonal to cubic phase. For the solid electrolytes, the composition of 90 mol% of ZrO_2 and 10 mol% of Y_2O_3 (YSZ) is the most suitable for different purposes [208–210].

To prepare a membrane oxygen electrode, the usual oxygen electrodes of the first or second kind are placed in the inner space of the solid electrolyte membrane. In the case of the gas oxygen electrode, a constant partial pressure of O_2 is created by passing air or pure oxygen through the inner space of the electrode. In the case of a metal-oxide electrode the "metal–metal oxide" mixture placed in the test-tube serves as a buffer mixture supporting a very low but constant pressure of oxygen; the use of powder mixtures of Ni|NiO, Cu_2O|CuO, Ti|TiO_2 and Pb|PbO [209] for this purpose has been reported in the literature. There are some data about the use of an inner silver electrode [178], whose basic scheme is as follows:

$$Pt|Ag|AgCl + Na_2CO_3 + chloride\ melt. \qquad (2.4.10)$$

This electrode is similar to the glass electrodes used for measurements in aqueous and aprotic solutions. According to the authors, it possesses good reversibility to oxide ions.

The electrochemical processes taking place at a membrane oxygen electrode can be divided into two stages. The first of them is the electrode process itself: either the reduction of gaseous oxygen at the metal surface or the reduction of a metal oxide according to such an equation:

$$MeO + 2\bar{e} \rightleftharpoons Me^0 + O^{2-}. \tag{2.4.11}$$

The second process is the transfer of the formed oxide ion through the solid electrolyte membrane. The potentiometric cells with membrane oxygen electrodes should be characterized by the slope of the $E-pO$ plot equal to $1.15RT/F$ if the electrode processes correspond to equations (2.4.3) and (2.4.8). The transport number of O^{2-} ions is an important criterion of the membrane electrode reversibility—for good solid electrolytes it is close to unity (1).

Membrane oxygen electrodes are the most convenient oxide-ion sensors for measurements of ionic melt basicity, since their use is not connected to pollution by foreign admixtures of the melt studied, or with passing gaseous oxygen through the melt. This excludes the drawbacks that are characteristic of other kinds of oxygen electrodes.

However, the properties of solid electrolytes of which the membranes are made, impose some limitations on the temperature range where the membrane oxygen electrode can be used. Perfil'ev and Fadeev determined the lower threshold temperature for the reversible operation of the solid electrolyte membrane to be close to 500 °C [211], owing to oxide-ion conductivity of the membrane. At higher temperatures, the membrane oxygen electrode is considered to be reversible to oxide ions, although the transport number of oxide ions, e.g. in YSZ, achieves the value of unity (1) at temperatures of the order of 1000 °C, i.e. its conductivity becomes completely ionic.

We shall now consider the methods for investigating the reversibility of the oxygen electrode in ionic melts and features of their operation under experimental conditions.

2.4.1 POTENTIOMETRIC METHOD OF STUDY OF OXYGEN ELECTRODE REVERSIBILITY

2.4.1.1 Direct calibration

Direct calibration is the simplest investigational method and is used widely for molten salts. Subsequent calculations of the equilibrium parameters of the

Lux acid–base reactions require data on the dependence of e.m.f. of the cell used on the equilibrium oxide-ion activity (concentration) in the melt. The calibration of the potentiometric cell is performed as described below.

Several known weights of a strong Lux base are added consecutively to the melt studied, which is placed in the potentiometric cell shown in Fig. 2.4.2. After each addition, the equilibrium e.m.f. value is determined. This routine usually permits calibration of the electrochemical cell in the pO range from 1 to 4. Note that the greater the molar mass of the strong Lux base added, the higher the pO values that can be achieved using the weight calibration. Another variant of the calibration consists in the addition of weights of the previously prepared solution of the strong Lux base to the ionic solvent studied, which is quenched after preparation and added to the melt as a solid (powder). This allows one to widen the range of the calibration, and usually values of pO close to 5 are achieved. Nevertheless, as mentioned earlier, the pure ionic melt contains oxide ions, which appear owing to the dissociation of the traces of oxygen-containing impurities; so they are hardly removed from the melt. Therefore, the addition of a small weight of the solution of base, which makes the equilibrium concentration of oxide ions equal, or close to that of the "admixture" oxide ion (usually their weights are of the order of analytical balance sensitivity, ~ 0.001 g), is often useless.

Potassium and sodium hydroxides and sodium peroxide are conventionally used as strong Lux bases for the calibration of the potentiometric cells.

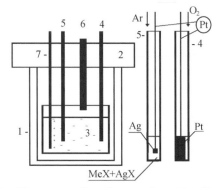

Fig. 2.4.2. The principle of the construction of a potentiometric cell with oxygen electrode: 1, cell with the cover (2); 3, crucible-container with the melt studied; 4, an oxygen electrode; 5, a reference electrode; 6, tube for adding weights of Lux acids or bases; 7, tube for passing the melt of inert gas (N_2, Ar, if necessary).

Their equivalent masses corresponding to the quantity of the Lux base whose complete dissociation results in the formation of 1 mol of oxide ions, are 80.0, 112.2 and 77.98 g, respectively. Other bases, e.g. BaO and Na_2CO_3, are rarely used for the calibration owing to some specific features of incomplete dissociation, which will be considered in the next part.

As a result of the calibration, a set of experimental points is usually obtained. These points are grouped in one or more sequences approximated by straight or curved lines, which can be explained reasonably. The linear plots obtained are treated by the least-squares method [212] to obtain the calibration e.m.f.–pO plots whose slopes give some information on the electrochemical processes taking place at the electrode surface (the number of electrons taking part in the elementary act of the reaction). These dependences are expressed and used for the calculations in the following form:

$$E = E_0(\pm S_{E_0}) + k(\pm S_k)\text{pO},\qquad(2.4.12)$$

where S_x are the standard deviations. Any non-linear or horizontal sections are usually explained by either the predominant effect of oxygen-containing admixtures (initial calibration section, at high pO) or by the melt saturation with respect to the used oxide-ion donor.

2.4.1.2 Indirect calibration of oxygen electrodes

Investigations devoted to the practical verification of the reversibility of oxygen electrodes in molten ionic liquids at high pO values are scant enough. Usually the electrochemical cell calibration is performed in the pO $= 1$–4 range, and subsequent calculations of the equilibrium oxide-ion molalities are performed on the assumption that the slope of the E–pO plot at higher pO remains constant.

Indeed, there are no factors limiting the reversible operation of gas oxygen electrodes at high and low pO values. Theoretically, if the concentration of the potential-determining particle in the solution is equal to zero, then the absolute magnitude of the potential of the corresponding electrode approaches infinity ($\pm\infty$). This conclusion has no physical sense; it means that the Nernst equation is applicable to the description of the electrochemical processes in the solution, if their concentration exceeds the certain limit. In practice, the deviations from the Nernst equation arise because of the effect of the binary electric layer formed at the electrode surface upon the concentration of extremely dilute solutions with low absolute concentrations of oxide ions [213].

The term "low absolute concentration" requires additional explanation. We would note that the term low absolute concentration is not to be confused with "low equilibrium concentration". The former signifies extremely dilute solutions with concentrations of the order of $10^{-7}-10^{-9}$ mol kg^{-1}, whereas the latter term denotes practically complete removal of ions or molecules from the reaction phase because of the formation of strong complexes and slightly soluble compounds.

The applicability of the Nernst equation for the description of electrode processes is limited only to low absolute concentrations of potential-determining particles since, in the general case, the ratio of the electrode surface to the volume of electrolyte solution is usually of the order of $S/V = 10-10^3$ m^{-1}. For the typical potentiometric cell used for measurement in molten salts, which is shown in Fig. 2.4.2, this value can be estimated as ≈ 10. In the case of low absolute concentration of the potential-defining particles, their concentration in the bulk of the solution is comparable with that in the binary electric layer at the electrode surface. Therefore, the formation of the binary electric layer is bound up with considerable changes of the bulk concentration, and the potential (e.m.f.) measured in such solutions will not correspond to the initial concentration of the potential-determining particles. In the case of low equilibrium concentrations this problem is not essential since, according to Le Chatelier's Law, removal of the potential-determining particles from the bulk of the solution, due to the formation of the binary electric layer at the electrode, causes a shift of the dissociation equilibrium towards the formation of these particles. The generally accepted Ag|AgCl electrode in aqueous solution can serve as a good illustration of all that we have said about the low equilibrium concentration.

The conventionally used preliminary purification of oxygen-less ionic melts from traces of oxide ions cannot reduce the absolute O^{2-} concentration to values lower than 10^{-5} mol kg^{-1}, which undoubtedly exceeds the threshold concentration of 10^{-7} mol kg^{-1}. The addition of even the strongest Lux acids causes only a reduction in the equilibrium concentration of O^{2-}, whereas the absolute concentration, which is the sum of the equilibrium concentrations of oxide ion and the conjugate base formed, remains unchanged.

Practical limitations of gas oxygen electrode reversibility in solutions of strong Lux acids may be connected either with the features of the behaviour of the inconsumable electrode material (Pt, Au) or with the dissolution (corrosion) of the solid electrolyte membrane material in the case where membrane electrodes are used.

A method of investigation of the oxygen electrode reversibility at high pO values in the melts based on alkali metal halides was reported in Ref. [214]. It is based on the construction of E–pθ plots using the data on the Lux acid–base processes of undoubted stoichiometry. The most convenient case is realized if the stoichiometric coefficients of the acid–base reactions are equal to 1. The investigations of the overwhelming majority of the Lux acid–base equilibria show that, as a rule, the neutralization of oxoacids in ionic melts is a multi-step process accompanied with the simultaneous existence of more than two acid–base forms (i.e. acids and conjugate bases) in the studied solutions; it is especially characteristic of homogeneous reactions with the participation of oxygen-containing acids (PO_3^-, V_2O_5, etc.). In contrast, the heterogeneous acid–base reactions resulting in precipitation or dissolution of MeO-type oxides in the melts:

$$1MeO \downarrow \rightleftharpoons 1Me^{2+} + 1O^{2-}, \tag{2.4.13}$$

can be characterized by the fact that all the stoichiometric coefficients are equal to 1 (unity). In the saturated solution of a metal oxide in molten salts the equilibrium is completely described by only two parameters: the equilibrium concentration of $[Me^{2+}]$ and that of $[O^{2-}]$.

Another essential advantage of the heterogeneous equilibria mentioned is the fact that, as a rule, the solutions containing an excess of metal cations are characterized by sufficiently high pO values. In this case, a wider pO range becomes available for the examinations.

This method may be substantiated in the following way.

In the saturated solutions of metal oxide of composition MeO, the molalities of the metal cations and oxide ions are mutually dependent and are bound with the solubility product, P_{MeO}:

$$P_{MeO} = m_{Me^{2+}} m_{O^{2-}}, \tag{2.4.14}$$

or

$$pP_{MeO} = -\log m_{Me^{2+}} + pO. \tag{2.4.15}$$

Taking into account the equilibrium molalities of the substances participating in the acid–base reaction, equation (2.4.15) can be rewritten in the form:

$$pO = pP_{MeO} + \log |m_{Me^{2+}}^0 - m_{O^{2-}}^0 + 10^{-pO}|$$

$$\approx pP_{MeO} + \log |m_{Me^{2+}}^0 - m_{O^{2-}}^0|, \tag{2.4.16}$$

since the equilibrium molality of oxide ions in the melt (10^{-pO}) is negligible compared to other terms under the logarithm sign.

Since the solubility product value (pP_{MeO}) is constant, the dependence of e.m.f. upon $-\log|m^0_{Me^{2+}} - m^0_{O^{2-}}|$ (hereafter this expression will be denoted as $p\theta$), plotted according to the potentiometric titration data obtained before reaching the equivalence point, is characterized by a slope equal to that of the E–pO calibration plot in absolute magnitude, but reversed in sign in the given pO range. To verify the correctness of the above-mentioned expression, we shall differentiate the equation for E against pO obtained from equation (2.4.16):

$$\frac{\partial E}{\partial pO} = \frac{\partial E}{\partial(pP_{MeO} + \log|m^0_{Me^{2+}} - m^0_{O^{2-}}|)}$$

$$= \frac{\partial E}{\partial(\log|m^0_{Me^{2+}} - m^0_{O^{2-}}|)} = -\frac{\partial E}{\partial p\theta}. \tag{2.4.17}$$

As this calibration plot cannot be obtained by the direct calibration method with additions of a strong Lux base, because of limitation of the lowest possible weight (it exceeds the minimal quantity, which can be weighted using an analytical balance) of the base and the existence of hardly removable oxygen-containing admixtures in the melt, the significance of the obtained indirect calibration data is very high. The results of investigations of oxygen-electrode reversibility at high pO values serve as substantiation of the correctness of thermodynamic parameters obtained with the use of this electrode in the stated pO region. Thus, the solubility data allow us to solve two problems simultaneously: to find the solubility product of the oxide studied and to check the oxygen function of the indicator electrode.

Certain limitations characteristic of this method are caused by dilution of the solution according to the consumption of the studied cation for the oxide precipitation. The dilution of the solution results in some increase of the solubility of the precipitated oxide, and the change in pP_{MeO} causes the increase of the E–$p\theta$ slope as compared to the Nernst one. The most typical E–$p\theta$ dependences for practically insoluble oxides are presented in Fig. 2.4.3. As a rule, the slopes of these linear dependences are close to $1.15RT/F$. The curves for the oxides possessing appreciable solubilities in ionic melts are exceptions to this regularity (see Fig. 2.4.3, dependence 4), since the initial

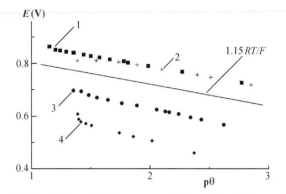

Fig. 2.4.3. Dependences of E vs pθ with an excess of the cations in the molten CsCl–KCl–NaCl eutectic at 600 °C: 1, Mg^{2+}; 2, Ni^{2+}; 3, Co^{2+}; 4, Mn^{2+}; the straight line slope corresponds to 1.15 RT/F.

section of this plot corresponds to the formation of a non-saturated solution, where the solubility product of the oxide is not achieved.

There exists another way for checking the reversibility of the indicator oxygen electrodes at high pO values. This method is connected with the calibration of the potentiometric cell in a reactive gas atmosphere—carbon dioxide being the most convenient for its creation since it possesses strong oxoacidic properties. Passing CO_2 through the ionic melts causes transformation of a certain number of oxide ions into carbonate ions and equilibrium (1.2.3) occurs under these conditions. The ratio of equilibrium molalities for the oxide and carbonate ions in all the solutions is constant in the given melt at constant temperature and CO_2 pressure. The balance of the initial concentration of oxide ion in the solution can be expressed as follows:

$$m^0_{O^{2-}} = m_{O^{2-}}(1 + K^{-1}_{(1.2.3)}).$$

(2.4.18)

From this equation it follows that the equilibrium molality of oxide ion is proportional to its initial molality, and the assumption that $pK_{(1.2.3)} > 3$ (which usually takes place in low-acidic melts) allows us to rewrite this equation as follows:

$$pO = pO^* + pK_{(1.2.3)},$$

(2.4.19)

and

$$\frac{\partial E}{\partial pO} = \frac{\partial E}{\partial (pO^* + pK_{(2.02)})} = \frac{\partial E}{\partial pO^*}. \tag{2.4.20}$$

Therefore, in the case considered, the slope of the $E-pO^*$ plot coincides with that of the calibration plot in the corresponding pO range, which can be determined with the help of equation (2.4.19). As an example, we consider the $E-pO^*$ dependence obtained in the KCl–LiCl eutectic melt at $600\,^{\circ}\text{C}$ (Fig. 2.4.4). The calibration plot consists of two sections with slopes of $1.15RT/F$ (at high pO values) and $2.3RT/F$ (at low ones) that is typical of the reversibility of the $Pt(O_2)|YSZ$ gas membrane electrode in molten halides of alkali- and alkaline-earth metals. As was mentioned earlier, passing CO_2 through the melt results in an increase in the e.m.f. due to the fixation of oxide ions and the formation of CO_3^{2-}. Subsequent additions of the base lead to a decrease in the e.m.f. The slope of the plot obtained in such a manner coincides (in e.m.f.) with the corresponding section of the calibration plot constructed from the data obtained in an argon atmosphere.

A rather high magnitude of the equilibrium constant of carbonate ion dissociation in the KCl–LiCl melt favours the shift of the indirect $E-pO^*$ plot (see Fig. 2.4.4) to that region of the $E-pO$ diagram where the upper section of the standard $E-pO$ calibration plot is located. However, it is possible to check the reversibility of indicator oxygen electrodes even at higher pO (e.m.f.) values that can be achieved by the addition of smaller quantities of a strong Lux base. The example presented demonstrates the potentialities of the so-called "carbonate" method for obtaining data on the reversibility of oxygen electrodes at high pO values in melts–solvents.

After the description of the most usable methods of investigation of oxygen electrode reversibility in molten salts, we shall turn to a consideration of the experimental results in specific melts.

2.4.2 EXPERIMENTAL RESULTS

2.4.2.1 Oxygen-containing melts

Shams El Din *et al.* reported potentiometric investigations of the reversibility of $Pt(O_2)$ gas oxygen electrode in molten KNO_3. Na_2O_2 was

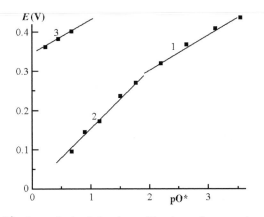

Fig. 2.4.4. E–pO^* plots obtained in the calibration of a membrane oxygen electrode Pt(O_2)|YSZ in the KCl–LiCl eutectice melt at 600 °C: 1,2, dependences obtained in argon atmosphere and 3, plot obtained in CO_2 atmosphere.

used for the calibration as a strong Lux base [88, 90–92]. The slope of the E–pO dependences obtained in the nitrate melt at 350 °C is approximately equal to 2.3RT/F, a value twice as much as the magnitude predicted by equation (2.4.3). This deviation of the slope of the E–pO plot may be explained by the so-called "peroxide function" of the gas oxygen electrodes in basic solutions, i.e. at relatively low temperatures and pressures of oxygen in the atmosphere over the melt, the following reduction process takes place at the electrode surface:

$$O_2 \uparrow + 2\bar{e} \rightleftharpoons O_2^{2-}. \tag{2.4.21}$$

Frederics, Temple and Trickett performed calibrations of the Pt(O_2) oxygen electrode with additions of Na_2CO_3 in the KNO_3–$NaNO_3$ equimolar mixture at 250–320 °C [215, 216]. It is well known that this Lux base is incompletely dissociated even in high-temperature ionic melts based on low-acidity alkali metal halides [197]. However, the investigators, using the data obtained by Kust [217, 218] in the said melt assumed that the complete transformation of dissolved carbonate ion into the oxide one, according to equation (2.3), took place under these experimental conditions. The slope of the E–pO^* calibration plot obtained at 300 °C is equal to 0.0294 ± 0.0008 V, that is somewhat lower than the theoretical slope of 1.15RT/F (0.0284 V). The slope close to 1.15RT/F is observed in other experiments performed in the

290–320 °C temperature range. Hence, from the obtained results it can be concluded that Pt(O_2) gas oxygen electrode shows a good reversibility to oxide ions, and that the potential-determining process at the electrode corresponds to equation (2.4.3).

Results similar to those described earlier were obtained by Zambonin *et al.* [219] who performed the calibration of gold and platinum gas oxygen electrodes in the KNO_3–$NaNO_3$ equimolar mixture at 250 and 350 °C, with additions of sodium carbonate as a Lux base. However, in contrast to the works by Frederics *et al.* [215, 216], the calibration was made in a reactive gas medium, the mixture of CO_2 and O_2 being used as the latter. Naturally, the gaseous phase possesses appreciable acidic properties and these results may be considered as indirect calibration. The slopes of E–log($[CO_2]/[CO_3^{2-}]$) plots for both oxygen electrodes, in this investigation, were stated to correspond to the $1.15RT/F$ one.

Schlegel and Uhr examined the behaviour of platinum and gold oxygen electrodes in $[Cr_2O_7^{2-}]/[CrO_4^{2-}]$ buffer solutions in the KNO_3–$NaNO_3$ eutectic melt in the temperature range 250–280 °C [220]; the examined buffer solutions are referred to as acidic ones since they contain excess of $Cr_2O_7^{2-}$ ion. Under the experimental conditions, the slope of the E–pO calibration plot is close to $1.15RT/F$. Also, the platinum oxygen electrode was found to be reversible to O^{2-} ions even in the absence of gaseous oxygen in the atmosphere over the melt. Proceeding from this peculiarity, the authors made conclusions about the formation of oxide films on the surface of platinum and other noble metals, transforming these gas electrodes to those of the second kind (metal-oxide ones). For example, the gas platinum–oxygen electrode Pt(O_2) was supposed to work as a Pt|PtO one. After the removal of the oxide film from the electrode surface by passing gaseous hydrogen through the melt, the electrodes were found to lose the sensitivity to oxide ions in the melts. It is interesting that the potential-determining reactions of the gas gold oxygen electrode at different temperatures are characterized by $z = 2$, which corresponds to the E–pO slope equal to $1.15RT/F$, in contrast to the platinum–oxygen electrode. Since chromate/dichromate buffer solutions are strongly acidic, the corresponding pO values belong to the acidic region and the data obtained are referred to indirect calibration.

Keenan and Williamson studied the use of the gas platinum oxygen electrode in the KNO_3–$NaNO_3$ equimolar mixture at 350 °C with an oxygen-free atmosphere over the melt [221]. The investigation was performed by the calibration of the electrode with quantities of carbonate in a carbon dioxide atmosphere, i.e. in an acidic medium. The slope of the $E-pO^*$ plot was determined from the calibration results to be equal to $2.3RT/(1.86F)$, and for the $E-\log p_{CO_2}$ a similar value was calculated as $2.3RT/(1.92F)$. The good reversibility on a platinum electrode to oxide ions, even in the absence of oxygen in the atmosphere over the melt, allows us to conclude that this electrode is classified among the metal-oxide ones, and that under O_2 pressure it possesses the mixed function. The formation of oxide films on the surfaces of metallic electrodes made of noble metals in contact with high-temperature ionic melts has been observed in some other investigations, e.g. see Refs. [222, 223]. The formation of oxide films at a platinum surface was noted for cells constructed from a YSZ plate with platinum electrodes applied to both sides [224].

Zambonin and Signorile reported the calibration of $Pt(O_2)$ and $Au(O_2)$ oxygen electrodes in the equimolar KNO_3-NaNO_3 mixture at 503 K [225, 226]. Sodium hydroxide was used as a strong Lux base. For the case of the gas platinum–oxygen electrode, the calibration's $E-pO^*$ dependence consists of two sections characterized by slopes of $1.15RT/F$ at low concentrations of base, and $2.3RT/F$ with titrant additions corresponding to moderately concentrated solutions of O^{2-}. These calibration plots are described by S-shaped curves, where the mentioned sections having slopes of $1.15RT/F$ and $2.3RT/F$ are resolved. These considerations deserve a particular diagram (see Fig. 2.4.5). The dissociation of sodium hydroxide according to equation (1.2.4) under the mentioned conditions is essentially incomplete. The slope of the $E-\log[OH^-]$ calibration plot corresponds to $2.3RT/F$ which the authors explain by the "hydroxide" function of the gold and platinum oxygen electrodes (since these plots coincide, both electrodes can be considered inconsumable), i.e. by the following electrochemical process:

$$\frac{1}{2}O_2 + H_2O + 2\bar{e} \rightleftharpoons 2OH^-. \tag{2.4.22}$$

At low concentrations of hydroxide ions ($10^{-6}-10^{-4}$ mol kg^{-1}), the slope of the $E-pO$ plot for the gold oxygen electrode is equal to $1.15RT/F$. The authors consider that the stabilization of oxide ions in the nitrate melts is caused by the presence of traces of water and silica, and that the latter dissolves in the melt owing to the interaction of the nitrate melt with the Pyrex glass from which the container for the melt is made. This means that the "stabilization" of oxide ions is nothing but the formation in the melt of silicate and hydroxide ions. If such a stabilization is absent in solutions, then oxide ions are oxidized by

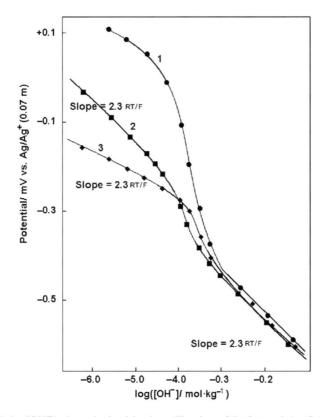

Fig. 2.4.5. E–log[OH$^-$] plots obtained in the calibration of Pt(O$_2$) and Au(O$_2$) gas oxygen electrodes in the molten KNO$_3$–NaNO$_3$ equimolar mixture at 503 K (in oxygen atmosphere): 1,2, plots for platinized (1), and smooth-platinum (2) and gold (3) indicator electrode, by P.G. Zambonin and G. Signorile [225].

gaseous oxygen or nitrate ions, which results in the formation of peroxide ions in the basic solutions. The peroxide ions are subject to oxidation into other peroxide species, e.g. superoxide:

$$O_2^{2-} + O_2 \rightleftharpoons 2O_2^-. \tag{2.4.23}$$

As for the mechanism of the Pt(O$_2$) gas electrode's work in the nitrate melt, the authors of Ref. [225] found it to become metal-oxide owing to the formation of a platinum oxide film on the electrode surface.

In his classic work [17], Lux used a gas gold oxygen electrode for oxide-ion activity measurements in the K$_2$SO$_4$–Na$_2$SO$_4$ equimolar mixture at 950 °C.

The Au(O_2) electrode was demonstrated to be reversible to oxide ions, and the slope of the E–pO plot was twice as high as the value predicted by the Nernst equation (2.4.2), i.e. that there was a one-electron potential-determining process at the electrode.

Investigations of the metal-oxide electrode reversibility seldom concern both oxygen-containing and oxygen-less ionic melts. There are some data on the work of these kinds of oxygen electrodes in molten KNO_3 at 350 °C.

Shams El Din and Gerges studied the reversibility of the metal-oxide electrodes Cu|CuO and Ag|Ag$_2$O in KNO_3 melt by the calibration of the electrode system with quantities of Na_2O_2 [89, 103]. Both the electrodes were found to be reversible to oxide ions according to the Nernst equation (2.4.8), and suitable as indicator electrodes for quantitative investigations of oxoacidity in nitrate melts.

Baraka *et al.* reported investigations of the reversibility of niobium (Nb|Nb$_2$O$_5$), tantalum (Ta|Ta$_2$O$_5$) and zirconium (Zr|ZrO$_2$) metal-oxide electrodes in molten KNO_3 at 350 °C [110–112]. The basic distinction of these metal-oxide electrodes from those mentioned above, e.g. in the previous paragraph, consists in the addition of oxides possessing appreciable acidic properties in the melt as sources of Zr^{IV}, Nb^{V} and Ta^{V} potential-defining particles. Nevertheless, all the mentioned electrodes demonstrate good reversibility to oxide ions, the slopes being equal to 0.068 V (Nb|Nb$_2$O$_5$), 0.065 V (Ta|Ta$_2$O$_5$) and 0.067 V (Zr|ZrO$_2$), whereas the, theoretical slope of $1.15RT/F$ at 350 °C is equal to 0.0618 V. The lower pO limits of the applicability of such, "acidic" electrodes and distortions in pO connected with the fixation of some quantity of oxide ion by the powder of the acidic oxide, were not determined.

The most significant data concerning the reversibility of the oxygen electrodes of different kinds in oxygen-containing melts are presented in Table 2.4.1. The results presented there allow us to conclude that gas oxygen electrodes are usually characterized by the slope $2.3RT/F$ with direct calibration. This is caused by redox interaction of gaseous oxygen with oxide ions dissolved in the melts, resulting in the formation of peroxide ions as the main potential-defining ions in the oxygen-containing melts. In contrast, the slopes obtained by calibration in acidic media are close to $1.15RT/F$ due to instability of the peroxide ions under the said conditions, so that the oxide ions become the main potential-defining ions. These electrodes were found to lose their sensitivity to oxide ions after removal of metal-oxide films from the surface of the electrode material. As for metal-oxide electrodes, they show a

TABLE 2.4.1
Approximate values of z for different electrodes reversible to oxide ions in some oxygen-containing melts

T (°C)	Electrode	Base	$\approx z$	References
KNO_3				
350	$Pt(O_2)$	Na_2O_2	1	[93–96]
	$Nb\|Nb_2O_5$		2	[110]
	$Ta\|Ta_2O_5$			[111]
	$Zr\|ZrO_2$			[112]
	$Cu\|CuO$			[94, 108]
	$Ag\|Ag_2O$			[108]
$KNO_3 - NaNO_3$				
250–320	$Pt(O_2)$	NaOH	1	[225]
	$Au(O_2)$	NaOH	1; 2	[225]
250	$Pt(O_2)$	Na_2CO_3	2	[215, 219]
		$Cr_2O_7^{2-}/CrO_4^{2-}$		[220]
$K_2SO_4 - Na_2SO_4$				
800	$Pt(O_2)$	Na_2O	1	[17]

good reversibility to oxide ions, with the E–pO slope equal to $1.15RT/F$, which agrees with equation (2.4.8). Irreversible work of the said electrodes has not yet been observed, which testifies to extremely low solubilities of the oxides used in the studied molten ionic media, at least at 350 °C.

2.4.2.2 Melts based on alkali metal halides

The Lux acid–base reactions in alkali metal melts are often investigated with membrane oxygen electrodes of various kinds. Other kinds of oxide-reversible electrodes are seldom used, and only for some specific cases. As for the reference electrodes, the most convenient are lead $Pb\|Pb^{2+}$ (Pb + a solution of $PbCl_2$ in the melt studied: this electrode is only used for studies in such chloride melts as the KCl—LiCl eutectic and the KCl–NaCl equimolar mixture) and silver $Ag\|Ag^+$ (Ag + a solution of silver halide in the melt studied) electrodes. The latter is not a "silver halide" electrode since, in contrast to aqueous solutions, silver halides are low-melting substances (melting points 455 °C for AgCl, 424 °C for AgBr and 554 °C for AgI) and there is no silver halide precipitate in the electrode space. Moreover, they are mixed with liquid alkali metal halides forming solutions close to ideal ones.

What we have said above especially concerns molten potassium halides, since the ionic radii of Ag^+ and K^+ are close.

Characteristic features of the construction of the reference electrodes mentioned consist in the materials serving as current taps. Platinum is used with silver reference electrodes, but cannot be used for lead electrodes since the lead forms alloys with Pt and the current taps are destroyed by such interactions. This requires the use of iron or tungsten wires as intermediate current-tap materials. The potentials of the mentioned reference electrodes are highly reproducible, and they were extensively and thoroughly studied over a wide range of temperatures and concentrations in potentiometric cells of the following kinds:

$$C(Cl_2 \uparrow)|melt + AgCl(or\ PbCl_2)|Ag(or\ Pb). \qquad (2.4.24)$$

The corresponding plots for chloride melts can be found, e.g. in Refs. [227, 228] that gives the possibility of expressing the obtained values referring to the chlorine electrode, which is generally accepted as a reference electrode in molten chlorides (with potential considered to be zero at all temperatures). The silver electrode in the KCl–NaCl eutectic melt mentioned in Ref. [227] is characterized by $\varphi = -0.939$ V against the chlorine electrode, and similarly the transport number of silver through the walls of the tube-container is negligible.

(i) KCl–LiCl (0.41:0.59) eutectic

Potentiometric investigations of the reversibility of gold and platinum gas oxygen electrodes in the KCl–LiCl (0.41:0.59) melt at 450 °C were reported in Ref. [229]. Lithium oxide was used as a strong Lux base for the calibration. Change in the partial pressure of oxygen in the electrode space causes changes in the e.m.f., which are half as large as the value of 0.036 V predicted on the basis of the Nernst equation (2.4.2) (the slope of experimental $E - \log p_{O_2}$ was found to be 0.02 V). From this result, Wrench and Inman concluded that these gas electrodes were irreversible to oxide ions according to the theoretical process (2.4.1). The $E - \log m_{Li_2O}$ plots for both gas oxygen electrodes are practically coincident and their slopes are close to $2.3RT/F$. Therefore, both electrodes are actually inconsumable and the process of reduction of gaseous oxygen to peroxide ions at the electrode surface takes place, i.e. the peroxide function which means the potential-determining process (2.4.21). The subsequent iodometric analysis of the quenched sample of the melt confirmed the presence of peroxide ions in the molten medium. At the same time, the plots of $E - \log m_{KOH}$ obtained with the use of KOH as a strong Lux base for gas gold and platinum oxygen electrodes possess different slopes. In the former case, it is close to $1.15RT/F$ and in the second one, it equals $2.3RT/F$. However, there is no explanation for this difference in Ref. [229].

Laitinen and Bhatia studied the behaviour of some metal-oxide electrodes in the KCl–LiCl eutectic melt at 450 °C [230]. The nickel metal-oxide electrode (Ni|NiO) was shown to possess bad reversibility to oxide ions in the melt, the E–pO slope being 0.053 V, in contrast to the theoretical one of 0.072 V. The calibration plots are characterized by considerable scattering of the experimental points. The reversibility of the Cu|Cu$_2$O metal-oxide electrode is in good agreement with the corresponding Nernst equation (2.4.6). Nevertheless, the very high solubility of cuprous oxide in the chloride melts makes this electrode practically impossible to use for oxoacidity investigations over a wide pO range, especially in acidic solutions. As for Pt|PtO and Pd|PdO metal-oxide electrodes, it should be noted that their work agrees with the Nernst equation (2.4.6) and can be used as indicators over a wider pO range than that of the Cu|Cu$_2$O oxygen electrode. A certain drawback of the mentioned metal-oxide electrodes is the relatively slow achievement of equilibrium conditions in these heterogeneous systems.

Landresse and Duyckaerts reported the results of a potentiometric investigation of the membrane oxygen electrode Ni|NiO|YSZ in the KCl–LiCl eutectic at 660 °C [231]. The slope of the E–pO calibration plot is approximately equal to $1.15RT/F$. The results of this study were used to estimate the solubility product of uranium oxide in the chloride melt.

The work of some of us (Cherginets, Banik and Rebrova) is devoted to studies of the reversibility of various oxygen electrodes such as Pt(O$_2$)|YSZ and Ni|NiO, in the KCl–LiCl eutectic melt at 400, 500, 600 and 700 °C. The potentiometric cell for the measurements at 400 and 500 °C consisted of the lead reference electrode (Pb|PbCl$_2$, 0.05 mol kg^{-1}) and one of the indicator oxygen electrodes [232]:

$$Pt|Pb|Pb^{2+}, KCl-LiCl\|KCl-LiCl, O^{2-}|O^{2-}\text{-electrode.} \qquad (2.4.25)$$

Li$_2$O, NaOH and Na$_2$CO$_3$ were used as Lux bases. The calibration of the cell with an oxygen electrode was performed in the pO range from 1 to 3. The obtained results are presented in Fig. 2.4.6.

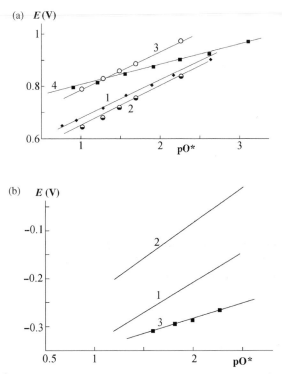

Fig. 2.4.6. E–pO^* calibration plots in the KCl–LiCl eutectic melt at 500 °C: (a) Pt(O$_2$)|YSZ, 1, NaOH; 2, Li$_2$O, 3; NaOH (($p_{H_2O} = 0.0096$ atm.); 4, Na$_2$CO$_3$; (b) Ni|NiO, 1, Na$_2$O$_2$; 2, Na$_2$O$_2$ ($p_{H_2O} = 0.0236$ atm.); 3, Na$_2$O$_2$ (400 °C).

The e.m.f. values of the cell with the membrane oxygen electrode (2.4.25) are positive, which corresponds to the running of the following potential-defining process in the cell

$$\frac{1}{2}O_2 + Pb^0 \downarrow \rightleftharpoons Pb^{2+} + O^{2-} \tag{2.4.26}$$

from the left to the right side. The cell with the metal-oxide nickel electrode in the studied range of oxide ion molalities is characterized by negative e.m.f. values. This means that the reaction which defines these e.m.f. values

$$NiO \downarrow + Pb^0 \downarrow \rightleftharpoons Ni^0 \downarrow + Pb^{2+} + O^{2-} \tag{2.4.27}$$

in the cell runs from the right to the left.

The E–pO^* plots of the $Pt(O_2)|YSZ$ membrane electrode at 500 °C (Fig. 2.4.6a) in the case of the use of strong Lux bases (NaOH or Li_2O) for the calibration are characterized by slopes close to $2.3RT/F$ − 0.174 ± 0.03 and 0.157 ± 0.004 V, respectively, compared to the theoretical one of 0.154 V. The presence of water vapour in the atmosphere over the melt does not result in an appreciable change of the E–pO^* slope. If Na_2CO_3 is used as a Lux base then the value of the E–pO^* slope is half as large as those for the above-mentioned strong Lux bases: 0.078 ± 0.004 V. This is caused by incomplete dissociation of the base in the chloride melt according to equation (1.2.3) and this means that the actual pO values are somewhat greater than the pO^* ones. So, the calibration plot obtained with the use of sodium carbonate is referred to the higher pO range. This allows us to elucidate the reversibility of the $Pt(O_2)|YSZ$ membrane electrode at high pO values in this melt: the potential-determining process is characterized by $z = 2$, and the slope of the E–pO plot is close to the theoretical one (0.077 V). The E–pO^* plots for the $Ni|NiO$ metal-oxide electrode in this melt at 400 and 500 °C are presented in Fig. 2.4.6b. The slopes of the calibration plots are close to that predicted by the Nernst equation for the metal-oxide electrodes (2.4.6), namely: the slopes at 500 °C are equal to 0.089 ± 0.010 V in solutions of Li_2O, and 0.095 ± 0.006 V in Na_2O_2 solutions. The slope at 400 °C is 0.062 ± 0.012 V, which practically coincides with the theoretical $1.15RT/F$ (0.067 V). There is no appreciable change in the E–pO^* slope in the atmosphere, which agrees qualitatively with the data obtained using the $Pt(O_2)|YSZ$ membrane oxygen electrode. It should be emphasized that the sign of deviations in the slopes as compared with the theoretical one are different: for the use of the membrane oxygen electrode there is a small decrease in the slope in the wet atmosphere from that obtained in a dry atmosphere, whereas measurements with the metal-oxide nickel electrode show a small increase in the slope in the wet atmosphere as compared with dry argon. Consequently, in contrast to the data of Laitinen and Bhatia [230], the practical examinations of the metal-oxide nickel electrode show that it is well reversible with respect to oxide ions in the molten KCl–LiCl eutectic and can be used for quantitative potentiometric investigations, mainly in basic media at temperatures below 500 °C, i.e. under the conditions where the membrane oxygen electrodes become hardly applicable. In basic solutions, the process of NiO dissolution is suppressed by excess of the Lux bases.

The reversibility of the $Pt(O_2)|YSZ$ membrane oxygen electrode with respect to oxide ions was examined at 600, 700 and 800 °C, with KOH being used as a strong Lux base for the potentiometric cell calibration. Since the temperatures of the experiments were relatively high, the silver electrode was used as a reference and the potentiometric cell scheme was as follows:

$$Pt|Ag|Ag^+(0.5 \text{ mol kg}^{-1}), KCl-LiCl\|KCl-LiCl, O^{2-}|$$
$$YSZ|Pt(O_2). \tag{2.4.28}$$

The E–pO plot of the cell with the gas membrane oxygen electrode at 600 °C is presented in Fig. 2.4.4, sections 1 and 2 [187]. This plot contains an inflection point at pO $\sim$ 2, which is characteristic of such membrane electrodes. The positive e.m.f. values of the cell (2.4.28) correspond to running the following potential-determining reaction:

$$\frac{1}{2}O_2 + 2Ag^0 \downarrow \rightleftharpoons 2Ag^+ + O^{2-} \tag{2.4.29}$$

from the left to the right.

Thus, the presence of the inflection point in the calibration E–pO plots is a distinctive feature of the work on the gas membrane oxygen electrode in high-temperature ionic melts. More exactly, there are two linear sections, with slopes corresponding to values of z equal to 1 and 2. This result, noted at first in our paper [233], showed that the electrode process at the gas membrane oxygen electrode was essentially dependent not on peculiarities of the assumed potential-determining process with the participation of the given Lux base (as was considered before), but on the equilibrium concentration of oxide ion created by dissociation of this base in the ionic melt.

In the case given, at low equilibrium molalities of oxide ions corresponding to pO values exceeding 2, the calibration E–pO plot is characterized by a slope equal to 0.094 ± 0.008 V. This is in good agreement with the theoretical slope of 0.086 V ($1.15RT/F$), although a little higher. The increase of the equilibrium O^{2-} molality to the values corresponding to pO magnitudes lower than 2 results in the appearance of the inflection in the calibration E–pO plot, and the section obtained at lower e.m.f. values is characterized by a slope which is approximately $2.3RT/F$, namely, 0.154 ± 0.009 V (against the theoretical slope of 0.173 V).

The described behaviour of the gas membrane oxygen electrode agrees qualitatively with the work on conventional gas oxygen electrodes in the nitrate melts considered above at different concentrations of oxide ions.

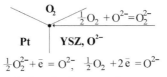

Fig. 2.4.7. Schematic diagram of work of gas membrane oxygen electrode Pt(O$_2$)|YSZ.

An explanation of the results is given by the author of this book in Ref. [234] as follows.

The use of a gas oxygen electrode in the inner electrode space for pO measurements requires a supply of gaseous oxygen to the surface of the metallic electrode as shown in Fig. 2.4.7. The solid-electrolyte membrane (e.g. YSZ) serves as a medium for the supply (conducting) of oxide ions, which are formed by the electrochemical processes that take place at the metal surface. In this case, the equilibrium

$$\frac{1}{2}O_2 + O^{2-} \rightleftharpoons O_2^{2-} \qquad (2.4.30)$$

takes place at the Pt|O$_2$|YSZ interphase boundary. Naturally, with low equilibrium oxide-ion concentrations in the melt (and consequently in the solid electrolyte) this equilibrium is shifted to the left. Therefore, at low oxide-ion concentrations in melts, the potential-defining reaction corresponds to the theoretical electrode process of a gas oxygen electrode described by equation (2.4.1). The reaction is also characterized by a higher oxidation potential compared to the reaction where peroxide ions take part. An increase of the equilibrium molality of oxide ions in molten medium results in a rise in their concentration at the "solid electrolyte |metal |gaseous oxygen" interphase boundary, and the concentration of peroxide ions formed according to reaction (2.4.30) approaches a certain value at which the electrode process (2.4.1) is blocked by peroxide ions located at the interface boundary. The electrode process in this case can be described by equation (2.4.5) and the corresponding chemical process in the potentiometric cell will be as follows:

$$\frac{1}{2}O_2^{2-} + Ag^0 \downarrow \rightleftharpoons Ag^+ + O^{2-}. \qquad (2.4.31)$$

It is obvious that at positive values for the e.m.f. of the cell (2.4.28) this equilibrium is shifted to the right.

Hence, the character of the reversibility of membrane oxygen electrodes with respect to oxide ions in ionic melts is dependent on the shift of the equilibrium in reaction (2.4.30). This reaction "switches on/off" one or another process for reduction of oxygen species present at the electrode to oxide ions. In this case, the peroxide function of the membrane oxygen electrode differs essentially from that of the conventional inconsumable gas oxygen electrode. To confirm or to disprove the proposed scheme, investigations of the membrane oxygen electrode are to be performed under conditions which provide an undoubted shift of reaction (2.4.30) towards the side of the formation of products, or the initial substances.

The results presented above give the possibility of confirming the correctness of the scheme shown in Fig. 2.4.7. We shall compare the calibration $E-pO$ dependences for the gas membrane oxygen electrode at 500 and 600 °C. From general considerations we should conclude that peroxide ions are more stable at 500 °C than at 600 °C and, therefore, the inflection point of the $E-pO$ calibration plot should be shifted to the high pO section. Actually, although this inflection exists (as confirmed by the calibration plot obtained using weights of Na_2CO_3 as a weak Lux base) it cannot be observed at the direct calibration at 500 °C. The inflection is easily determined in the calibration plot obtained at 600 °C (see Fig. 2.4.4), since under these conditions the stability of O_2^{2-} is lower and the section with a slope of $2.3RT/F$ is shifted towards lower pO values (≈ 2).

The behaviour of a $Pt(O_2)|YSZ$ membrane oxygen electrode at 700 °C was studied in Refs. [162, 189]. The calibration $E-pO$ plots obtained, one of which is presented in Fig. 2.4.8, clearly demonstrate the drawbacks of the direct calibration method at low initial molalities of oxide ions.

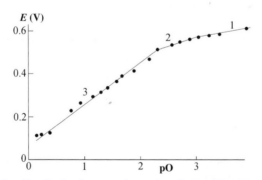

Fig. 2.4.8. Calibration $E-pO$ plot for potentiometric cell (2.4.28) with a membrane oxygen electrode in the KCl–LiCl eutectic melt at 700 °C: 1, the section where the effect of oxygen admixtures is predominant; 2, section with 3, section with $z = 1$.

The initial weight of the Lux base (KOH) corresponding to pO values considerably higher than 3 results in the appearance of a section with a slope appreciably less than $1.15RT/F$. This section is characterized by comparable concentrations of the introduced Lux base and the oxygen admixtures in the melt. In the second section, the effect of oxygen admixtures on the total oxide ion concentration is negligible and the resulting plot has a slope of 0.109 ± 0.01 V (at 700 °C $1.15RT/F$ is equal to 0.0965 V). This section allows us to calculate the concentration of oxygen-containing admixtures in the initial melts: it can be estimated as $(1.95 \pm 0.5)10^{-4}$ mol kg^{-1}. The weight of the base corresponding to pO values lower than 2 gives rise to the third section of the calibration plot, characterized by a slope of 0.190 ± 0.018 V $(2.3RT/F = 0.193$ V) corresponding to a one-electron process at the membrane electrode.

The calibration plot for the membrane oxygen electrode at 800 °C has no characteristic features, in contrast with similar plots obtained at 600 and 700 °C; the inflection point is located at pO ≈ 2 and the slope at low pO values is appreciably less than $2.3RT/F$. The latter fact is seemingly caused by the elevated acidity of the Li$^+$-containing melt in comparison, for example, with the KCl–NaCl equimolar mixture.

The parameters of the calibration E–pO plots (2.4.12) for different kinds of oxygen electrodes obtained in the KCl–LiCl eutectic melt at different temperatures are displayed in Table 2.4.2.

(ii) Molten KCl–NaCl equimolar mixture and CsCl–KCl–NaCl eutectic

These chloride melts are referred to as ionic media with low-acidic properties. As seen from the previous Part, they can be used as a zero (reference melts) for the construction of the general acidity scale of high-temperature ionic melts. The behaviour of most oxygen electrodes was studied in the molten KCl–NaCl, which is generally accepted as the "classic" molten medium.

One of the first investigations of the Pt(O$_2$) electrode as an indicator for the detection of O^{2-} in the KCl–NaCl eutectic melt at 700 °C was performed by Littlewood and Argent [235]. They studied the behaviour of the said oxygen electrode in an atmosphere containing HCl, H$_2$O and O$_2$ with different oxide-ion concentrations in the melt. The changes of oxygen pressure were found to affect the activity of O^{2-} in the melt slightly, whereas the effect of water and HCl was greater. The Pt(O$_2$) oxygen electrode was found to become a redox

TABLE 2.4.2
Parameters of the calibration E–pO plots (2.4.12) in the KCl–LiCl eutectic melt at different temperatures

Electrode	Donor of O^{2-}	E^0 (V)	k (V)	$\sim z$	pO^*
400 °C					
Ni|NiO	Na_2O_2	-0.408 ± 0.020	0.062 ± 0.012	2	1–3
KCl–LiCl, 500 °C					
Pt(O_2)|YSZ	Li_2O	0.544 ± 0.010	0.174 ± 0.030	1	1–4
	NaOH	0.484 ± 0.007	0.157 ± 0.040	1	1–4
	Na_2CO_3	0.723 ± 0.008	0.078 ± 0.004	2	2–4
Ni|NiO	Li_2O, Na_2O_2	-0.525 ± 0.015	0.089 ± 0.010	2	1–3
Ni|NiO	Na_2O_2, p_{H_2O}	-0.449 ± 0.011	0.095 ± 0.06	2	1–3
600 °C					
Pt(O_2)|YSZ	KOH	0.116 ± 0.021	0.094 ± 0.008	2	>2.3
		0.027 ± 0.011	0.154 ± 0.009	1	<2.3
700 °C					
Pt(O_2)|YSZ	KOH	0.191 ± 0.020	0.109 ± 0.010	2	>2
		0.020 ± 0.030	0.190 ± 0.018	1	<2
800 °C					
Pt(O_2)|YSZ	KOH	0.099 ± 0.022	0.109 ± 0.012	2	>2
		-0.052 ± 0.015	0.174 ± 0.008	1	<2

one if its potential against the standard chlorine electrode reaches the value of -0.44 V.

The platinum gas oxygen electrode reversibility in this melt was studied by Shapoval, Grishchenko and Zarubitskaya for subsequent calculations of the acid–base equilibrium parameters. Direct calibration of the potentiometric cell in the KCl–NaCl eutectic melt at 700 °C was performed with weights of sodium peroxide and sodium tungstate [148, 149]. The E–log $m_{O_2^{2-}}$ plots were characterized by a slope of 0.0917 V in both cases, and this was close to the value predicted by equation (2.4.3) for a temperature of 700 °C (0.0965 V). Later the same group found that the increase of Na_2O_2 concentration in the KCl–NaCl melt over 3×10^{-2} mol kg^{-1} resulted in a gradual change of slope of the calibration plots from the value corresponding to $1.15RT/F$ to that of the second section which is close to $2.3RT/F$. The origin of this slope consists in the peroxide function already mentioned. The results show that increasing oxide ion concentrations in the chloride melt cause the formation of stable peroxide ions, as a result of interaction between dissolved oxide ions with gaseous oxygen supplied into the electrode space which is in contact with the studied melt. The reversibility of the $Pt(O_2)$ gas electrode was shown to be disturbed by the presence of the redox forms of tungsten compounds.

The direct calibration of the potentiometric cell with the $Pt(O_2)$ electrode, using sodium carbonate as a Lux base, was stated in Refs. [139, 141] to lead to an E–pO plot with a slope of 0.155 V, which was intermediate between the values of $1.15RT/F$ and $2.3RT/F$.

Cherginets and Banik reported the direct calibration of the gas platinum oxygen electrode in the KCl–NaCl eutectic melt at 700 °C [233]. The calibration was made with weights of NaOH in the pO range from 1.20 to 2.91 (Fig. 2.4.9a, dependence 1). The slope of the mentioned calibration plot was calculated to be 0.176 ± 0.037 V. This agrees with the Nernst slope for the electrode process accompanied by the transfer of one electron ($2.3RT/F$). As with the above-considered results by Shapoval, such a value of the slope is explained by the potential-determining reaction (2.4.21) taking place at the

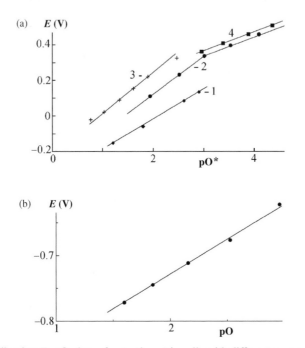

Fig. 2.4.9. Calibration E–pO plots of potentiometric cells with different oxygen electrodes in the KCl–NaCl eutectic melt at 700 °C: (a) $Pt(O_2)$: 1, NaOH; $Pt(O_2)|ZrO_2$; 2, NaOH, Ar; 3, BaO_2, Ar; 4, K_2CO_3,Ar; (b) Ni|NiO: NaOH, Ar.

inconsumable electrode in basic solutions, i.e. by the peroxide function. It should be noted that the e.m.f. values of the potentiometric cell are changed in sign. That means that the chemical reactions in the electrochemical cell:

$$Pt|Ag|Ag^+, KCl-NaCl\|KCl-NaCl, O^{2-}|YSZ|Pt(O_2). \qquad (2.4.32)$$

which can be described in such a manner

$$\frac{1}{2}O_2\uparrow + Ag^0\downarrow \rightleftharpoons Ag^+ + \frac{1}{2}O_2^{2-} \qquad (2.4.33)$$

runs from left to right at higher pO values (exceeding 2.5), and the decrease in pO below 2.5 results in a reversal in the direction of the reaction (2.4.33).

The potentiometric investigation of the metal-oxide nickel electrode in the KCl–NaCl melt at 700 °C shows that this electrode is reversible to oxide ions (see Fig. 2.4.9b) and the slope in the pO range from 1 to 3 is estimated to be equal to 0.109 ± 0.010 V. This is in a good accordance with the $1.15RT/F$ values characteristic of this kind of oxygen electrode (Table 2.4.3).

The reversibility of the $Pt(O_2)|ZrO_2(CaO)$ membrane oxygen electrode in molten KCl–NaCl has been studied in a number of works. Thus, the authors of Refs. [65, 236, 237] reported the application of the electrode with a $ZrO_2(CaO)$ solid electrolyte membrane for the potentiometric investigation of some heterogeneous acid–base equilibria and the construction of the row of cation acidity in the KCl–NaCl melt.

TABLE 2.4.3
Parameters of the calibration E–pO plots (2.4.12) of cells with different oxygen electrodes in the KCl–NaCl eutectic melt

Electrode	Donor of O^{2-}	E^0 (V)	k (V)	$\sim z$	pO^*
KCl–NaCl, 700 °C					
$Pt(O_2)|YSZ$	NaOH	0.000 ± 0.005	0.114 ± 0.010	2	>3
		-0.307 ± 0.030	0.216 ± 0.030	1	<3
$Pt(O_2)$	NaOH	-0.258 ± 0.070	0.176 ± 0.037	1	1.2–3
$Pt(O_2)$	Ni^{2+}/NiO	0.601	0		>6.8
		-0.424 ± 0.015	0.149 ± 0.003	1.5	5.6–6.8
		-0.333 ± 0.056	0.182 ± 0.020	1	2–3.3
$Ni|NiO$	NaOH	-0.947 ± 0.020	0.109 ± 0.010	2	1–3
$Ni|NiO$	Na_2O_2, p_{H_2O}	-0.139 ± 0.013	0.156 ± 0.006		2–3
KCl–NaCl, 800 °C					
$Pt(O_2)|YSZ$	KOH	0.226 ± 0.082	0.128 ± 0.031	2	>3
		-0.371 ± 0.013	0.176 ± 0.005	1	<3

The slopes of the calibration E–pO dependences obtained in the absence of gaseous oxygen in the atmosphere over the melt are approximately equal to the theoretical one (0.965 V). In particular, for the investigated Lux bases they are estimated to be 0.101 ± 0.007 V for Na_2CO_3 and 0.113 ± 0.002 V for SrO and NaOH. With a reactive atmosphere over the melt consisting of oxygen or air, the slope of the calibration E–pO plot becomes twice as high as $1.15RT/F$ in solutions of the strong Lux bases. As for the weak bases, it has been stated that passing oxygen containing gases through the melt does not cause an appreciable increase in the slope of the E–pO* calibration plot. For example, the calibration plot obtained with weights of SrO in air is characterized by a slope of 0.109 ± 0.013 V that remains close to $1.15RT/F$.

The calibration of a potentiometric cell with the $Pt(O_2)$ membrane electrode with the use of KOH, in the range of pO values from 1 to 4, results in a slope value of 0.0995 V [236]. However, a $Pt(O_2)|ZrO_2(CaO)$ gas membrane oxygen electrode should exhibit the peroxide function in the stated pO range, and, therefore, the correctness of the data presented is very doubtful.

Combes $et\ al.$ reported the use of $Ni|NiO|\ ZrO_2(CaO)$ membrane electrode for investigating oxoacidic processes in the molten KCl–NaCl equimolar mixture at 1000 K [154, 238–240]. This electrode is made as a fine mixture of Ni and NiO powders placed in the test-tube manufactured of the solid electrolyte ceramics. Pizzini and Morlotti gave the proportion of Ni:NiO to be 90:10%. This mixture is preliminarily sintered at high temperatures and then used as an electrode material [241]. Its use requires the application of a high-input-impedance electrometric amplifier (with an internal resistance of the order of 10^9 Ω). Preliminary calibration of the cell with the oxygen electrode was performed with BaO weights in the pO range from 1.15 to 2, i.e. the base was added to the formation of a saturated solution of BaO in the KCl–NaCl melt. The slope of the said E–pO depende nce is equal to 0.100 V. The use of Na_2O or electrochemically generated oxide ions for the calibration, instead of BaO, resulted in E–pO plots possessing slopes close to $1.15RT/F$, namely, 0.097 ± 0.002 V.

The potentiometric investigations of the $Pt(O_2)|YSZ$ membrane oxygen electrode at 700 and 800 °C by the calibration using various Lux bases have been reported in some of our papers. The potentiometric cell used for the studies in the molten KCl–NaCl eutectic consisted of the oxygen electrode of the type under study and the silver reference one [233]. The oxide-ion donors were BaO_2, NaOH, and sodium and potassium carbonates. The calibration was made in the pO range from 1 to 4. The E–pO calibration plots for the potentiometric cell including the $Pt(O_2)|YSZ$ oxygen electrode are presented in Fig. 2.4.9a. The data of this figure show that the use of strong bases for the calibration gives slope-values close to $2.3RT/F$ at low pO values, and the plots obtained using carbonate ions are characterized by a slope of approximately $1.15RT/F$ (0.101 ± 0.010 V). This agrees qualitatively with similar data obtained in the molten KCl–LiCl eutectic. The measurements made over a wide range of base concentrations (NaOH) result in the typical two-sectional plot consisting of two linear sections: the first of them is located at pO values exceeding 3, with the slope value of 0.114 ± 0.010 V which is somewhat greater than $1.15RT/F$. The increase of oxide ion concentrations above 10^{-3} mol kg^{-1} (corresponding to pO < 3) gives rise to a line with the slope of 0.216 ± 0.030 V—a value which exceeds the corresponding theoretical slope of $2.3RT/F$, also. Elevated concentrations of gaseous oxygen in the atmosphere over the melt cause some reduction in the slope of the E–pO plot; it becomes equal to 0.156 ± 0.021, which is intermediate between $1.15RT/F$ and $2.3RT/F$. The dependence obtained with the use of BaO_2 as an oxide ion donor (Fig. 2.4.9a, curve 3) is seen to be sufficiently shifted to the ordinate axis, although the slope value 0.209 ± 0.032 V remains in good agreement with $2.3RT/F$. This testifies to incomplete dissociation of BaO, formed by decomposition of BaO_2 in the unsaturated solutions:

$$BaO_2 \rightarrow BaO(Ba^{2+} + O^{2-}) + \frac{1}{2}O_2 \uparrow . \qquad (2.4.34)$$

The decomposition of pure barium peroxide is known to begin at 600 °C and dissolution decreases this temperature [114].

Moreover, the behaviour of the $Pt(O_2)|YSZ$ membrane oxygen electrode has been studied by the potentiometric method in dry and wet atmospheres over the studied chloride melts. The results show that the E–pO* slope in a dry atmosphere (0.168 ± 0.015 V) is somewhat greater than the corresponding value in wet argon containing approximately 0.01 atm. of H_2O (0.156 ± 0.006 V). The E–pO calibration plot for the $Pt(O_2)$ gas oxygen electrode (Fig. 2.4.9a, dependence 1) is located above the corresponding dependence for the $Pt(O_2)|YSZ$ membrane oxygen electrode (Fig. 2.4.9a, dependence 2). This fact is very important for the practice of oxoacidity investigations, since it permits determination of the integrity of the solid electrolyte membrane on the basis of e.m.f. data only. So, the shift of the e.m.f. value downwards points to the existence (or appearance) of cracks in the membrane. Because of these cracks, the melt studied is in immediate contact with the inner gas electrode.

The data represented by dependence 4 (Fig. 2.4.9a) show that carbonate ions are incompletely dissociated according to equation (2.3), even in diluted solutions, since the corresponding calibration $E-pO^*$ plot is located above the similar plot obtained with the use of strong bases (KOH or NaOH). The calibration $E-pO$ plot for the $Pt(O_2)|YSZ$ electrode, obtained at 800 °C, has no appreciable distinction from the plots constructed at 700 °C; it consists of two linear sections with the inflection point located at $pO \approx 3$. The parameters of the calibration $E-pO$ plots obtained in the molten KCl–NaCl eutectic melt are contained in Table 2.4.3.

The indirect calibration method allowed the practical determination of the upper pO limits of the reversible work, and consequently, the applicability of different kinds of oxygen electrodes for measurements of oxide ion activities in molten salts. We have obtained some interesting results connected with the reversibility of $Pt(O_2)$ gas oxygen electrode and $Pt(O_2)|YSZ$ membrane electrode toward oxide ions at pO values observed in the titration of strong cation acids in the molten KCl–NaCl eutectic at 700 °C [233, 234]. We shall consider the data for the $Pt(O_2)$ gas oxygen electrode (Fig. 2.4.10). The investigation of the reversibility was performed according to the techniques described on p. 160, using the data from potentiometric titration of Ni^{2+} cations with weights of NaOH. From Fig. 2.4.10a it follows that at the beginning of the titration the $Pt(O_2)$ oxygen electrode is not sensitive to changes of the equilibrium oxide ion concentration (curve 2), whereas the $Pt(O_2)|YSZ$ membrane electrode maintains good reversibility to O^{2-} ions under these conditions (curve 1). When the pO value approaches 6.8 (from the membrane electrode data), the gas electrode is "switched on" and starts to work reversibly to oxide ions [see Fig. 2.4.10b (plot 2)]. For the $Pt(O_2)|YSZ$ membrane oxygen electrode, the data from indirect calibration shown in Fig. 2.4.10b (plot 1) show that it remains reversible to oxide ions even at the their equilibrium concentrations taking place in acidic solutions containing an excess of Ni^{2+} cations. The value of the $E-pO$ slope calculated before reaching the equivalence point is equal to 0.114 ± 0.006 V; after reaching the equivalence point it becomes twice as high as the value obtained before the equivalence point, namely, 0.229 ± 0.020 V. These estimates are in good

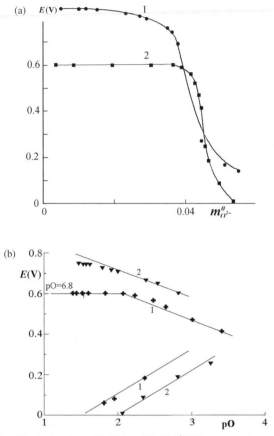

Fig. 2.4.10. Results of investigation of Pt(O_2) and Pt(O_2)|YSZ oxygen electrode reversibility in the molten KCl–NaCl eutectic at 700 °C: (a) potentiometric curves: 1, Pt(O_2)|YSZ; 2, Pt(O_2); (b) E–pθ dependences: 1, Pt(O_2)|YSZ; 2, Pt(O_2).

agreement with the data for the membrane oxygen electrode obtained by the direct calibration method (see p. 179). Similar investigations performed in solutions of other cation acids, such as Co^{2+} and Mg^{2+}, in the molten KCl–NaCl also confirm this conclusion. This remains correct for the use of the Pt(O_2)|YSZ membrane electrode for studies of oxoacidity in ionic melts of other anion compositions (bromides and iodides).

Returning to the data presented in Fig. 2.4.10b, we should mention that the loss of Pt(O_2) gas oxygen electrode's reversibility at pO > 6.8 is caused by the actual working of this electrode as a metal-oxide one even at high pO values.

It means that after immersing the $Pt(O_2)$ electrode in the solutions studied, the following chemical and electrochemical processes occur:

$$Pt^{2+} + 2\bar{e} \rightleftharpoons Pt^0 \downarrow, \qquad (2.4.35)$$

$$Pt^{2+} + O^{2-} \rightleftharpoons PtO \downarrow. \qquad (2.4.36)$$

Literature data confirm the fact of the formation of stable oxide films at the surface of platinum metal in ionic melts in oxygen-containing atmospheres, and hence the possibility of realization of the above-mentioned processes. Therefore, while the oxide film exists on the platinum surface, the solubility product of PtO in the melt is achieved, and the electrode becomes reversible to O^{2-}. After complete dissolution of this film in acidic solutions in ionic melts, which terminates at pO > 6.8:

$$PtO \downarrow + Ni^{2+} \rightleftharpoons NiO \downarrow + Pt^{2+} \qquad (2.4.37)$$

the $Pt(O_2)$ gas oxygen electrode transforms into a metal one, since at the surface of this electrode only the potential-determining process (2.4.35) takes place. Since the concentration of Pt^{2+} ions after complete dissolution of PtO remains constant, a plateau arises in the indirect-calibration-dependence, unless there are no additions of strong complexing agents to the solution studied.

The direction of reaction (2.4.37) is confirmed by the results obtained by Laitinen and Bhatia [230], who found that the solubility of NiO in the KCl–LiCl eutectic melts was considerably lower than that of PtO. Of course, such a sequence of solubilities remains unchanged in other melts.

When the values of pO become lower than 6.8, owing to the neutralization process proceeding, the $Pt(O_2)$ electrode begins to "feel" changes of the equilibrium oxide ion concentration and the slope of the $E-p\theta$ dependence is estimated to be 0.149 ± 0.015 V, which is not equal to $1.15RT/F$ (Table 2.4.3), the value of z being close to 1.5. A possible cause of this effect is the simultaneous running of potential-determining processes with the transfer of one and two electrons per elementary act of the reactions at the electrode surface. These processes may be the discharge of Pt^{2+} and Pt^+ ions that co-exist in the melt under the experimental conditions. Another possible reason consists in the competition of the peroxide function of the gas platinum oxygen electrode with the metal-oxide one.

After reaching the equivalence point, the excess of oxide ions is accumulated in the melt and this causes the slope of the $E-pO$ plot to equal

0.182 ± 0.020 V, which is a little lower than $2.3RT/F$. The electrochemical process that takes place at the electrode surface is just the pure, peroxide function of the gas electrode at low pO values, since the excess of the oxide ions in the melt favour the formation of O_2^{2-} owing to the interaction (2.4.4) that occurs upon bubbling of gaseous oxygen through high-temperature ionic melts. From the theoretical viewpoint there is no upper pO limit to the reversible work of the gas oxygen electrodes. However, what we have said above leads to the conclusion that there exist conditions limiting the reversible work of each particular electrode. These are bound up with characteristic features of the running of the electrochemical process (2.4.1) at the surface of this electrode.

For the membrane oxygen electrodes, their range of reversible work is depicted by the rhombus in the e.m.f.–pO diagram constructed on the basis of the theoretical analysis of Tremillon *et al.* (Fig. 2.4.11) [180]. From this diagram it follows that the pO limit of its reversible work is close to 13 (molar fractions); the value calculated in molalities lies within $11-12$. The lower pO limit is caused by corrosion and destruction of the membrane material in basic melts (formation of zirconates). As for e.m.f. limits, they are dependent on the oxidation of oxide ions to gaseous oxygen (the upper limit) and the reduction of ZrO_2 to zirconium oxides of lower oxidation degrees (the lower limit).

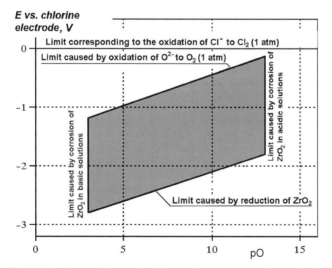

Fig. 2.4.11. The limits of reversible work of the membrane oxygen electrodes in chloride melts (KCl–NaCl eutectic) melt at 1000 K (pO values expressed in molar fractions), by R. Combes, J. Vedel and B. Tremillon [180].

Recently we have managed to check practically the position of the upper pO limit (acidic boundary) of reversible work of membrane oxygen electrodes for the case of $Pt(O_2)|YSZ$ [242]. The upper limit of the reversible work of the membrane oxygen electrode should correspond to the equilibrium concentration of oxide ions in the melt saturated by one of the two oxides (ZrO_2, Y_2O_3) of the solid electrolyte membrane. The material (alumina) of the crucible possesses a solubility in the melts that is considerably lower than those of ZrO_2 or Y_2O_3 because of common regularities connecting the solubility with the cation radius and the melting point. This means that if we study the solubility of an oxide whose cation radius is smaller than that of Y^{3+}, the initial section of the titration curve should include a plateau similar to that observed in Fig. 2.4.10, dependence 2. To perform the said study we chose $CrCl_3$, since it is sufficiently stable in the KCl–NaCl equimolar mixture at 700 °C (its melting point is 1192 °C [121]) in contrast to the more available ferric chloride $FeCl_3$ (boiling point close to 315 °C [121]), which sublimes under the experimental conditions, and the $FeCl_3$ remaining in the ionic solution is transformed into $FeCl_2$. The anhydrous chromium(III) chloride was obtained for the following titrimetric examination by the carbohalogenation method, i.e. by heating Cr_2O_3 in CCl_4 vapour. The description of the corresponding carbohalogenation can be found in Ref. [243]:

$$Cr_2O_3 + 3CCl_4 \uparrow \rightleftharpoons 2CrCl_3 + 3COCl_2 \uparrow . \qquad (2.4.38)$$

The data for the potentiometric titration of Cr^{3+} cations with weights of NaOH are shown in Table 2.4.4. The corresponding curve of the potentiometric titration depicted in Fig. 2.4.12 is characterized by the existence of a plateau in the initial section of the curve, located at the pO value approximately equal to 11 (section 1). This means that measurements of concentrations lower than 10^{-11} mol kg^{-1} cannot be realized in chloride melts using the membrane oxygen electrode According to the titration with Lux base additions, the equilibrium concentration of oxide ions increases. When the value of O^{2-} concentration exceeds a threshold concentration, a gradual decrease of the instrumentally measured e.m.f. value (and of pO) as the base's weight rise is observed. Therefore, we may conclude that the range of oxide-ions equilibrium concentrations available for potentiometric investigations using the membrane oxygen electrode with YSZ solid electrolyte has an upper limit corresponding to a pO value of 11.

TABLE 2.4.4

Results of the potentiometric titration of $CrCl_3(\sim 0.02$ mol $kg^{-1})$ with NaOH in the KCl–NaCl eutectic melt at 700 °C

$m_{O^{2-}}^0$	E (V)	pO	$\bar{n}$	$pP_{Cr_2O_3}$
0.000	1.191	11.04	–	–
0.001	1.195	11.07	0.05	36.44
0.004	1.197	11.09	0.14	36.55
0.008	1.199	11.11	0.32	36.73
0.011	1.190	11.03	0.46	36.60
0.015	1.180	10.94	0.61	36.47
0.022	1.147	10.65	0.88	35.92
0.023	1.130	10.50	0.92	35.53
0.024	0.802	7.62	0.97	26.97
0.025	0.742	7.10	1.01	25.46
0.035	0.095	2.26	1.17	–

The solubility-product values calculated on the basis of the stoichiometry of the following reaction:

$$Cr_2O_3 \downarrow \rightleftharpoons 2Cr^{3+} + 3O^{2-}, \tag{2.4.39}$$

should be attributed to the process of dissolution of zirconium dioxide rather than to the dissolution of Cr_2O_3 or Y_2O_3, whose solubility products are considerably lower, see Refs. [244, 245]. Since the formation of Zr^{4+} under

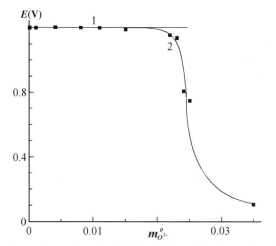

Fig. 2.4.12. Dependence of e.m.f. vs the initial molality of the base–titrant in the titration of $CrCl_3$ (~ 0.02 mol kg^{-1}) with NaOH weights in the molten KCl–NaCl eutectic at 700 °C.

the experimental conditions is less probable than that of ZrO^{2+}, the solubility products were calculated from the equation:

$$ZrO_2 \downarrow \rightleftharpoons ZrO^{2+} + O^{2-}, pP_{ZrO_2}, \qquad (2.4.40)$$

The average value of the solubility product of ZrO_2 was found to be $pP_{ZrO_2} = 12.62$. This is in good agreement with the data obtained by Komarov and Krotov [246], who estimated the solubility of ZrO_2 in the molten KCl–NaCl equimolar mixture at 1000 K as $\sim 10^{-5}$ mol kg^{-1}. Indeed, according to our data, the concentrations of ZrO^{2+} and O^{2-} in the saturated solution should be close to 5×10^{-7} mol kg^{-1}. Thus, the magnitude obtained by the authors of Ref. [246] should be considerably higher, owing to the somewhat higher temperature and the additional concentration of non-dissociated ZrO_2 in the said melt.

All that we have mentioned above allows us to assume that the upper limit of the reversible work of the oxygen electrode with a solid electrolyte membrane made of YSZ is connected with the chemical stability of the material of the membrane, which loses the zirconium dioxide both in strongly acidic and strongly basic media.

Although the molten CsCl–KCl–NaCl eutectic is one of the most convenient melts for investigations of various properties, owing to its relatively low melting point together with weak acidic properties, the reversibility of the oxygen electrode in this melt has not been studied so far. Studies performed at the Institute for Single Crystals concerns, among other aspects, the work of the $Pt(O_2)|YSZ$ membrane oxygen electrode at temperatures 600, 700 and 800 °C.

The potentiometric cell construction was similar to that in equation (2.4.28): NaOH and KOH were used as Lux bases for obtaining calibration plots. The behaviour of the membrane oxygen electrode agrees qualitatively with the results obtained in studies of other alkali metal halide melts (Fig. 2.4.13). All the calibration plots are characterized by an inflection point whose position shifts to higher pO values as the melt temperature lowers. At 700 and 800 °C changes in the character of the potential-determining process are observed at pO = 3, at 600 °C the inflection takes place at pO = 4 (Table 2.4.5). It is obvious that such a shift is caused by the peroxide function of the gas membrane oxygen electrode.

The parameters of the E–pO calibration plots in the molten CsCl–KCl–NaCl eutectic, obtained at different temperatures, allows confirmation of the

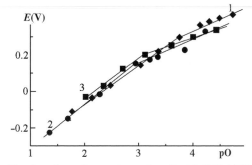

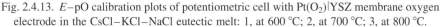

Fig. 2.4.13. E–pO calibration plots of potentiometric cell with Pt(O$_2$)|YSZ membrane oxygen electrode in the CsCl–KCl–NaCl eutectic melt: 1, at 600 °C; 2, at 700 °C; 3, at 800 °C.

above conclusion concerning the existence of two kinds of reversibility of the membrane oxygen electrode in ionic melts.

(iii) Other melts based on alkali metal halides

Rybkin and Seredenko employed the Pt(O$_2$)|ZrO$_2$(CaO) gas membrane oxygen electrode for measurements of oxobasicity in molten potassium chloride at 800 °C and in caesium iodide at 650 °C [247]. The calibration of the electrochemical cell in a KCl melt allowed them to obtain slope values equal to 0.213 ± 0.006 V when Na$_2$O$_2$ was used as an oxide-ion donor, and 0.106 ± 0.004 V when K$_2$CO$_3$ was used as a Lux base. The former value is approximately equal to $2.3RT/F$, whereas the latter is close to $1.15RT/F$.

TABLE 2.4.5
Parameters of E–pO calibration plots (2.4.12) for Pt(O$_2$)|YSZ membrane oxygen electrode in the CsCl–KCl–NaCl eutectic melt at different temperatures

O^{2-} donor	E^0 (V)	k(V)	$\sim z$	pO*
CsCl–KCl–NaCl, 600 °C				
KOH	0.005 ± 0.016	0.089 ± 0.004	2	>4
	-0.475 ± 0.010	0.207 ± 0.004	1	<4
CsCl–KCl–NaCl, 700 °C				
KOH	-0.151 ± 0.010	0.106 ± 0.020	2	>3
	-0.538 ± 0.026	0.235 ± 0.010	1	<3
CsCl–KCl–NaCl, 800 °C				
KOH	0.099 ± 0.022	0.109 ± 0.012	2	>2
	-0.052 ± 0.015	0.174 ± 0.008	1	<2

This points to the dependence of the slope value of the strength of the Lux base used as an oxide-ion donor. For molten CsI at 650 °C, it should be noted that the calibration of both Cs_2CO_3 and KOH resulted in a slope of 0.092 ± 0.006 V, which agrees with the slope of $1.15RT/F$ at this temperature. Concerning both of these melts, in the opinion of the authors, the use of sodium peroxide for the calibration results in the accumulation of O_2^{2-} ions in the chloride melt, and because of this the slope of the $E-pO$ plot approaches $2.3RT/F$. However, the experimental conditions (inert gas atmosphere) usual for such a routine favour the decomposition of peroxide ions in these melts and they decompose to O^{2-} with the evolution of oxygen. Rybkin and Banik calibrated the same electrode in an NaI melt at 700 °C. The calibration was performed over $-\log[CO_3^{2-}]$ a range from 1 to 4; Na_2CO_3 was used as a Lux base [248]. The slope of the $E-pO$ calibration plot was found to be 0.100 V, which is close to the $1.15RT/F$ predicted by equation (2.4.3). These investigations are the final ones among the pioneer works devoted to the application of membrane oxygen electrodes for investigations of the oxoacidic properties of molten halides.

Our results concern the reversibility investigations of the $Pt(O_2)|YSZ$ membrane oxygen electrode in molten individual alkali metal halides of potassium and caesium at 800 °C and of sodium halides at 830 °C [183, 197, 249]. The reversibility of the membrane oxygen electrode and its applicability for measuring the content of O^{2-} in molten NaI and CsI have significant practical applications, and it was examined over a wide temperature range from the melting point of the corresponding halide to approximately 800 °C [249]. For sodium halides these studies were performed up to 830 °C, which is somewhat higher than the melting point of NaCl. All the results obtained are contained in Table 2.4.6.

It may be assumed that for all the studied ionic melts, the behaviour of the $E-pO$ calibration—dependence remains unchanged; it has two sections with the slopes corresponding to values $z = 1$ and 2. The inflection point in the $E-pO$ plots obtained in molten potassium halides at 800 °C is located at the pO value equal to 3; the slope values at high pO practically coincide with $1.15RT/F$ (0.106 V), whereas in the strongly basic section the slopes of the dependences are somewhat lower than the theoretical one ($2.3RT/F$), which is equal to 0.212 V (see Table 2.4.6).

By analysing the $E-pO$ dependences in molten caesium halides at 800 °C, it is easily seen that these dependences are two-sectional with the inflection points located at pO ≈ 3. Even at high pO values (low oxide-ion

TABLE 2.4.6

Parameters of E–pO calibration dependences (2.4.12) for potentiometric cells with $Pt(O_2)|YSZ$ membrane oxygen electrode in melts of individual alkali metal halides

Melt, T (°C)	O^{2-} donor	E^0 (V)	k(V)	$\sim z$	pO*
KCl, 800	KOH	-0.162 ± 0.016	0.109 ± 0.008	2	>3
		-0.498 ± 0.006	0.206 ± 0.002	1	<3
KBr, 800	KOH	-0.283 ± 0.019	0.109 ± 0.003	2	>3
		-0.758 ± 0.006	0.231 ± 0.002	1	<3
KI, 800	KOH	-0.484 ± 0.056	0.112 ± 0.021	2	>3
		-0.645 ± 0.014	0.188 ± 0.009	1	<3
CsCl, 800	KOH	0.102 ± 0.010	0.091 ± 0.023	2	>3
		-0.752 ± 0.051	0.270 ± 0.017	1	<3
CsBr–KBr, 700	KOH	-0.172 ± 0.010	0.110 ± 0.010	2	>3
		-0.454 ± 0.007	0.204 ± 0.004	1	<3
CsBr, 800	KOH	0.142 ± 0.045	0.129 ± 0.012	2	>3
		-0.622 ± 0.060	0.321 ± 0.025	1	<3
CsI, 650	KOH	-0.202 ± 0.031	0.105 ± 0.009	2	>3
		-0.373 ± 0.010	0.165 ± 0.005	1	<3
CsI, 700	KOH	-0.152 ± 0.020	0.114 ± 0.006	2	>3
		-0.320 ± 0.008	0.193 ± 0.005	1	<3
CsI, 750	KOH	-0.103 ± 0.032	0.092 ± 0.010	2	>3
		-0.268 ± 0.020	0.154 ± 0.009	1	<3
CsI, 800	KOH	-0.145 ± 0.032	0.102 ± 0.011	2	>3
		-0.280 ± 0.018	0.158 ± 0.009	1	<3
NaCl, 830	NaOH	-0.127 ± 0.072	0.106 ± 0.024	2	>2
		-0.356 ± 0.014	0.204 ± 0.009	1	<2
NaBr, 830	NaOH	-0.078 ± 0.041	0.091 ± 0.015	2	>2
		-0.265 ± 0.002	0.160 ± 0.001	1	<2

concentrations) appreciable deviations of the slopes from $1.15RT/F$ are observed. As a rule, the slope values at low pO considerably exceed the, theoretical slope. Although for CsI this value is equal to 0.158 V and corresponds to $z = 1.35$, for CsCl and CsBr melts the corresponding values are equal to 0.270 V ($z = 0.79$) and 0.321 V ($z = 0.66$) [197]. The caesium bromide melt was found to be more "capricious" than other Cs^+-based melts.

It is interesting that the E–pO dependences in the CsBr–KBr (2:1) bromide melt at a lower temperature (700 °C) are characterized by the slope of 0.204 ± 0.004 V, which agrees with the 2.3 slope RT/F. Therefore, it is reasonable to assume that the addition of potassium ions to the CsBr melt favours stabilization of the ionic solvent. The same observation may be made for the substitution of the bromide melt anion: there are no appreciable anomalies in the behaviour of the membrane oxygen electrode in the caesium iodide melt (however, the iodide melt possesses stronger reducing properties).

The melts of sodium halides were studied at 830 °C, which is slightly higher than the melting point of NaCl, the most refractory halide. The latter's melting point is within 801–808 °C, from the data of several authors. The E–pO calibration plots for these melts consist of two sections and the inflection points are located near pO = 2. These values are appreciably lower than those for the potassium and caesium halides considered above. The observed deviation may be explained by two causes. The first is the fact that the oxoacidic properties of molten sodium halides are *a priori* stronger than those of the potassium and caesium salts, and hence the stability of peroxide ions in Na^+-based ionic media should be appreciably lower. However, the alkali metal halides formed by the same cations differ considerably in oxoacidic properties. The other, and more reasonable, explanation consists in the observation that alkali metal oxides (excluding Li_2O and Na_2O) are unstable at elevated temperatures. They are subject to disproportionation with the formation of the corresponding alkali metal peroxide dissolved in the melt at temperatures exceeding 400 °C [250]:

$$2K_2O \rightarrow 2K + K_2O_2, \tag{2.4.41}$$

and this trend strengthens from potassium oxide to caesium oxide. It is quite natural that the occurrence of a similar equilibrium shifted to the right favours the formation of peroxide ions in alkali metal halides containing strong Lux bases. The existence of stable O^{2-} ions in the melts results in "switching on" of the peroxide function of the membrane oxygen electrodes in strongly basic media; the potential-determining process is described by the slope, $z = 1$ or $2.3RT/F$.

A characteristic feature of the E–pO calibration plots in the molten sodium halides is the reduced values of the slopes at low pO; they are considerably less than $2.3RT/F$, whereas at high pO the slopes are approximately equal to $1.15RT/F$.

It is very interesting to examine the behaviour of the membrane oxygen electrode with a lowered partial pressure of gaseous oxygen in the inner space of the electrode (inside the YSZ test-tube) in the melts. The peroxide function of gas oxygen electrodes (and of the gas membrane oxygen electrode, as well) is caused by the formation of stable peroxide ions at the interface boundary. Therefore, the reduction of the partial pressure of O_2 in the inner electrode space should result in a shift of the inflection point of the E–pO calibration plot to lower pO values. It follows from the above-considered investigations that the membrane oxygen electrode with an inner Ni|NiO metal-oxide electrode demonstrates reversibility with a one-slope plot, which is characterized by a slope close to $1.15RT/F$.

To prepare the membrane oxygen electrodes with a lowered partial pressure of gaseous oxygen, the equimolar mixtures of powders of metal with the corresponding oxide were placed into a test-tube made of YSZ solid electrolyte. The platinum tap was also inserted into the test-tube. Afterwards the test-tube was plugged with asbestos mixed with silicate glue. The partial pressure of oxygen diminishes in the following sequence: $Cu/Cu_2O >$ $Ni/NiO > Ti/TiO_2$. All the prepared electrodes are characterized by sufficient reversibility to oxide ions (see Fig. 2.4.14). As is seen from this

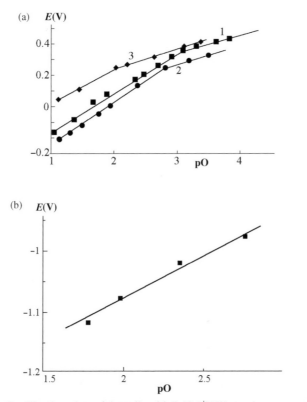

Fig. 2.4.14. $E-pO$ calibration plots of the cells with $Pt(O_2)|YSZ$ membrane oxygen electrodes with different partial pressures of O_2 in the inner electrode space in molten CsI at 700 °C: (a) 1, O_2; 2, Cu|Cu$_2$O; 3, Ni|NiO; (b) Ti|TiO$_2$.

figure, Pt|Cu/Cu$_2$O|YSZ and Pt|Ni/NiO|YSZ membrane oxygen electrodes, like the gas membrane oxygen electrode (under a pressure of O$_2$ equal to 1 atm.), are characterized by two-sectional E–pO calibration plots consisting of two linear sections of slope 1.15RT/F and 2.3RT/F. At the same time, the calibration of the Pt|Ti/TiO$_2$|YSZ membrane oxygen electrode yields a linear plot without inflection; the slope of this corresponds to a value 1.15RT/F. The inflection point for the cell with the membrane electrode filled with Cu + Cu$_2$O mixture is located near pO $\sim$ 3; the use of Ni + NiO mixture shifts the inflection point to pO $\approx$ 2. Hence, a reduction in the partial pressure of oxygen in the inner electrode space causes a shift of the inflection point up to its disappearance, which can be successfully explained in terms of the peroxide function of the gas oxygen electrodes. The stability of peroxide ions is known to reduce as the partial pressure of oxygen decreases, or the melt-temperature rises. Then, the inflection point and the section of the calibration plot corresponding to the slope 2.3RT/F should shift to lower pO values—as actually takes place.

The principal features of reversibility that are characteristic of metal-oxide and gas oxygen electrodes in molten alkali metal nitrates remain unchanged when these electrodes are used for measurements in molten alkali metal halides. The slope of the E–pO calibration plot of the cells with the membrane oxygen electrodes depends considerably on the means used for the creation of the O$_2$ partial pressure in the inner electrode space. This may be either by passing gas (oxygen or air) through the inner electrode space, or placing the "metal-oxide" mixture into the solid electrolyte membrane. The E–pO slope for the Ni|NiO|YSZ membrane electrode has been shown in a number of studies to close to 1.15RT/F. At the same time, the passing of oxygen through the inner electrode space with a platinum tap (electrode Pt(O$_2$)|YSZ) results in the two-sectional calibration plot, with slope 2.3RT/F at low pO values and slope 1.15RT/F at high pO, or with the use of weak Lux bases for the calibration. The Pt(O$_2$)|YSZ membrane oxygen electrode was shown in our investigations to be characterized by differing slopes, not only in the solutions of the various Lux bases but also in those of the same base but with different initial base concentrations. The latter is explained by the formation of peroxide ions at the "platinum | oxygen | solid electrolyte" interface boundary.

It should also be mentioned that practically all the calibration plots obtained with the use of different membrane electrodes are characterized by values of z, which are close but statistically different (being lower) from the corresponding integer numbers. This may be assumed to be one of the principal features characterizing the behaviour of oxygen electrodes with a solid electrolyte membrane made of YSZ ceramics. As a rule, the slopes of E–pO plots correspond to the theoretical $2.3RT/F$ and $1.15RT/F$ ones to within an accuracy closer than 10–15%.

2.4.2.3 Melts based on alkali- and alkaline-earth halides

Some papers have been devoted to studies of the behaviour of the membrane oxygen electrode with "inner" electrodes (i.e. placed into solid electrolyte test-tubes) $Ag|Ag^+$, $Ni|NiO$ and $Pt(O_2)$.

Castrillejo *et al.*, reported the investigation of the reversibility of the $Ag|AgCl,Na_2CO_3|YSZ$ membrane electrode in an equimolar molten $CaCl_2$–KCl mixture at 575 °C [178]. This electrode was shown to be sufficiently reversible to oxide ions, and the E–pO slope-value was close to $1.15RT/F$ (0.084 V). Calibration under CO_2 pressure (1 atm.) allowed the value of the dissociation constant of CO_3^{2-} (2.3) to be estimated as $10^{-1.4}$.

We have performed investigations of the behaviour of $Pt(O_2)|YSZ$ membrane oxygen electrode in the ionic melts based on alkali metal chlorides doped with alkaline earth chloride ($CaCl_2$, $SrCl_2$, $BaCl_2$). The concentration of the latter was approximately 25 mol% at 700 °C, and all the compositions corresponded to eutectic mixtures. The results of the calibration are presented in Table 2.4.7 and shown in Fig. 2.4.15. From the data presented it is seen that strengthening of the acidic properties of the cations in the sequence Ba^{2+}– Sr^{2+}–Ca^{2+} causes a gradual shift of the E–pO calibration plots to lower pO values and, finally, the inflection point disappears in the Ca^{2+}-containing

TABLE 2.4.7

Parameters of the E–pO calibration plots (4.12) for the cells with $Pt(O_2)|YSZ$ membrane oxygen electrode in melts based on alkali and alkaline earth metal chlorides at 700 °C

Melt	E^0 (V)	k (V)	$\sim z$	pO^*
$CaCl_2$–KCl	0.182 ± 0.009	0.087 ± 0.013	2.22	pO > 1.3
$SrCl_2$–KCl–$NaCl$	0.002 ± 0.014	0.144 ± 0.012	1.34	pO < 2.5
	0.103 ± 0.048	0.105 ± 0.016	1.83	pO > 2.5
$BaCl_2$–KCl–$NaCl$	-0.112 ± 0.012	0.140 ± 0.006	1.38	pO < 2.7
	-0.020 ± 0.010	0.112 ± 0.005	1.72	pO > 2.7
$BaCl_2$–KCl	-0.298 ± 0.006	0.159 ± 0.010	1.21	pO < 2.7
	-0.163 ± 0.010	0.116 ± 0.003	1.66	pO > 2.7
KCl–$LiCl$	0.020 ± 0.030	0.190 ± 0.018	1.02	pO < 2
	0.191 ± 0.020	0.109 ± 0.010	1.77	pO > 2

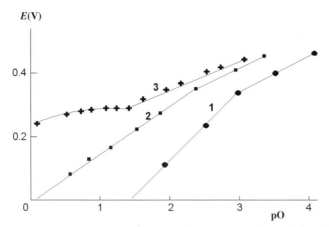

Fig. 2.4.15. E–pO dependence of Pt(O$_2$)|YSZ membrane oxygen electrode in molten BaCl$_2$–KCl–NaCl (1), SrCl$_2$–KCl–NaCl (2) and CaCl$_2$–KCl (3) melts at 700 °C.

chloride melts. It is more correct to say that the saturation of CaCl$_2$–KCl melts by CaO is achieved at an oxide ion concentration considerably lower than that corresponding to the inflection point of the E–pO calibration plot for a hypothetical solvent without the upper limit of basicity.

Another peculiarity of the alkali metal chlorides doped with alkali metal chlorides appears in the values of the slopes of the peroxide section; they are appreciably lower than 2.3RT/F. The behaviour of the Pt(O$_2$)|YSZ membrane oxygen electrode in chloride melts based upon alkali- and alkaline-earth metal chlorides is similar to that in the KCl–LiCl. In this case, the changing position of the inflection point of the E–pO calibration plot agrees with the positions of these melts doped with acidic cations (Ba–Sr–Li–Ca) in the general oxoacidity scale.

2.4.2.4 KCl–NaCl–NaF eutectic

Investigations of the Pt(O$_2$)|YSZ membrane oxygen electrode in the KCl–NaCl–NaF mixed melt at 700 °C allowed a very interesting result to be obtained concerning the problem of making measurements of the oxide-ion concentrations in fluoride melts using the equipment available for ionic melts based on other halide ions. The E–pO calibration plot for the Pt(O$_2$)|YSZ oxygen electrode in this melt differs basically from those considered above (see Fig. 2.4.16). First, the dependence obtained is not linear, but it looks like a

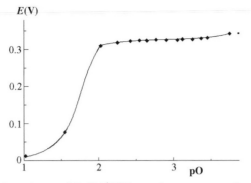

Fig. 2.4.16. E–pO dependence of Pt(O_2)|YSZ membrane oxygen electrode in the molten KCl–NaCl–NaF eutectic at 700 °C.

potentiometric titration curve. The most probable reason for such a behaviour of the membrane oxygen electrode is corrosion of polyvalent cations from the solid electrolyte ceramics or constructional materials (alundum, melted Al_2O_3). To establish the cause of such unusual behaviour of the E–pO calibration plot, samples of the molten KCl–NaCl and KCl–NaCl–NaF eutectics were extracted from the melts, after being held in the alundum crucibles immersed in the melt YSZ test-tubes for 2 h, and quenched to room temperature. Also, samples were obtained of the salt mixtures prepared from the pure reagents without melting.

Quantitative chemical analysis of the obtained melt samples showed the absence of Zr and Y in all the tests. Aluminium was found only in samples of the mixed chloride–fluoride mixture that had preliminarily been fused and kept for 2 h in an alundum crucible. The concentration of aluminium was estimated to correspond to the molality of AlO^+ ions in the said melt, of the order of 1×10^{-2} mol kg^{-1}. This is in good agreement with the data given in Fig. 2.4.16, where an e.m.f. drop similar to the drop at the equivalence point is observed at a concentration of O^{2-} close to 1.5×10^{-2} mol kg^{-1}.

Aluminium cations are well known to form very strong fluoride complexes [251], and, therefore, fluoride melts interact to an appreciable degree with the Al_2O_3-containing constructional materials. This leads to dissolution of alumina in molten mixtures containing F$^-$, followed by the formation of fluoro-complexes. Addition of the Lux bases favours the interaction's running in the reverse direction, i.e. it results in precipitation of Al_2O_3 from its solutions in the fluoride melt. These ideas allow us to assume that the following process in the melts can cause destruction of the alundum equipment:

$$Al_2O_3 \downarrow + 2nF^- \leftrightarrow 2AlOF_n^{-(n-1)} + O^{2-}. \qquad (2.4.42)$$

This reaction is shifted appreciably to the right, compared with the interactions with other halide ions. This fact is explained in the context of the HSAB concept: the formation of complexes of Al with Cl^-, Br^-, or I^- ions is less favourable than of the complex fluoroaluminates, since fluoride ion is the hardest halide base and Al^{3+} ion is referred to as the strongest hard acids. The F^- and O^{2-} ions seem to possess closely similar hard basic properties in molten salts.

Thus, owing to the high equilibrium concentration (of the order of 10^{-1} mol kg^{-1}) of admixtures (Al, O^{2-}) in fluoride melts prepared or used for measurements in Al_2O_3-based containers, serious quantitative investigations of oxoacidity cannot be realized in such media if the constituent parts of the equipment are made of this constructional material.

2.4.3 CONCLUSIONS

The potentiometric studies of the behaviour of $Pt(O_2)$ and $Pt(O_2)|YSZ$ gas oxygen electrodes show that the $E-pO$ calibration plots for these electrodes consist of two linear sections with different slopes approximately equal to $1.15RT/F$ [at low equilibrium oxide-ion concentrations, which correspond to the generally accepted process of gaseous oxygen reduction, by equation (2.4.1)], and $2.3RT/F$ in concentrated solutions of strong Lux bases. The fact that the latter slope is twice as large as the value predicted by equation (2.4.1) may be explained by the peroxide function of gas electrodes and this peroxide function differs for the above-mentioned electrodes. In the case of the $Pt(O_2)$ gas oxygen electrode the peroxide function consists in the formation of stable peroxide ions in the solution, owing to interaction with gaseous oxygen of oxide ions dissolved in the melt, and the electrode process is described by equation (2.4.21). In contrast, the reason for the peroxide function in the case of the $Pt(O_2)YSZ$ membrane electrode is the formation of stable peroxide ions at the interface boundary, "platinum | solid electrolyte| gaseous oxygen", but not in the melt studied. Therefore, the electrode process at this electrode differs somewhat from that occurring at the $Pt(O_2)$ electrode, and is described by equation (2.4.5). The experimentally obtained values of slopes of the E–pO calibration plots for the case of the membrane oxygen electrode are higher by 10–15% than the corresponding theoretical magnitudes. The duration of the e.m.f. stabilization for the membrane oxygen electrode is relatively short in acidic solutions (20–30 min). In basic media it increases to 1–1.5 h, owing to gradual dissociation of the strong Lux bases used (NaOH, KOH or peroxides).

The temperature rise, increase of the melt acidity or reduction of the gaseous oxygen pressure in the inner electrode space are established to shift the inflection point of the two-sectional calibration plots of the cells with the $Pt(O_2)YSZ$ membrane oxygen electrode to the low pO region, since all these factors decrease the peroxide ion's stability at elevated temperatures.

Investigations of the behaviour of the $Pt(O_2)YSZ$ membrane oxygen electrode in solutions possessing highly acidic properties show that its acidic limit of reversible work is caused by corrosion of the material of the solid electrolyte membrane, owing to considerable increase of ZrO_2 and Y_2O_3 solubilities in media possessing enhanced affinity to oxide ions. The upper acidic limit estimated from the titrimetric data corresponds to pO values close to 11 in the molten KCl–NaCl eutectic at 700 °C.

As for the $Pt(O_2)$ gas oxygen electrode, it should be noted that its upper acidic limit of reversible work is caused by the "metal-oxide" nature of this electrode, which actually works as a Pt|PtO one. This means that an oxide film is formed on the surface of this electrode and after its complete dissolution in acidic media the electrode loses its reversibility to O^{2-}. The upper acidic limit of the reversible work of the $Pt(O_2)$ electrode in the KCl–NaCl equimolar mixture was found to be near pO = 6.8. The decrease of pO to lower values makes this electrode sensitive to oxide ions. However, even in this case, the slope of the E–pO calibration plot deviates appreciably both from $1.15RT/F$ and $2.3RT/F$. In the pO range from 1 to 4 this electrode is reversible to oxide ions, owing to the peroxide function, with the slope of the E–pO plot being equal to $2.3RT/F$. This allows us to conclude that the most reliable results can be obtained by using this electrode for measurements of the oxide ion concentrations in basic solutions, although the period of the e.m.f. stabilization is too long.

As found in our investigations, the Ni|NiO metal-oxide electrode is well reversible to oxide ions in all the halide melts studied. The E–pO calibration plot is linear, with a slope corresponding to $1.15RT/F$ at all the temperatures studied. At elevated temperatures (above 500 °C), such electrodes can only be used for studies of oxoacidity in basic solutions (dissociation of strong and weak Lux bases). A reduction in the experimental temperature widens the acidic limit of the reversible work of these electrodes to high pO values, and they become the only electrodes suitable for obtaining reliable results in molten nitrate melts at temperatures of 200–400 °C (e.g., the Cu|CuO metal-oxide electrode).

Part 5. Investigations of Dissociation of Lux Bases in Ionic Melts

As repeatedly mentioned in the previous parts, the correctness of the constants obtained for the Lux acid–base equilibria depends essentially on the correctness of the assumption about the completeness of dissociation under the experimental conditions of the strong Lux bases, which are used for potentiometric titration of the studied acids and for calibration of the potentiometric cell with the indicator oxygen electrode. For the latter factor, the accordance to the corresponding Nernst equation of the potential-determining process at the indicator electrode is less important than the completeness of the oxide-ion donor's dissociation with the formation of O^{2-}.

Undoubtedly, the oxides formed by most acidic metal cations contained in the ionic solvent should be appropriate for the oxoacidity studies as the strongest Lux bases. However, the syntheses used for obtaining the said oxides are often laborious [250]. Moreover, their storage over a long period and their use for weight titration are very serious problems. Therefore, investigators are mostly forced to use other bases whose aqueous solutions display strongly basic properties. They are guided by the assumption that such bases will undergo complete dissociation when dissolved in an ionic melt. Besides this, the use of some Lux bases without any examination of their basic properties has been reported in certain literature sources and the choice of the base is not substantiated at all.

Alkali metal hydroxides (NaOH and KOH), peroxides (Na_2O_2) and carbonates (Na_2CO_3 and K_2CO_3) are often used as standard strong Lux bases in molten salt media. *A priori* the criterion of a strong basicity of aqueous solutions of these substances allows us to consider them to be completely dissociated in melts at high temperatures and in an inert atmosphere in the potentiometric cell. Naturally, the presence of water vapour, oxygen and carbon dioxide in the atmosphere over the melt suppresses the dissociation of these bases. These conditions are created deliberately for investigation of the dissociation processes under the equilibrium conditions. This branch of studies is closely connected with another one, of immediate practical importance. This concerns investigations of the interactions between reactive gas atmosphere of acidic or basic character, with ionic melts—and, particularly with molten alkali metal halides. We shall now consider the behaviour of the Lux bases in molten salts in this context.

2.5.1 REACTIONS OF IONIC MELTS WITH GASES OF ACIDIC OR BASE CHARACTER

Gases in the atmosphere over ionic melts can interact:

- with the main components of the ionic liquids and
- with the admixtures dissolved in them.

The reactions of the former kind (e.g. melt hydrolysis) mean the presence of an active gaseous component (water vapour), which serves as a Lux base and is the source of oxide ions arising in the melt. In contrast, the gaseous reagent, which takes part in reactions of the second kind demonstrates acidic properties by fixing (in some cases, with subsequent removal from the melt) or destroying various oxygen impurities that enter the melts by various means. These reactions also occur during the process of purification of molten alkali metal halides by halogenating agents.

The fusion of individual metal halides and of their mixtures is accompanied, as a rule, by high-temperature hydrolysis (recently, the term, "pyro-hydrolysis" has been used), which occurs due to the presence of traces of water vapour in the atmosphere over the liquid salts. These traces arise by evaporation of water from hygroscopic salts or are present in the inert gas, which has been insufficiently dried before use.

The presence of basic admixtures in ionic liquids after melting their component causes many unpleasant consequences. The wide use of Li^{+}-containing chloride melts as supporting electrolytes for high-temperature chemical-power sources shows that oxygen-containing impurities cause considerable increase of the solubility of the free alkali metal owing to the formation of sub-ions of composition Li_2^{+}. This leads to immediate transfer of the alkali metal from the anode to the cathode, where it destroys the electrode material. For example, in rechargeable Li–Ni power sources such an interaction occurs according to the following equation:

$$2Li_2^{+} + NiCl_2 \downarrow \rightarrow Ni^0 \downarrow + 4Li^{+} + 2Cl^{-}. \qquad (2.5.1)$$

This reaction is not accompanied with the appearance of electric current in the outer cell, worsens the technical characteristics and shortens the lifetime of the power sources. Nevertheless, apart from the work of Lysy and Duyckaerts [252], there are no data on investigations of the behaviour of oxygen-containing impurities in the KCl–LiCl eutectic melt.

It should be added that the oxygen-containing admixtures are extremely undesirable in molten fluorides, which are used as scintillation materials. It is surprising that molten fluorides have an enhanced affinity for oxygen, although this cannot be expected from the values of the redox potentials of a "halogen–halide ion system". The removal of the mentioned impurities from the fluoride melts is a very difficult problem. Other practically important media where oxygen impurities are undesirable are the molten sodium and caesium iodides, which are used for the growth of large scintillation detectors. Oxygen-containing impurities can play two roles in these crystals. It has been shown that doping of pure CsI with CO_3^{2-} leads to the formation in the crystal lattice of centres of blue luminescence and that the energy transfer from the lattice to these centres is sufficiently effective [253, 254]. In contrast, oxygen admixtures in the crystals doped by other activators cause the formation of traps and quenching centres.

Following our consideration of some cases of the negative effect of oxygen impurities on the properties of melts and products made from them, we shall now discuss the investigations of pyrohydrolysis.

2.5.1.1 High-temperature hydrolysis of melts based on alkali metal halides

The acid–base reactions of this kind are studied mainly for individual alkali metal halides and some of their most used mixtures.

Investigation of the pyrohydrolysis of solid and fused NaCl in the temperature range 600–950 °C was reported by Hanf and Sole [255], who proposed the so-called dynamic method for this purpose. This consists in passing (bubbling) a wet inert gas (nitrogen) through the layer of the solid salt or the melt, which gives rise to the process of pyrohydrolysis:

$$NaCl_{s,l} + H_2O \uparrow \rightarrow NaOH_l + HCl \uparrow . \qquad (2.5.2)$$

The HCl produced from the melt or solid is absorbed by water and its quantity determined by conventional acid–base titration. The treatment of the experimental results is as follows. The expression for the rate of HCl-production may be presented as:

$$dn/d\tau = vp_{HCl}/p, \qquad (2.5.3)$$

where p is the total pressure, p_{HCl} is the "equilibrium" partial pressure of HCl, and v is the rate of the wet gas supply.

Using this relationship, the equilibrium constant of equation (2.5.2) can be expressed as:

$$K = \gamma n p_{HCl}/(N p_{H_2O}),\qquad(2.5.4)$$

where γ is the activity coefficient of NaOH in solid or liquid NaCl. The integration of equation (2.5.3), taking into account equation (2.5.4), leads to the following time-dependence of the number of moles of HCl:

$$n = [2vN p_{H_2O} K/(p\gamma)]^{1/2} \tau^{1/2},\qquad(2.5.5)$$

which is linear in the coordinates $n-\tau^{1/2}$ as shown in Fig. 2.5.1. One should expect this dependence to pass through the coordinate origin ($\tau = 0$, $n = 0$), but this is not the case. The reason for such a deviation from linearity is that the actual interaction (2.5.2) is complicated by diffusion of NaOH in sodium

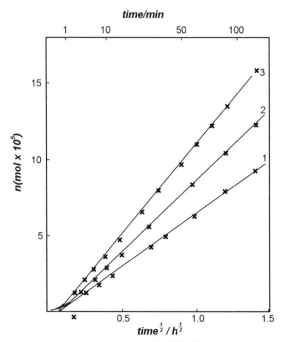

Fig. 2.5.1. Effect of temperature on the hydrolysis of NaCl above its melting point at 850 °C (1), 900 °C (2) and 950 °C (3). (By Hanf and Sole [255].)

chloride—especially in the solid-phase pyrohydrolysis. Taking this circumstance into account, the pyrohydrolysis model was modified in Ref. [255] by introducing a correction denoted as λ:

$$n = [2vNp_{H_2O}K/(p\gamma)]^{1/2}(\tau + \lambda(e^{-t/\lambda} - 1))^{1/2}, \tag{2.5.6}$$

In essence, this is the time required for minimization of the diffusion limitations, and this modification of the model allowed the plot to be made linear by choosing an appropriate λ value.

The obtained experimental data testify that high-temperature hydrolysis is negligible, although the rate of yielding HCl from the treated salts was somewhat higher than the value expected, owing to dissolution of NaOH in sodium chloride. The experimentally obtained plots of the dissociation constant against the inverse temperature are presented in Fig. 2.5.2. These plots show breaks at the melting point of NaCl, and the slopes of the straight-line plots differ somewhat.

Proceeding from the obtained results, the authors calculated the activity coefficients for NaOH in NaCl below and above its melting point. They found that these values exceeded 1 at lower temperatures and were below 1 in liquid NaCl. The dependences of the activity coefficients against the inverse temperature are expressed by the following equations:

$$\log \gamma_{NaOH} = 0.392 - 608\, T^{-1}, \quad T < T_{mp}, \tag{2.5.7}$$

$$\log \gamma_{NaOH} = -0.580 - 1038\, T^{-1}, \quad T < T_{mp}, \tag{2.5.8}$$

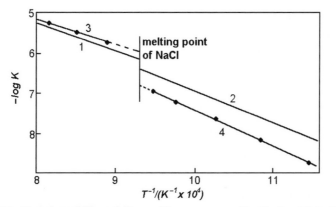

Fig. 2.5.2. Variation of K/γ and K_{theor} with temperature. (By Hanf and Sole [255].)

It is obvious that the main disadvantage of the method proposed by Hanf and Sole [255] is the assumption that the partial pressure of HCl in the gas mixture is equal to the equilibrium one. Detailed study of the description of the experiment, however, shows that the actual HCl pressure is appreciably lower than the equilibrium one, since pyrohydrolysis occurs in the system, although its rate is not high.

Rybkin and Nesterenko used the same experimental routine to investigate the processes of high-temperature hydrolysis and oxidation of solid and fused sodium iodide [256, 257]. According to the authors' estimation, the enthalpies of the following reactions

$$NaI + H_2O \rightleftharpoons NaOH + HI, \tag{2.5.9}$$

$$NaI + 0.5O_2 \rightleftharpoons 0.5Na_2O_2 + 0.5I_2, \tag{2.5.10}$$

are 32.8 ± 2 and 13.9 ± 1 kJ mol^{-1} in the case of solid NaI, whereas for molten NaI they are 30.6 ± 2 and 9.3 ± 2 kJ mol^{-1}, respectively. Consequently, as claimed by the authors [256, 257], sodium iodide is more prone to oxidation than to pyrohydrolysis. This applies to both the solid powder and the fused salt. Sodium peroxide formed in the oxidation process is stated to be insoluble in crystalline sodium iodide. Nevertheless, the conclusions made cannot be considered certain, since they are based on the values of the enthalpies and not on the magnitudes of the Gibbs energies of reactions (2.5.9) and (2.5.10).

A simple but reliable experimental routine was used by Combes, Vedel and Tremillon to investigate the pyrohydrolysis of the KCl–NaCl equimolar mixture [240]. For this purpose, a mixture of HCl and H$_2$O obtained by passing inert gas through aqueous solutions of HCl of certain concentrations was supplied into the inner space of the potentiometric cell with an indicator oxygen electrode (Ni|NiO|YSZ). The measurements of the equilibrium oxide-ion concentrations, performed by the potentiometric method, permitted calculation of the equilibrium constants of reaction (3.6), and determination of their thermal dependence:

$$pK_{(3.6)} = 55.3 \times 10^3 T^{-1} - 40.2. \tag{2.5.11}$$

This correlation shows that at 1000 K the value of pK is equal to 15.1, and hence that at this temperature, equilibrium (1.3.6) is shifted to the right.

The routine above was used to study the processes taking place in the pyrohydrolysis of the KCl–LiCl eutectic at 500 °C [232] and the equilibrium constant of reaction (1.3.6) was found to correspond to $pK = 9.77 \pm 0.4$. The results obtained in Ref. [232] mean that the pyrohydrolysis of the melts, with the formation of both O^{2-} and of OH^- ions, is thermodynamically unfavourable.

$$2H_2O + 2Cl^- \rightleftharpoons 2HCl + (OH)_2^{2-}, \qquad (2.5.12)$$

The experiment gives evidence that it is completely suppressed in the presence of the Lux bases in concentrations providing an equilibrium O^{2-} concentration of the order of 10^{-3} mol kg^{-1}.

Smirnov, Korzun and Oleynikova investigated the pyrohydrolysis of individual alkali metal halide melts, excluding the lithium and rubidium salts [258] over a wide temperature range. The equilibrium on which the calculations are based on is:

$$MX + H_2O \uparrow \rightleftharpoons MOH + HX \uparrow, \qquad (2.5.13)$$

where "M" and "X" denote alkali metal and halogen, respectively. To determine the hydrolysis constants, the following potentiometric cell with the liquid junction was used:

$$(H_2O + H_2)Au|NMOH + (1 - N)MX_1|MX_1|C(X_2). \qquad (2.5.14)$$

The e.m.f. value of the cell (2.5.14) may be expressed via the equilibrium concentration of the reagent as:

$$E = E^0 + \frac{RT}{F} \ln \frac{P_{H_2}^{1/2} P_{X_2}^{1/2}}{P_{H_2O}} a_{MOH}. \qquad (2.5.15)$$

On the basis of the potentiometric measurements the authors of Ref. [258] calculated the equilibrium constants for all the studied alkali metal halides from the following formula:

$$\ln K = \frac{F}{RT} E - \frac{\Delta G_{HX}^0}{RT} + \ln \frac{P_{H_2}^{1/2} P_{X_2}^{1/2}}{P_{H_2O}} a_{MOH}. \qquad (2.5.16)$$

The logarithms of the conditional equilibrium constants of pyrohydrolysis of all the studied alkali metal halides (2.5.13) are shown in Table 2.5.1, they are considerably lower than zero. This means that, as with the chloride melts, other alkali metal halides such as the bromides and iodides are not prone to

TABLE 2.5.1

Coefficients of dependences of $\ln K$ of reaction (2.5.13) against the inverse temperature: $\ln K = \ln K_0 + kT^{-1}$, for molten alkali metal halides

	Na^+		K^+		Cs^+	
	$\ln K_0$	$-k$	$\ln K_0$	$-k$	$\ln K_0$	$-k$
Cl^-	9.31 ± 1.1	$25{,}294 \pm 1297$	8.37 ± 0.59	$26{,}796 \pm 670$	8.01 ± 0.52	$27{,}173 \pm 579$
Br^-	3.51 ± 0.56	$19{,}786 \pm 657$	5.60 ± 0.68	$25{,}661 \pm 785$	5.63 ± 0.77	$26{,}588 \pm 865$
I^-	4.05 ± 0.73	$20{,}315 \pm 841$	5.65 ± 0.76	2612 ± 816	5.30 ± 0.57	2699 ± 650

Smirnov et al. [258].

hydrolysis under the standard conditions, which means, in particular, that the pressures of the gases be 1 atm.). The equilibrium constants for molten NaCl are in good agreement with the results obtained by Hanf and Sole [255].

The disposition of the molten alkali metal halides to the pyrohydrolysis (2.5.13) has been shown in Ref. [258] to decrease together with the increase of radius of both alkali metal cation and halide ion. Apart from the above-said constants, the authors of Ref. [258] estimated the activity coefficients of the solutions of OH^- ions in molten alkali metal halides to approach unity (1), i.e. the properties of these solutions are close to ideal ones. Qualitatively the conclusions concerning the activity coefficients agree with the data of Hanf and Sole [255]. Nevertheless, the activity coefficients of hydroxides at the melting points of alkali metal halides are close to unity (1) and then they are reduced by a factor of two or three.

2.5.1.2 Purification of halide ionic melts from oxide-ion admixtures

Investigations of the processes of alkali metal halide purification from oxygen-containing admixtures are mainly of applied importance. They are performed on the melts widely used for various scientific and industrial purposes. Different halogenating agents are often used for the removal of oxygen-containing admixtures, among which hydrogen halides and free halogens should be mentioned [259–267]. The interactions of such reagents with oxide ions in melts are represented by the following principal schemes:

$$2HX \uparrow + O^{2-} \rightleftharpoons H_2O \uparrow + 2X^-, \tag{2.5.17}$$

$$X_2 \uparrow + O^{2-} \rightleftharpoons 0.5O_2 \uparrow + 2X^-, \tag{2.5.18}$$

where X denotes halogen. For reaction (2.5.17), it should be noted that the water formed as one of the reaction products as the conjugate Lux base, in turn, shows pronounced acidic properties, fixing oxide ions into hydroxide ions; this interaction is superimposed

$$H_2O + O^{2-} \rightleftharpoons 2OH^-, \tag{2.5.19}$$

on equation (2.5.17) and retards the purification process. Naturally, the efficiency of such a disturbance of the purification is dependent on the hygroscopicity of the halides: the more hygroscopic the initial salt, the stronger the extent of holding water by the corresponding melt. The termination of the hydrogen halide's action on the melt may result in the starting of the reverse process of hydrolysis, since the reactions in equation (2.5.17) are seen from the previous section to be reversible. Another factor which causes the shift of reaction (2.5.17) to the left is the thermal instability of HBr and HI,

$$2HX \uparrow \xrightarrow{t} H_2 \uparrow + X_2 \uparrow, \tag{2.5.20}$$

and the removal of hydrogen halide from the reaction sphere shifts the equilibrium to the side of its formation.

Novozhilov *et al.* studied the solubility of HCl in molten alkali- and alkaline earth metal chlorides [268, 269] and found the solubility of HCl in these melts to obey Henry's law. The thermal dependences of the solubility (the Henry coefficient values) of HCl in molten alkaline earth metal chlorides, which are exposed to the pyrohydrolysis, are presented in Fig. 2.5.3. The dependences are close to linear an0d their treatment by the least-squares method allows the calculation of the thermodynamic parameters of HCl dissolution in the chloride melts. These values are presented in Table 2.5.2.

A comparison of the entropy values with those of dissolution of argon and CO_2 in molten fluorides [270] shows that they are very close. This means that the process of HCl dissolution in molten chlorides runs according to the physical mechanism, i.e. the dissolved molecules occupy voids in the melt structure. A small reduction in the gas solubility on moving from $CaCl_2$ to $BaCl_2$ melts is explained by an increasing surface tension in the said sequence of molten halides.

The dissolution process of HCl in molten alkali metal halides is accompanied by the formation of H-bonds, whose energies (kJ mol^{-1}) increase in the sequence Na (21.9)—K (25.1)—Rb (27.2)—Cs (31.0) [269]. The strength of H-bonds in molten chlorides is dependent on the polarizing

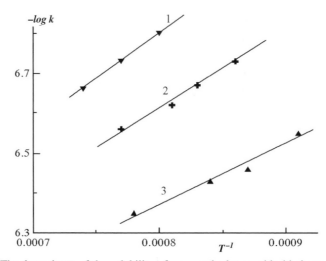

Fig. 2.5.3. The dependence of the solubility of gaseous hydrogen chloride in molten alkaline earth metal chlorides against the inverse temperature: 1, $BaCl_2$; 2, $SrCl_2$; 3, $CaCl_2$; from the data of Ref. [269].

action of the melt-cation. The H-bond becomes more stable if the polarizing action of the cation on chloride ion reduces, which favours the formation of "$Cl–H\cdots Cl^-$" associates. It should be noted that in molten alkaline earth metal chlorides, the absorption bands corresponding to the formation of H-bonds have not been found.

The production of magnesium requires reliable methods for the preparation of anhydrous magnesium chloride and, therefore, the related processes attract the attention of researchers. A thermochemical consideration of the dehydration of $MgCl_2 \cdot n\, H_2O$ and the analysis of possible ways for its purification from water and oxide-ion pollutants were performed by Kipouros and Sadoway [271]. An experimental study of the dehydration of mixtures of $KMgCl_3 \cdot 2H_2O$ and $MgCl_2 \cdot 2H_2O$ with KCl under different pressures of HCl (up to 2.5 MPa) and rates of heating was realized by Salyulev and

TABLE 2.5.2
Thermodynamic parameters of HCl dissolution in molten alkali-metal chlorides

Melt	T (K)	ΔH (kJ mol^{-1})	S (J mol^{-1} K^{-1})
$BaCl_2$	1097	42.15	32.6
$SrCl_2$	1198	37.96	31.4
$CaCl_2$	1295	27.95	25.5

Kaluzhnikova [272]. Heating of the mixtures under HCl pressures results in the formation of transparent melts, and then exposure for some minutes at 720–820 K leads to practically complete evaporation of water from the melts. Pressures of HCl exceeding 1.5 MPa suppress the formation of a precipitate of MgO in the process of dehydration. However, reduction of the pressure of the chlorinating agent leads to boiling of the melts and the appearance of an MgO suspension in them. This is explained by the decomposition of magnesium hydroxychloride according to the following reactions,

$$Mg(OH)Cl \rightarrow MgO \downarrow + HCl \uparrow, \tag{2.5.21}$$

$$2Mg(OH)Cl \rightarrow MgCl_2 + MgO \downarrow + H_2O \uparrow. \tag{2.5.22}$$

and traces of water are observed in the cooled parts of the experimental apparatus. If the treatment of the melts with HCl is repeated twice or thrice, the solid phase is not formed after the action of the acidic gas ceased. What we have said above concerns the melts having a high concentration of $MgCl_2$. The presence of KCl in the mixtures causes the formation of compounds in the solid state and of strong halide complexes of magnesium in melts. Chloride ions in such compounds replace water from the coordination sphere around the Mg^{2+} ion, and this substitution of coordinated particles facilitates the process of $MgCl_2$ dehydration and weakens the running of the pyrohydrolysis. Another explanation consists in the decrease in $MgCl_2$ activity in the melt, which shifts the equilibrium of the following reaction

$$MgCl_2 + H_2O \rightleftharpoons MgO \downarrow + 2HCl \uparrow, \tag{2.5.23}$$

to the left. By taking into account all that has been discussed above, one may make the following conclusion. The formation of complex metal–ammonium halides facilitates the obtaining of oxygen-free halides after calcination of the initial hydrate, and the use of NH_4X (since the radii of K^+ and NH_4^+ ions are close) as an easily removable component for the synthesis of anhydrous halides is the most convenient way for their synthesis.

Novozhilov reported the polythermal investigation of HCl solubility in the melts of $KCl-MgCl_2$ system [273]. The pure individual halides and melts containing 13, 33, 50, and 73 mol% of $MgCl_2$ were studied. The dependences of the Henry coefficient on the inverse temperature are linear (see Fig. 2.5.4),

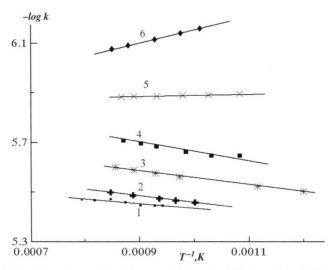

Fig. 2.5.4. The dependence of the Henry coefficient for HCl in melts of KCl–MgCl$_2$ system against the inverse temperature: 1, KCl; 2, KCl—MgCl$_2$ (13 mol% of MgCl$_2$); 3, KCl–MgCl$_2$ (33 mol% of MgCl$_2$); 4, KCl–MgCl$_2$ (50 mol% of MgCl$_2$); 5, KCl–MgCl$_2$ (73 mol% of MgCl$_2$); 6, MgCl$_2$.

which allows us to use the following equation

$$-RT \log k = \frac{\Delta H}{2.303RT} - \frac{\Delta S}{2.303R}, \qquad (2.5.24)$$

for calculation of the thermodynamic characteristic of the solutions of HCl in the said chloride melts. The data obtained are presented in Table 2.5.3 together with the parameters of the said polythermal plots:

$$-\log k = -\log k_0 + aT^{-1}. \qquad (2.5.25)$$

In the authors' opinion, the thermodynamic data presented show that the whole of the concentration region studied can be treated from the viewpoint of the formation of MgCl$_4^{2-}$ coordination compounds in the melts. Dissolution of hydrogen chloride in the melts is not accompanied by an appreciable interaction with the melt components (KCl and MgCl$_2$). Therefore, the work of the formation of the void in the melts should be the main factor, which defines the behaviour of the temperature dependence. The changes in the deviation of the derived ΔH and ΔS values from the additive magnitudes possess an extremal character; the maximum deviations are observed at an

TABLE 2.5.3
Thermodynamic parameters and coefficients describing the dissolution of HCl in melts of the KCl–MgCl$_2$ system

Concentration of MgCl$_2$ (mol%)	0	13	33	50	73	100
ΔH (kJ mol^{-1})	5.6	3.93	2.67	1.34	9.45	18.7
ΔS (J mol^{-1} K^{-1})	−4.4	−6.48	−9.58	−12.88	−9.2	−5.02
$-\log k_0$	5.63 ± 0.03	5.72 ± 0.01	5.83 ± 0.01	5.97 ± 0.04	5.83 ± 0.01	5.64 ± 0.01
a	−201 ± 37	−266 ± 6	−272 ± 9	−309 ± 40	51.6 ± 4	51.1 ± 6

MgCl$_2$ concentration near 33 mol%. The ratio KCl:MgCl$_2$ at this point of the phase diagram allows us to divide it into two regions, which are characterized by an excess of KCl and MgCl$_2$ against K$_2$MgCl$_4$ composition. The excess quantity of MgCl$_2$ as compared to K$_2$MgCl$_4$ melt causes a breakdown of the KCl structure, which leads to a reduction of "free" chloride ions in the melt and, hence, the fraction of the Cl–H$\cdots$Cl$^-$ bonds.

Nevertheless, there exists a simpler explanation of the observed facts. The character of the changes in the Henry coefficient differs with the variation of MgCl$_2$ concentration in the melts. KCl is known to form one stable and congruently melting compound of composition KMgCl$_3$. It is interesting that the excess of the individual components against this composition defines the behaviour of the thermal dependence with the slope, which is characteristic for the excess substance. Indeed, the Henry coefficient of HCl for alkaline earth metal chlorides decreases with the temperature, whereas the changes observed in alkali metal chlorides are the inverse (see Fig. 2.5.4). Therefore, the plots 1,2,3 are similar to those of alkali metal chloride, and plots 5 and 6 are typical for molten individual alkaline earth metal chlorides. As for the equimolar KCl–MgCl$_2$ mixture (congruently melted compound), its behaviour is similar to that of melts containing excess of KCl.

The effectiveness of the purification of molten lithium halides by hydrogen halides is appreciably lower than mentioned above [232]. This is explained by the fact that molten lithium halides keep the water, which can dissolve in these melts in considerable quantities. Even mixed-halide mixtures retain this property, e.g. the molten KCl–LiCl eutectic keeps the dissolved water strongly at temperatures of the order of 400 °C [179], and bubbling of dry HCl during an hour does not result in complete removal of H$_2$O.

Purification of substances that are prone to decomposition with chlorine evolution is especially effective at elevated temperatures since, as shown by Ryabukhin and Bukun [274], the solubility of chlorine in alkali metal chloride

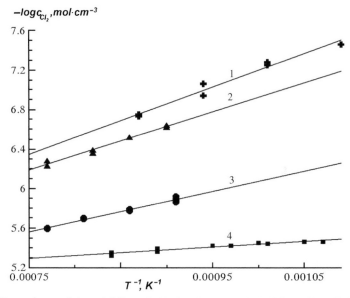

Fig. 2.5.5. Dependence of the solubility of chlorine (at $p_{Cl_2} = 1$ atm.) in molten alkali metal chlorides against the inverse temperature $(-\log c_{Cl_2,MeCl} = f(T^{-1}))$:1, LiCl; 2, NaCl; 3, KCl; 4, CsCl.

melts increases with the temperature (see Fig. 2.5.5). The corresponding dependences $(-\log c_{Cl_2,MeCl}(mol\ cm^{-3}) = f(T^{-1}))$ can be approximated by the following formulae:

$$-\log c_{Cl_2,LiCl} = 3.79(\pm 0.24) + 3.41(\pm 0.25) \times 10^3 T^{-1}, \qquad (2.5.26)$$

$$-\log c_{Cl_2,NaCl} = 3.98(\pm 0.14) + 2.94(\pm 0.16) \times 10^3 T^{-1}, \qquad (2.5.27)$$

$$-\log c_{Cl_2,KCl} = 4.02(\pm 0.09) + 2.04(\pm 0.10) \times 10^3 T^{-1}, \qquad (2.5.28)$$

$$-\log c_{Cl_2,CsCl} = 4.86(\pm 0.05) + 0.57(\pm 0.04) \times 10^3 T^{-1}. \qquad (2.5.29)$$

From these dependences, it follows that an increase of the radius of the alkali metal cation results in decrease (in absolute magnitude) of the slopes. The free terms of the three first dependences have close values. It is interesting that there is no reason to divide the said dependences into two groups (Li and Na vs K and Cs) on the basis of the slope data, although the dissociation of carbonate ion (see p. 237) allows one to make such a division.

Therefore, the equilibrium concentration of Cl_2 increases with temperature and the equilibrium (2.5.38) is shifted to the right according to Le Chatelier's rule.

As for the effectiveness of purification of halide melts with free gaseous halogens, this is defined by the standard redox potentials of the corresponding $C(X_2)$ halogen electrodes, as compared with that of the oxygen electrode, and reduces in the sequence chlorine–bromine–iodine. Similarly, the action of the latter halogen on the oxide-ion-containing melt may result in an apparent removal of the oxygen-containing admixtures; however, the total concentration of oxygen remains unchanged owing to the following process of halogen disproportionation:

$$3I_2 \uparrow + 3O^{2-} \rightleftharpoons IO_3^- + 5I^-. \tag{2.5.30}$$

The iodate ions formed are stable enough in molten media and can return oxygen upon the addition of substances of acidic character to the iodide melts, e.g.

$$IO_3^- + 5I^- + 3Mg^{2+} \rightleftharpoons 3I_2 \uparrow + 3MgO \downarrow, \tag{2.5.31}$$

and the real equilibrium-concentration of the Lux acid (Mg^{2+}) in the melt will be appreciably lower than the expected one.

One of the most useful applications of melt treatment by halogens is the dissolution of free metals. Such a treatment is especially convenient for the synthesis of derivatives of certain noble metals, which are hardly dissolved in the generally accepted superacidic media at room temperature. Thus, Sinitsyn *et al.* investigated the chlorination of metallic iridium in molten alkali metal chlorides [275]. The chlorination process leads to the formation of a mixture of Ir^{III} and Ir^{IV} in a 2:1 molar ratio: the chlorides are immediately fixed into the complexes, Me_2IrCl_6 and Me_3IrCl_6 that prevents evaporation of the iridium chlorides from the melts (the degree of removal does not exceed 0.5%). Stabilization of Ir^{IV} is caused by the formation of clusters containing $Ir^{III} \cdots Ir^{IV} \cdots Ir^{III}$ fragments and, therefore, the ratio of the oxidation degrees of Ir is independent of the melt composition. The composition of the polynuclear complex shows that it should be equal to 2:1.

The purifying action of ammonium halides (NH_4X) may be explained, in the simplest manner, by their thermal decomposition (sublimation) with the formation of the corresponding hydrogen halides, which act as halogenating agents [276–278]. If the quantity of the hydrogen halide is excessive, the remaining NH_4X precipitates as a solid phase on the cool parts of the reaction vessels. The excess of ammonia equivalent to the quantity of consumed HX is removed in a flow of inert gas or by pumping out the reactor. This means that the reversible reaction

$$NH_4X \uparrow = NH_3 \uparrow + HX \uparrow \qquad (2.5.32)$$

is used for the purification.

The considerations above lead to the conclusion that ammonium halides should be recommended, mainly, as halogenating agents only for drying and sintering of the initial salts. Their use for the industrial purification of melts is hardly expedient since there are considerable problems with providing a persistent flow of NH_4X through the melt, whereas HX and volatile organic halides (carbohalogenation) have undoubted advantages in this respect.

Nevertheless, the use of NH_4X for purification of solid salts is substantiated by the fact that it forms complex halides with salts of multivalent cations. As an example, we may consider the process for obtaining anhydrous $MgCl_2$. Dissolution of MgO in HCl, followed by the evaporation of some water results in the crystallization of $MgCl_2 \cdot 6H_2O$; this salt is well known to lose HCl and H_2O at $\sim 400\,°C$, with MgO remaining:

$$MgCl_2 \cdot 6H_2O \xrightarrow{400°C} MgO + 2HCl \uparrow + 5H_2O \uparrow, \qquad (2.5.33)$$

and drying of the hydrate in a flow of HCl is not fruitful. This is explained by oxygen entering from water into the coordination shell of the Mg ion and the concurrence of Cl^- and stronger hard acid, O^{2-}. Addition of NH_4Cl to the hexahydrate results in the formation of an intermediate adduct of composition $(NH_4)_2MgCl_4 \cdot n\,H_2O$, in which the oxygen is displaced by chloride ions in the coordination sphere. This complex salt is easily dehydrated, even in an inert gas or in air. After complete sublimation of NH_4Cl, anhydrous $MgCl_2$ remains in the reaction vessel and may be smelted into a compact mass with a melting point near $742\,°C$. Similar considerations apply to anhydrous rare earth halides considered to be promising new scintillation materials [279–282], which are obtained in a similar way. It is interesting to note that the applicability of such an approach may be estimated from the phase diagrams of $KX–MeX_n$. The K^+ cations are known to possess an ionic radius close to that of NH_4^+ and the formation of the stable complex compound K_mMeX_{n+m} points to the fact that the corresponding $(NH_4)_mMeX_{n+m}$ complex should be stable also, and NH_4X can be recommended for dehydration of the multivalent metal halide. The pair K_2MgCl_4 (congruently melted compound)$-(NH_4)_2MgCl_4$ may be demonstrated as a successful example. An essential advantage of this method of purification is the removal of iron admixtures from the purified melts: the iron is removed as $FeCl_3$ in the process of sublimation and condenses on the cooled parts of the reactors together with excess of NH_4Cl.

The ability of NH_4Cl to substitute oxide ions in compounds by chloride ions is used for the chemical synthesis of chlorides by treatment of oxide raw material in molten salts. For example, Tselinckii and Samokhvalova report the obtaining of copper chlorides in a molten salt mixture, which is a very convenient way of synthesizing copper halides [283]. The heating of CuO-containing raw material with ammonium halides at 150–450 °C leads to the formation of low-melting eutectics of composition CuX_n-NH_4X. This is accompanied by the evaporation of water and ammonia from the melt:

$$CuO + 2NH_4X \rightarrow CuX_2 + 2NH_3 \uparrow + H_2O \uparrow, \qquad (2.5.34)$$

$$6CuO + 6NH_4X \rightarrow 3Cu_2X_2 + N_2 \uparrow + 4NH_3 \uparrow + 6H_2O \uparrow . \quad (2.5.35)$$

Then the molten mixture can be obtained as an anhydrous melt and, after cooling, as a solid salt. This routine allows the obtention of a substance containing 99.6–99.9 mass% of the main reagent, which is better than that prepared using traditional synthesis routines. Moreover, ammonium halides are extremely convenient for the preliminary purification of halide melts before their use in various kinds of investigations; the addition of crystalline NH_4X results in a considerable reduction in the concentration of oxygen-containing impurities, down to $10^{-5}-10^{-6}$ mol kg^{-1} with respect to equilibrium O^{2-} molality [214]. We use this method for oxide solubility studies in which not only the melt but also the added metal halide is subjected to this treatment, which often allows to "feel for" unsaturated solution sections in the titration curves.

Many investigators consider different variations of carbohalogenation as the most convenient way for removing oxygen-containing impurities from molten ionic halides [284–291].

The halogenating action of generally accepted chlorinating agents, such as CCl_4, $COCl_2$, Cl_2, and $C + Cl_2$ mixture (where carbon is introduced into the melt in the form of acetylene soot), is substantiated by the thermodynamic calculations of the following reactions [284]:

$$CCl_4 \uparrow + 2O^{2-} \rightarrow CO_2 \uparrow + 4Cl^-, \tag{2.5.36}$$

$$COCl_2 \uparrow + O^{2-} \rightarrow CO_2 \uparrow + 2Cl^-, \tag{2.5.37}$$

$$C + Cl_2 \uparrow + O^{2-} \rightarrow CO \uparrow + 2Cl^-. \tag{2.5.38}$$

As shown by thermodynamic analysis, the effectiveness of all the above purification processes (2.5.36)–(2.5.38) is practically the same as is seemingly caused by the formation of gases that are inert in relation to halide melts (CO and CO_2) [285]. However, reaction (2.5.36) possesses an appreciable advantage, since its application is not connected with the formation of considerable quantities of extremely toxic compounds (Cl_2, $COCl_2$) during the purification, in contrast with reactions (2.5.37) and (2.5.38). Therefore, practically all the studies applied for the removal of oxygen-containing impurities from chloride melts are just connected with the use of CCl_4 [286–290]. The purification plant comprises a saturator with liquid CCl_4, an inlet for inert gas, and an outlet for the gas saturated with CCl_4.

The process of purification of alkali metal chlorides runs via the following stages:

- melting of the initial salt mixture in an atmosphere of inert gas and CCl_4, accompanied with the formation of a black carbon suspension because of the pyrolysis of organic admixtures; sometimes a layer of carbon floats up and aggregates at the melt surface,
- treatment of the melt by bubbling of a saturated vapour of CCl_4 in an inert gas flow, the carbon suspension gradually disappears owing to chlorination, which occurs according to the following reaction:

$$CCl_4 \uparrow + C + 2O^{2-} \rightarrow 2CO \uparrow + 4Cl^-. \tag{2.5.39}$$

  The usual treatment temperature is near $800\,°C$, and its duration is approximately 1–2 h. This depends on the melt acidity (a shorter time is required for KCl—NaCl, and a longer one is needed for hygroscopic chlorides).
- The final stage is accompanied with the absence of wetting; the liquid does not wet the walls of the quartz container, and the oxygen-containing admixture at this stage can be estimated by values of 10^{-4} mol kg^{-1}.

Aleksandrov *et al.* [292] reported a study of the processes of rare-earth metal oxide chlorination by CCl_4 at 200–840 °C. A quartz container does not react with CCl_4 up to 750 °C, but heating to higher temperatures causes a decrease in the container's mass owing to the formation of volatile compounds of silicon according to the following reaction:

$$SiO_2 + 2CCl_4 \rightarrow SiCl_4 \uparrow + 2COCl_2 \uparrow .$$ (2.5.40)

As for rare-earth metal oxides, it should be noted that their chlorination was performed with a consumption of a quantity of CCl_4 4–5 times as much as required according to the stoichiometry. The thermal dependence of the rate of chlorination passes through a maximum, which is caused by the peculiarities of mass transfer in the treated samples. A decrease in the rate of chlorination of solid samples with temperature is caused by their coagulation and reduction of the specific surface area. Chlorination of the oxide materials at temperatures above the melting points of the corresponding chlorides is limited by the solubility of CCl_4 and products of its thermolysis in the formed melts, and by their diffusion in the liquid phase. The degree of conversion of the studied rare-earth metal oxides into chlorides, at the temperature corresponding to the maximal rate of chlorination, does not exceed 85–95%.

Carbohalogenation of bromide and iodide melts is very effective, also. For this purpose, the use of C_2H_5Br, $CHBr_3$, or CBr_4 [291] for bromide melts and of C_2H_5I [285] for iodide melts is recommended. An essential advantage of the carbohalogenation process consists in the considerable reduction of the concentrations of transition-metal cations in the purified melts, especially as it concerns the use of polyhalogen-derivatives. However, the process of purification with the use of organic halides with a low content of halogen results in accumulation of appreciable quantities of carbon (soot) in the melt. When the purified melt is used for crystal growth, the suspended particles are rejected by the solid–liquid interface. However, carbon suspensions are not desirable components of the molten medium and, therefore, the use of the monohalide derivatives cannot be considered as a convenient way for carbohalogenation. The most suitable volatile organic halides should be characterized by a C:X ratio less than 1:2 (CH_2X_2, CHX_3, CX_4). For the case of bromides, the possible way uses mixtures of bromine with volatile halocarbons. In this case, the substances evaporate in an inert gas flow according to their saturated vapour pressures. This provides the action of $C + Br_2$ mixture on the melt.

From the point of view of chemical equilibrium, carbohalogenation results in the formation of CO_2 which, in turn, reacts with oxide ions and transforms them into sufficiently stable carbonates, thereby retarding the main process of the purification. Naturally, carbonate is the main oxygen-containing admixture remaining in the halide mixture obtained after the carbohalogenation.

Eckstein, Gross and Rubinova reported the use of halide derivatives of silicon for the purification of alkali metal halide melts from oxygen impurities [293] by means of the interaction of SiX_4 with the mixtures by the reaction:

$$SiX_4 \uparrow + 2O^{2-} \rightarrow SiO_2 \downarrow + 4X^-.$$ (2.5.41)

Silicon dioxide, one of the products of this interaction, is insoluble in pure alkali metal halides and separates from the molten medium owing to the difference in densities. Thermodynamic analysis of the processes of molten iodide purification with different halogenating agents shows that their effectiveness reduces in the sequence $SiI_4 > HI > I_2$ [294]. An obvious advantage of silicon halides for the purification of halide melts used for single-crystal growth is the fact that their use does not result in the appearance of additional impurities in the purified melts, since these processes are usually performed in quartz (SiO_2) vessels-reactors.

The described purification methods are effective enough in an over-whelming majority of cases, but nevertheless each method has its own specific impurities remaining in the melt after the purification. For example, the use of HCl or NH_4Cl is connected with the appearance of water traces, whereas CCl_4 and other similar halogen derivatives result in the formation of carbonate. On the other hand, the purification by HCl or NH_4Cl starts even with mild heating (up to $200-300\,°C$) that avoids corrosion of the quartz container when used. On the contrary, CCl_4 affects the oxygen impurities starting only from the temperatures of the pyrolysis of organic compounds, i.e. $400-450\,°C$, and in this case, corrosion of the quartz container becomes vital. This means that the features of the subsequent application of a given ionic melt requires the choice of a specific method of purification of this melt from oxygen impurities, taking into account its advantages and drawbacks.

2.5.2 BEHAVIOUR OF LUX BASES IN IONIC MELTS

The investigation of the equilibria occurring in the dissociation processes of oxide ion donors (Lux bases) in ionic melts is of great practical importance both for experimental studies of oxoacidity and for various technological purposes.

The investigations of oxygen-electrode reversibility and oxoacidity reactions in molten salts require preliminary choice of a "standard" oxide-ion donor (Lux base), which can be considered to be completely dissociated under the experimental conditions. The correctness of this last assumption affects essentially the results of the investigations performed with the use of the said Lux bases. Detailed consideration of the levelling of acidic (basic) properties in ionic melts based on alkali metal halides leads to the conclusion that alkali metal oxides are the strongest bases, and the completeness of their dissociation in the melts based on the corresponding salts is undoubted. Nevertheless, the practical use of these oxides is considerably complicated by their chemical and thermal instability. In particular, K_2O, Rb_2O and Cs_2O are subject to disproportionation at temperatures above 400 °C, which result in the formation of a free alkali metal and the corresponding peroxide. Therefore, alkali metal peroxides (O_2^{2-}) and hydroxides (OH^-), which are more stable during storage, are used widely for the investigations; sometimes barium oxide (BaO) is used as a Lux base. In some special cases, carbonate compounds (Na_2CO_3) are employed for potentiometric titration and calibration. The dissociation processes with the participation of two latter bases run according to the following equations:

$$O_2^{2-} \rightleftharpoons \frac{1}{2}O_2 \uparrow + O^{2-}, \qquad (2.5.42)$$

$$BaO \rightleftharpoons Ba^{2+} + O^{2-}. \qquad (2.5.43)$$

Regarding the above dissociation reactions, it should be noted that the process of BaO dissociation results in the accumulation of foreign Ba^{2+} ions in the melts; this, in turn, changes the chemical properties of the melt-solvent. Also, BaO possesses a limited solubility in molten alkali metal halides and its dissociation is essentially incomplete even in unsaturated solutions. Even if we neglect the incomplete dissociation, there are other limitations on the practical use of this base for studying oxoacidic properties of ionic melts. It cannot be used in the following cases:

- for studies of the solubilities of alkaline earth metal oxides, which form solid solutions with BaO;
- for the investigation of the Lux acid–base equilibria with the participation of oxoanions forming compounds with Ba^{2+} ions that are

insoluble in the melts: $(Ba_3(PO_4)_2$, $BaCrO_4$, etc. can be mentioned as examples;

– for studies in bromide and iodide ionic melts where the incompleteness of the acid–base dissociation strengthens considerably.

The behaviour of carbonates, hydroxides and peroxides in ionic melts should be considered separately.

The matter consists in the fact that although alkali metal hydroxides are classified among strong bases, they are nevertheless distilled without decomposition at temperatures near 1300 °C. Sodium and potassium carbonates melt without decomposition at 854 and 890 °C, respectively, and the partial pressure of CO_2 over the said melts is low enough even at 1000 °C. Thermal dissociation of Na_2O_2 starts at temperatures close to 460 °C, whereas K_2O_2 melts without appreciable decomposition at 490 °C [114, 121]. Hence, these bases should be sufficiently stable under the experimental conditions provided in most experimental oxoacidity studies (200–800 °C).

On the other hand, dissolution of the said bases in ionic melts in an inert atmosphere should cause their complete dissociation according to Le Chatelier's Law, since the partial pressures of conjugated acidic gases (e.g. H_2O, CO_2, O_2) are clear to be zero and the constituent cations of the melt fix the formed oxide ions. The degree of a Lux base's decomposition is defined by the equilibrium concentration and acidity of the constituent cations of the melt, the temperature, the solubility of the conjugate acid–gas in the melt, and its concentration as a component in the inert gas over the melt.

Owing to the action of a number of factors and difficulties in estimating their effect on the completeness of dissociation of the said bases, the corresponding data obtained from thermodynamic calculations are not in agreement with their actual behaviour. Therefore, the desirability of experimental investigations of the Lux-base dissociation can scarcely be exaggerated.

2.5.2.1 Sodium peroxide, Na_2O_2

As mentioned above, this substance is more stable in long-term storage than are the oxides of alkali metals. Reference data show that heating pure Na_2O_2 to temperatures above 460 °C results in its decomposition to sodium oxide [114, 121]. Dissolution of Na_2O_2 in molten salts leads to decomposition at appreciably lower temperatures. Nevertheless, in molten nitrates possessing strong oxidation properties it is stable enough. Flinn and Stern performed barometric and potentiometric investigations of the reaction of oxide ions with nitrate melt [295]. A weight of an oxide ion dissolved in a basic medium in the presence of gaseous oxygen in the atmosphere over the melt is completely transformed into O_2^{2-}. In contrast, an acidic medium favours complete breakdown of peroxide ions and fixation of the formed oxide ions by the acid. X-Ray examinations of peroxide solutions in molten nitrates show that the saturation of KNO_3-NaNO_3 with peroxide ions results in precipitation of Na_2O_2 but not of K_2O_2 [296], which agrees with the above-described phenomenon of the levelling of oxobasic properties in molten ionic media. Potentiometric investigations of the dissociation of O_2^{2-} ion in nitrate melts demonstrate that at $200-350\,°C$ it is the main form of the existence of oxygen ions in basic media [297–300], since oxide ions are oxidized by the anion of the melt (NO_3^-) according to the following reaction:

$$NO_3^- + O^{2-} \rightarrow O_2^{2-} + NO_2^-.$$
(2.5.44)

Moreover, the formation of other species of oxygen ions, such as superoxide, etc. is possible under these conditions. Of course, the use of the gas oxygen electrode for potentiometric studies requires the presence of gaseous oxygen in the atmosphere over the melt, which favours the transformation of oxide ions into different peroxide compounds. Nevertheless, in an inert atmosphere thoroughly purified from oxygen the difference between the said oxide-ion donors may be very considerable—this specially concerns relatively low temperatures, as the decomposition of pure alkali metal peroxides starts in the temperature range $\sim 400-500\,°C$. Despite the above-mentioned, the $[O^{2-}]:[O_2^{2-}]$ concentration ratio under a constant pressure of oxygen should remain constant. This assumption is confirmed by investigations of metal-oxide electrode reversibility in nitrate melts. These show that all the said electrodes are reversible to those oxide ions which remain under equilibrium conditions with the different peroxide species in the melt.

The decrease in the basicity of ionic melts owing to both the strengthening of the acidic properties of constituent cations of the melt [300] and the decreasing equilibrium concentration of oxygen ions [101, 220] reduces peroxide-ion stability; the O^{2-} ion becomes the main form of existence of oxygen ions under these conditions. The addition of sodium peroxide to molten $LiNO_3$ has been found [301] to result in a yield of 0.5 mol of gaseous oxygen per 1 mol of the added Na_2O_2. Subsequent electro-chemical investigations demonstrated the complete absence of O_2^{2-} ions in the melt.

The stability of the different oxygen species in molten carbonates was studied by Cassir *et al.* [302]. Theoretical predictions made by the authors and the experimental results obtained are in good agreement and show that in $Na_2CO_3-K_2CO_3$, $Li_2CO_3-Na_2CO_3$, $Li_2CO_3-K_2CO_3$ and $Li_2CO_3-Na_2CO_3-K_2CO_3$ melts, the peroxide species can only be stabilized in basic media. Superoxide species, unstable in lithium-containing carbonate, can be stabilized in the $Na_2CO_3-K_2CO_3$ melt even under slightly basic conditions. This means that, even in the case of such basic molten media as alkali metal carbonates, the constituent cation acidity defines the stability of peroxide ions; its decrease is shown in Ref. [302] to favour the formation of different peroxide species.

In order to study the features of Na_2O_2 in the molten KCl–LiCl eutectic at 400 and 500 °C, a potentiometric investigation with the use of Ni|NiO metal-oxide electrode was performed [232]. The E–pO calibration plot obtained with the addition of O_2^{2-} to the melt was found to be coincident with the corresponding plot for NaOH base in a dry inert atmosphere. However, owing to the existence of a certain concentration of removed carbonate impurity in Na_2O_2, the plot was located at higher e.m.f. values than the calibration plot obtained with the use of NaOH or Li_2O. This fact may be explained by the low degree of CO_3^{2-} dissociation, but not by a weak basicity of O_2^{2-} ions.

Deanhardt and Stern investigated the stability of O_2^{2-} ion in higher-temperature chloride and sulphate melts, such as NaCl (at 830 °C) and Na_2SO_4 (at 920 °C) [244, 245]. They showed that peroxide ions could be formed in appreciable concentrations only in the atmosphere of free oxygen over the melt, whereas an inert atmosphere resulted in complete destruction of O_2^{2-} to oxide ions. The values of the equilibrium constants of (0.42) in molten NaCl and Na_2SO_4 at 1100 K were estimated in Refs. [244, 245] to be within 0.05–0.07 for the NaCl melt and 0.03–0.05 for the Na_2SO_4 one.

Similar conclusions can be drawn after a detailed analysis of the data in papers [303–307], whose authors investigated the interactions of gaseous oxygen with molten alkali metal chlorides by the potentiometric method using a $Pt(O_2)$ gas-oxygen electrode. The potentiometric cell consisted of the pair "indicator oxygen electrode–chlorine reference electrode". The studies show that the following reactions of chloride-melt oxidation with oxygen:

$$O_2(gas) + 4MCl(melt) \rightleftharpoons 2M_2O(melt) + 2Cl_2(gas), \qquad (2.5.45)$$

$$O_2(gas) + 2MCl(melt) \rightleftharpoons M_2O_2(melt) + Cl_2(gas), \qquad (2.5.46)$$

$$O_2(gas) + MCl(melt) \rightleftharpoons MO_2(melt) + \frac{1}{2}Cl_2(gas) \qquad (2.5.47)$$

are shifted towards the initial reagents to such an extent that the interaction products are formed in chloride melts in disappearing concentrations, which cannot be determined by the traditional methods of chemical analysis [306]. It should be noted that Na_2O is the main product of the oxidation of NaCl by gaseous oxygen, whereas the concentrations of different peroxide species are lower (by a factor of approximately 10). The said peroxide species (M_2O_2 and MO_2) are found to predominate over M_2O compounds in molten potassium and caesium chlorides. This may be explained by the fact that K_2O, Rb_2O and Cs_2O are prone to disproportionation at temperatures exceeding 400 °C. The stated trend strengthens as the cation radius of alkali metals increases when passing from the potassium compound to the caesium one. Elevation of the melt temperature results in strengthening of the affinity of the chloride melts for oxygen, i.e. the oxidation processes become more favourable. The increasing ability of alkali metal halides to form peroxide species in the process of their oxidation can be explained by the decreasing oxoacidic properties of the cations in the sequence Na−K−Rb−Cs.

Hence, it can be concluded that peroxide ions are thermodynamically unstable in high-temperature melts of alkali metal halides. However, in the presence of O_2 in the atmosphere over the halide melts they may exist in sufficient concentrations, even at temperatures considerably exceeding that of decomposition of pure alkali metal peroxides.

The above-mentioned considerations show that sodium peroxide cannot be considered a strong base at temperatures below 400 °C, even in an oxygen-free atmosphere because of incomplete dissociation according to equation (2.5.42). The practical impossibility of purifying this Lux base from carbonate ions imposes additional limitations on its use for oxoacidity studies in non-oxidizing melts and at relatively low experimental temperatures.

2.5.2.2 Alkali metal carbonates, Me₂CO₃

In many papers the acid–base dissociation of carbonate ion according to equation (2.3) is considered to be incomplete, excluding ionic melts based on lithium halides. Besides the melt acidity, the temperature of ionic melts affects considerably the process of CO_3^{2-} dissociation.

Frederics *et al.* studied the behaviour of carbonate ion in molten KNO_3– $NaNO_3$ equimolar mixtures at 250 and 320 °C [215, 217]. They concluded that the dissociation of CO_3^{2-} in the said melt was appreciably incomplete, and that the oxide ions formed were stabilized by fixation in pyronitrate complexes by the solvent anion. In contrast, Desimoni, Sabbatini and Zambonin, who performed potentiometric studies in molten KNO_3–$NaNO_3$ gave evidence of incomplete dissociation of the carbonate ion according to equation (2.3) in the specified ionic melt [219]. An extremely simple and reliable experimental routine was used by Zambonin to study carbonate-ion dissociation in a nitrate melt [308]. The author analyzed cooled samples of carbonate ion solutions in the nitrate melt with the use of the standard technique of titration with carbonate–hydroxide and carbonate–hydrogencarbonate [309]. As is known, the acid–base titration of aqueous carbonate solutions with strong acids is a two-stage process. The first stage is the transformation of carbonate into bicarbonate:

$$CO_3^{2-} + H^+ \rightleftharpoons HCO_3^-, \qquad (2.5.48)$$

and the second stage ends by the formation of carbonic acid, or hydrate of carbon dioxide:

$$HCO_3^- + H^+ \rightleftharpoons H_2CO_3(CO_2 \uparrow + H_2O), \qquad (2.5.49)$$

If we have a solution of pure carbonate then the consumption of strong acid at each titration stage is the same, whereas a carbonate solution polluted with alkali metal hydroxide is characterized by a higher consumption of acid for the first stage of the neutralization than for the second one; the bicarbonate admixtures result in the inverse consumption of the acid for the stages mentioned.

In the case considered, only the appearance of surplus hydroxide can be expected in the case of appreciable dissociation of CO_3^{2-} ions in the nitrate melts with the yield of the formed CO_2 from the solution in the melt. However, the experiments found the consumption of the acid–titrant for each step of the titration to be the same and, hence, the dissolved carbonate ions are completely stable in nitrate melts formed by cations of weak acidity.

Bocage *et al.* used sodium carbonate as a strong Lux base for determination of the solubility products of CaO and MgO in the molten $BaCl_2-CaCl_2-NaCl$ eutectic mixture at 600 °C [185]. The determination was performed by the potentiometric titration technique. The dissociation process of Na_2CO_3 in the melt was assumed to be complete, without an experimental check of this assumption. It is quite natural that the sole fact of the presence in the chloride melt of Ca^{2+} cations possessing strongly acidic properties cannot be considered as exhaustive evidence of complete breakdown of CO_3^{2-} ion. Moreover, solid $CaCO_3$ decomposes to an appreciable degree only at temperatures near 825 °C.

The dependence of the dissociation constant (pK) of carbonate ion in the molten KCl–NaCl equimolar mixture was found by Combes *et al.* [310] to be expressed by the following equation,

$$pK_{(2.3)} = 1.32 \times 10^{-4}T^{-1} - 8.4, \tag{2.5.50}$$

which made it possible to perform the estimation at a temperature of 1000 K: the pK value was equal to 4.8. In the opinion of the authors of Ref. [310], such a pK value had to prove complete dissociation of carbonate ion at melt temperatures exceeding 1000 K, since low partial pressures of CO_2 in the atmosphere over the melt provide complete transformation of CO_3^{2-} to O^{2-}. Nevertheless, potentiometric investigations of carbonate ion dissociation in the molten KCl–NaCl equimolar mixture at 700 °C show that the equilibrium position of the process (1.2.3) is not shifted completely to the right, even in the absence of CO_2 pressure over the melt [141]. The "formal" stability constant of CO_3^{2-}, implied by the authors to correspond to the concentration ratio CO_3^{2-}/O^{2-} was equal to 2.5 ± 0.5 (concentrations expressed as molar percent). In this study, the authors assumed that the constant partial pressure of CO_2 over the melt, even in a CO_2-free atmosphere, and the change of the initial CO_3^{2-} concentration, did not affect this pressure. Investigations performed in a similar way on the molten KCl–LiCl eutectic (at 500 °C) and equimolar KCl–NaCl mixture (at 700 °C) gave evidence that even in very dilute solutions, with concentrations of about 10^{-3} mol kg^{-1}, the degree of dissociation of CO_3^{2-} under the said conditions is not higher than 70% [232]. Therefore, the assumption about complete dissociation of the carbonate ions dissolved in alkali-halide melts heated up to 1000 K in a CO_2-free atmosphere [310] cannot be confirmed by experimental data.

A similar conclusion may be made based on the results of our investigations performed by calibration of the potentiometric cell using the indicator oxygen electrode with additions of the base studied (K_2CO_3) and concentrations of the base within the $pO^* = 3-4$ range. The comparison of the calibration-dependences with the similar plot obtained with the use of the standard strong base allowed calculation of the degree of dissociation of CO_3^{2-} ion under the experimental conditions. For this purpose we investigated the carbonate dissociation at temperatures of 690, 710, and 720 °C. The degree of carbonate-ion dissociation was found to increase with rise in the melt temperature; however, even at temperatures near 720 °C (993 K) it did not achieve 100%, although the concentration of the solution is very low. The same may be said about the data presented in Ref. [141]. Since the slope of the calibration $E-pO^*$ dependence for carbonate ions is appreciably less than that for the calibration plots with additions of strong bases (0.155 and 0.193 V, respectively), it is obvious that the degree of CO_3^{2-} dissociation decreases together with an increase in the initial base concentration. Assuming that the assumption of the authors of Ref. [141] about the constant partial pressure of CO_2 over the melt is true, it may be deduced that the formal stability constant

$$K' \rightleftharpoons \frac{[CO_3^{2-}]}{[O^{2-}]} \rightleftharpoons \frac{(1-\alpha)[CO_3^{2-}]_0}{\alpha[CO_3^{2-}]_0} \rightleftharpoons \frac{(1-\alpha)}{\alpha} = \text{const} \quad (2.5.51)$$

should actually be unchanged. The experiments performed give evidence that the opposite is the case. Hence, there is no constant effective concentration of CO_2 in the melt where the donor of CO_3^{2-} is dissolved under a CO_2-free atmosphere.

As seen, the studies of CO_3^{2-} dissociation under non-equilibrium conditions are connected with the solubility of the CO_2 formed in ionic melts. The greater the solubility value, the higher the degree of holding CO_2 by the ionic solvent that retarded the decomposition of carbonate solutions. To obtain some information on the solubility of carbon dioxide in molten alkali metal halides over a wide temperature range we shall consider the following works.

Novozhilov investigated the solubility of CO_2 in melts of individual alkali metal chlorides at various temperatures [311]. The dependences of the Henry coefficients against the inverse temperature were linear for all the chlorides

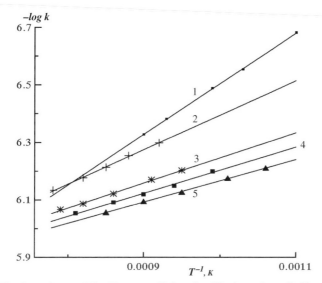

Fig. 2.5.6. The dependence of the Henry coefficient for CO_2 in molten alkali metal chlorides against the inverse temperature: 1—LiCl, 2—NaCl, 3—KCl, 4—RbCl, 5—CsCl.

studied (Fig. 2.5.6), and the CO_2 solubility increased somewhat on proceeding from LiCl to CsCl. Comparison of these results with the data on argon solubility in the specified melts shows that they are close, although by assuming the physical mechanism of the gas dissolution one could expect an appreciably lower solubility of CO_2 in the melts. This is explained by the fact that, although both gas molecules are non-polar (dipole moments equal to zero), CO_2 is prone to formation of complexes of composition such as:

$$
\begin{array}{c}
Cl^- \\
\vdots \\
Me^+ \cdots O = C = O \cdots Me^+ \\
\vdots \\
Cl^-
\end{array}
$$

with constituent parts of the ionic melts, which favour stabilization of the dissolved CO_2 molecules.

The coefficients of the obtained thermal plots and the thermodynamic characteristics of the process of CO_2 dissolution calculated according to equation (2.5.24) in molten alkali metal chlorides are presented in Table 2.5.4.

As seen from this table, ΔH and ΔS values decrease with the increase of the atomic number of the alkali metal, which is caused by the reduction of the energy necessary to create holes in the corresponding melt. In the case of CsCl

TABLE 2.5.4

Thermodynamic parameters and coefficients describing the dissolution of CO_2 in molten alkali metal chlorides

Melt	LiCl	NaCl	KCl	RbCl	CsCl
ΔH (kJ mol^{-1})	41.0	32.7	26.8	24.7	22.8
ΔS (J mol^{-1} K^{-1})	10.6	4.3	1.2	0.0	-1.17
$-\log k_0$	4.74 ± 0.02	5.20 ± 0.02	5.37 ± 0.02	5.40 ± 0.03	5.43 ± 0.01
a	-1762 ± 13	1198 ± 25	874 ± 24	803 ± 33	735 ± 13

some ordering of the melt-structure is observed that is detected by the small negative entropy value.

Bratland *et al.* reported the investigation of CO_2 solubility in molten alkali metal halides (NaCl, KCl, KBr, KI) by volumetric, thermogravimetic and freezing methods [270]. The determined dependences of the Henry coefficient against the inverse temperature are described by the following equations:

$$-\log k = 5.12 + 1280T^{-1}, \text{ for NaCl,} \tag{2.5.52}$$

$$-\log k = 5.28 + 980T^{-1}, \text{ for KCl,} \tag{2.5.53}$$

and

$$-\log k = 5.24 + 880T^{-1}, \text{ for KBr.} \tag{2.5.54}$$

The simplified hole model was shown to describe the data on CO_2 solubility in the alkali metal halide melts with good accuracy. The entropy changes in the process of dissolution are close to -1 J mol^{-1} K^{-1}, which agrees with the data of Novozhilov [311], and the solubility data obtained for the molten chlorides are in good agreement with Ref. [311]. An interesting fact was revealed— the solubility of CO_2 increased by four times upon the addition of a small concentration of Ni^{2+} ($\sim 10^{-3}$ mol kg^{-1}), introduced into molten NaCl as an admixture. A study of the kinetics of the dissolution process showed that the rate of CO_2 dissolution in alkali metal halide melts was defined by the rate of transfer of CO_2 from the gaseous phase into the liquid, but not by the diffusion and convection of the dissolved molecules in the melt.

Barbin *et al.* performed investigations of carbonate-ion stability in ionic melts of a KCl–NaCl system containing 30, 50, 70, and 90 mol% of NaCl in the 752–852 °C temperature range [312]. In order to study this equilibrium, the salt mixture with the addition of sodium carbonate in a concentration to provide a saturated solution of Na_2CO_3 in KCl–NaCl melt, was melted and heated up to the working temperature in a glass–carbon crucible, and kept in a CO_2 atmosphere for several hours. Under these conditions, the gas mixture over the melt was enriched with carbon monoxide (approx. 10 vol% of CO), formed as a result of the Boudoir reaction between the crucible walls and CO_2:

$$CO_2 \uparrow + C \downarrow \rightarrow 2CO \uparrow . \tag{2.5.55}$$

Since there was no direct reaction of CO with the chloride melt, its concentration in the gas mixture was neglected and corrections for the partial pressure of CO_2 were not made. The melt was analyzed periodically; for this purpose, several samples were taken off the melt, dissolved in water, and subjected to acid–base titration in aqueous solution, to determine the ratio of concentrations of Na_2O to Na_2CO_3 as described above. These studies provided the temperature dependences of the equilibrium constant of (1.2.3) in four melts of the KCl–NaCl system in the 1 temperature range 1025–1125 K, as follows:

$$\log K = 9.495 - 14,076T^{-1}, \quad 1053–1125 \text{ K},$$
$$30 \text{ mol.\% of NaCl}, \tag{2.5.56}$$

$$\log K = 10.059 - 14,553T^{-1}, \quad 1025–1125 \text{ K},$$
$$50 \text{ mol.\% of NaCl}, \tag{2.5.57}$$

$$\log K = 9.279 - 13,599T^{-1}, \quad 1025–1125 \text{ K},$$
$$70 \text{ mol.\% of NaCl}, \tag{2.5.58}$$

$$\log K = 8.302 - 12,406T^{-1}, \quad 1072–1125 \text{ K},$$
$$90 \text{ mol.\% of NaCl}. \tag{2.5.59}$$

Elevation of the temperature leads to an increase in the equilibrium constant of equation (1.2.3). Similar changes are observed when the molar fraction of Na^+ cations in the melts increases and the dependence of $\ln K$ against the inverse effective cation radius of the melts calculated according to the following equation:

$$r_{eff} = N_{Na^+} r_{Na^+} + N_{K^+} r_{K^+}, \tag{2.5.60}$$

is found to be practically linear. The comparison of the values obtained for the equilibrium constants, with the thermochemical calculations for the equilibrium between solid Na_2O and Na_2CO_3

$$\log K = 6.377 - 15{,}684T^{-1}, \quad 1000 - 1131 \text{ K, pure NaCl,} \quad (2.5.61)$$

showed that there was a considerable increase of K values as compared with the thermochemical data, and that the deviations became more substantial as the Na^+ concentration increased. For example, the estimation performed for the solid phase reaction at 1000 K gave the pK value to be 9.31, whereas the value determined practically for the molten KCl–NaCl equimolar mixture was pK = 4.49. Of course, the corrections taking into account the latent heat of melting of Na_2O and Na_2CO_3 give the pK value to be 4.93, which is in good agreement with the experimental data, and shows that the properties of the said solutions of Na_2CO_3 in the chloride melts are close to ideal. The practical significance of the work of Barbin *et al.* [312] lies in the possibility of calculating the dissociation constants of carbonate ion in mixtures of intermediate compositions based on the effective radius of the melt cation.

Polythermal investigations of equilibrium (1.2.3) in the molten CsCl–KCl–NaCl eutectics (600, 700, and 800 °C) and KCl–NaCl eutectics (700 and 800 °C), and in an individual KCl melt at 800 °C were reported in Ref. [313]. These studies were performed by consecutive calibration of the electrode system with addition of a strong Lux base in an argon atmosphere and in pure CO_2, as shown in Fig. 2.5.7. The data show that passing CO_2 through the melts causes a considerable increase in e.m.f. This means a considerable reduction of the equilibrium O^{2-} concentration owing to its fixation by CO_2 followed by the formation of carbonate. From the known slope of the calibration E–pO^* dependence one can estimate the pK value using the formula presented in Fig. 2.5.7. The equilibrium molality of

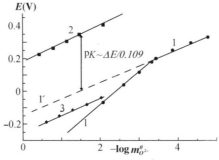

Fig. 2.5.7. E–pO^* dependence of cell with the $Pt(O_2)|YSZ$ membrane oxygen electrode in molten KCl at 800 °C: 1, KOH, Ar; 2, KOH, CO_2; 3, K_2CO_3, Ar.

TABLE 2.5.5
Experimental and calculated data for equilibrium (1.2.3) in molten KCl at 800 °C

pO^*	E (B)	pO	pK
2.09	0.405	5.25	3.16
1.47	0.345	4.70	3.23
1.14	0.306	4.34	3.20
0.79	0.263	3.94	3.15
0.44	0.227	3.61	3.17

The average value of pK = 3.18 ± 0.03.

oxide ions in the melts saturated with CO_2 can be neglected in comparison with the theoretically possible initial molality in the denominator of the following formula:

$$pK = -\log \frac{m_{O^{2-}} p_{CO_2}}{(m_{O^{2-}}^0 - m_{O^{2-}})}. \tag{2.5.62}$$

More thorough calculations show that the above-described express routine is accurate enough (see Table 2.5.5).

The dependence 3 in Fig. 2.5.7 was obtained by calibration of the potentiometric cell with K_2CO_3 additions, in an inert atmosphere (argon). It is easily seen that even in the absence of CO_2, carbonate-ion dissociation in the melts is not complete. It should also be added that as the increase of the initial concentration of the Lux base rises, its degree of dissociation falls. The pK values in Ref. [313] were calculated as being equal to 3.2 ± 0.1 (KCl at 800 °C); in the molten KCl–NaCl they are 5.1 ± 0.1 at 700 °C, and 3.6 ± 0.1 at 800 °C; in the molten CsCl–KCl–NaCl eutectic the corresponding values are 4.12 ± 0.1 at 600 °C, 3.31 ± 0.1 at 700 °C, and 2.87 ± 0.1 at 800 °C. The dependence of the pK of carbonate-dissociation in the molten CsCl–KCl–NaCl eutectic against the inverse temperature is linear and can be approximated by the following equation:

$$pK = -2.68(\pm 0.7) + 5900(\pm 600)T^{-1}. \tag{2.5.63}$$

Although the oxoacidic properties of the CsCl–KCl–NaCl and KCl–NaCl melts are practically the same, the pK values of equilibrium (1.2.3) in the former melt are considerably higher than in the latter, at the same temperature. The reason for this is as follows: the dissociation constant of carbonate ion in any melt is mainly defined by the stability of the most unstable individual carbonate formed by the constituent cations of the melt; the greater

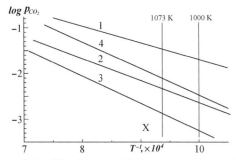

Fig. 2.5.8. Dependence of the CO_2 decomposition pressure of over individual alkali metal carbonates against the inverse temperature (from the data of Ref. [314]): 1, Li_2CO_3; 2, Na_2CO_3; 3, K_2CO_3; 4, Cs_2CO_3.

the decomposition pressure over the said carbonate, the higher the value of the carbonate-ion dissociation constant in the melt. We shall now explain this statement in detail.

The temperature dependences of the decomposition pressure of CO_2 over the individual alkali metal carbonates are presented in Fig. 2.5.8. As seen from these plots, the decomposition pressure of CO_2 over Cs_2CO_3 is approximately 10 times as high as the corresponding value for K_2CO_3, although the acidic properties of Cs^+ and K^+ cations were shown in Part 3 to be practically the same. The above considerations lead to the conclusion that reaction (1.2.3) cannot be used as a standard acid–base equilibrium to estimate the relative oxoacidic properties of melts of different cation composition.

The data presented in Fig. 2.5.8 allow us to establish that the thermal dependences for lithium and sodium carbonates (denoted by 1 and 2, respectively) are characterized by close slopes. The slopes of the dependences of decomposition pressure over K_2CO_3 (3) and Cs_2CO_3 (4) are close, too, but are considerably different from those for the pair Na–Li. A possible reason for such a distinction consists in the different abilities of the alkali metal oxides to disproportionate with the formation of peroxide ions according to reaction (4.41), which favours the shift of the equilibrium (1.2.3) to the right. Sodium and lithium oxides are stable enough and are not disposed to the formation of peroxides in oxygen-free atmosphere, whereas oxides of potassium, rubidium, and caesium are disproportionate to an appreciable degree at temperatures exceeding 400–500 °C.

The investigation of the behaviour of carbonate ions in the molten KCl–LiCl eutectic at 600, 700, and 800 °C was reported in Refs. [315, 316].

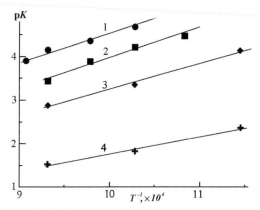

Fig. 2.5.9. Dependence of pK of reaction (1.2.3) against the inverse temperature in molten
NaI (1), CsI (2), CsCl–KCl–NaCl (3) and KCl–LiCl (4).

The experimental routine used was similar to that described in Ref. [313].
Carbonate ion, added to the melt in the form of K_2CO_3, was shown by the
experiments to be completely destroyed with the evolution of CO_2 in an inert
atmosphere. The values of pK of the equilibrium (1.2.3) were equal to
2.39 ± 0.08 at 600 °C, 1.81 ± 0.14 at 700 °C, and 1.53 ± 0.1 at 800 °C (see
Fig. 2.5.9). As in the case of the molten CsCl–KCl–NaCl eutectic, the thermal
dependence was practically linear and can be described by the following
equation:

$$pK = -2.30(\pm 0.6) + 4070(\pm 500)T^{-1}. \qquad (2.5.64)$$

Polythermal investigations of carbonate-ion stability in the technologically
important melts CsI and NaI in the temperature range from the melting point
to ~ 800 °C were performed in Refs. [249, 317]. The CsI melt was studied at
temperatures of 650, 700, 750 and 800 °C and the corresponding pK values
were equal to 4.47 ± 0.1, 4.23 ± 0.1, 3.89 ± 0.1 and 3.44 ± 0.1, respectively. Sodium iodide, which has a significantly higher melting point, was
investigated at temperatures of 700, 750, 800 and 830 °C that allowed the pK
values to be obtained as 4.68 ± 0.1, 4.39 ± 0.1, 4.17 ± 0.1 and 3.92 ± 0.1,
respectively. The dependences of pK against the inverse temperature for all
the studied melts are presented in Fig. 2.5.9. The dependences for the studied
iodide melts calculated in Refs. [17, 249] are:

$$K = -2.80(\pm 0.9) + 6770(\pm 850)T^{-1} \text{ for CsI} \qquad (2.5.65)$$

and

$$pK = -1.49(\pm 0.6) + 6010(\pm 600)T^{-1} \text{ for NaI.} \qquad (2.5.66)$$

The thermal dependences for all the studied melts, excluding KCl–LiCl, are characterized by practically coincident slopes of the order of 6×10^3, whereas for the molten KCl–LiCl eutectic the slope is lower by a factor of 1.5.

Apart from the above-mentioned investigations performed over sufficiently wide temperature ranges, there are some studies of carbonate-ion dissociation performed at a single temperature in alkali metal halide melts of the same cation composition and of individual alkali halides.

The investigation of equilibrium (1.2.3) in molten potassium halides at 800 °C was reported in Ref. [183]; the experimental routine being similar to that of Ref. [313]. The change of the pK value of equilibrium (1.2.3) in the sequence KCl–KBr–KI was found in Ref. [183] to be non-monotonous: 3.2 ± 0.2 (KCl) → 2.4 ± 0.2 (KBr) → 4.4 ± 0.2 (KI). The similar dependence in the sequence CsCl–CsBr–CsI was stated in a later study [197] to pass through a minimum in the bromide melt: 2.22 ± 0.06 (CsCl) → 1.20 ± 0.1 (CsBr) → 3.44 ± 0.1 (CsI). Despite the similar form of the dependences for potassium and caesium halides, the pK values obtained in the latter halides are lower approximately by one unit as compared with the corresponding potassium halides. Such a resemblance of these dependences allows us to assume that the dissociation-constant changes in the sequence are caused only by the difference in oxoacidic properties of the melts; the bromide melts possess the highest acidic properties and the iodide ones are the weakest Lux acids. To complete the set of such studies we investigated the behaviour of carbonate ion in molten sodium halides at 830 °C. In this case, the values of pK of the equilibrium (1.2.3) increase in the sequence NaCl (2.63 ± 0.1)– NaBr (2.88 ± 0.06)–NaI (3.92 ± 0.2) [198]. The experimental results for all the studied individual alkali metal halides are presented in Fig. 2.5.10.

In contrast to the potassium and caesium halide melts, a monotonous change of the dissociation constant of carbonate ion in the sequence of sodium halides was observed. This distinction has been explained on p. 147 by the different stabilities of the inner complexes formed by melt ions in individual molten alkali metal halides, and by the different character of their changes with the change in the melt anion. The stability of the complex changes greatly in the potassium and caesium "chloride–bromide–iodide" sequences (the minimum is observed in the bromide melts), whereas in the sodium halides the chloride complexes possess the lowest stability and the iodide

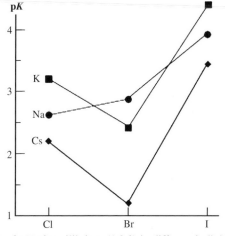

Fig. 2.5.10. The values of pK of equilibrium (1.2.3) in different individual alkali metal halide melts at 800 °C (for the sodium salts the data were obtained at 830 °C).

complexes are the most stable. The decrease in the halide-complex stability results in the accumulation of "free" alkali metal cations in molten media, which increases the melt acidity, and thus leads to a reduction of the carbonate-ion stability in the melt.

Assuming a pK value of 2 to be the threshold for the carbonate-ion stability in molten salts in a CO_2-free atmosphere, it may be concluded that in halide melts consisting, even partially, of lithium salts and more acidic multivalent cations (Ca^{2+}, Mg^{2+}, Ln^{3+}, etc.) carbonate ion is unstable and undergoes complete decomposition with the formation of the equivalent quantity of oxide ions in the melt. As for the other alkali metal halides studied, similar behaviour of CO_3^{2-} ions can be expected only for CsCl, CsBr, and KBr melts at temperatures exceeding 800 °C. Of course, instability of carbonate ion in the melts does not mean an automatic disappearance of the oxygen admixture from the melt, since oxide ions arise instead of CO_3^{2-} owing to their complete breakdown. This means that only for the melts characterized by a low pK value of the equilibrium (1.2.3), the carbohalogenation method of purification is the most suitable.

2.5.2.3 Alkali metal hydroxides, MeOH

There are a few works devoted to the quantitative investigation of hydroxide ion dissociation in high-temperature ionic melts. Zambonin reported a potentiometric investigation of NaOH dissociation in the molten equimolar KNO_3-NaNO_3 mixture at 500 K [46, 225]. The equilibrium constant of reaction (1.2.4) was found there to be close to 10^{-18}, which is evidence of sufficient stability of OH^- ions at the temperature of the experiment. While considering the equilibrium constant value, it should be concluded that hydroxide ions do not decompose even in an atmosphere free of water vapour, although there are no data on the stability of OH^- ion in a dry atmosphere in the cited papers.

Wrench showed that hydroxide ions did not undergo decomposition according to equation (1.2.4) in the molten KCl–LiCl eutectic mixture at 400 °C [318]. This is probably connected with strong fixation of water by this melt, and may be explained by the extremely high hygroscopicity of LiCl [179]. The elevation of the melt temperature to 500 °C results in complete decomposition of hydroxide ions and removal of water from the melt, as is confirmed by both polarographic [319, 320] and potentiometric [232] measurements.

Combes *et al.* reported investigations of the equilibrium constant of reaction (1.2.4) in the molten KCl–NaCl equimolar mixture [119, 180, 240]. The experimental routine consists in addition of a weight of O^{2-} donor to the melt; afterwards the equilibrium e.m.f. of the cell with the indicator oxygen electrode ($Ni|NiO|ZrO_2$) is measured and the e.m.f. values are determined at partial pressures of water vapour from 0.0013 to 0.011 atm. The calculations of the equilibrium constant of reaction (1.2.4), performed using the obtained e.m.f. values, show that in the temperature range 737–817 °C the dependence of the pK of the equilibrium (1.2.4) upon the inverse temperature is expressed by the following equation:

$$K_N = 8.15 \times 10^3 T^{-1} - 6.75. \tag{2.5.67}$$

Since the stoichiometry of the dissociation process was beyond doubt, there were no examinations over a wide range of oxide-ion concentrations.

Studies of the process of hydroxide ion dissociation performed over a wide range of initial concentrations of oxide ions and partial pressures of H_2O demonstrate that in molten alkali metal halides this process does not comply with the generally accepted equation (1.2.4) [232, 321]. The dissociation process of OH^-, both in the KCl–LiCl eutectic and in the molten KCl–NaCl equimolar mixture, was found to proceed according to the scheme:

$$(OH)_2^{2-} \rightarrow O^{2-} + H_2O \uparrow. \tag{2.5.68}$$

The method used to investigate hydroxide-ion dissociation is as follows. A weight of a strong Lux base (Li_2O, Na_2O_2, NaOH) is added to the melt studied and the equilibrium e.m.f. of the cell with an indicator oxygen electrode (Ni|NiO or $Pt(O_2)$|YSZ) is measured. Then the mixture of gases Ar + H_2O, obtained by bubbling the initial inert gas through either the saturated aqueous solution of NaCl or solutions of H_2SO_4 characterized by a constant pressure of water vapour at a given temperature, is passed thorough the studied melts. As a rule, passing of a wet gas causes a considerable increase in the e.m.f. due to a reduction of the equilibrium oxide-ion concentration in the melt. After stabilization of the e.m.f. value with the flow of the wet gas, dried inert gas is bubbled through the melt, which makes it possible to examine both the rate of the e.m.f. decrease and its final (equilibrium) value. Then the next portion of the oxide-ion donor is dissolved in the melt and the above-described sequence of measurements is repeated.

The described experimental routine used in our studies possesses some advantages as compared with the techniques described above, which have been used by other authors [119, 240]. The main advantage is that the investigations are performed on a wide range of oxide-ion donors, which allows more detailed information to be obtained about the dissociation process. The directed shift of the calculated values of the dissociation constant with changes in the initial concentration of strong Lux bases is the reason for reconsidering the reaction's stoichiometry. The potentiometric results obtained allow one to estimate the completeness of the dissociation of hydroxide ions obtained immediately in the melts from the strong bases (alkalimetal oxides or sodium peroxide). Finally, the analysis of the e.m.f. values before and after passing the wet gas through the melt allows one to determine the degree of the chloride melt's pyrohydrolysis in basic solutions.

Before describing the results of our experiment it should be noted that the e.m.f. value measured in the solutions of the initial Lux base at temperatures of 500 °C and higher coincides with the final e.m.f. value after passing the wet gas, both for the KCl–LiCl eutectic and the KCl–NaCl equimolar mixture. This shows that, in the melts studied, the hydroxide ions are completely dissociated with the formation of O^{2-} (according to the potentiometric results). The traces of water remaining in the melts after the bubbling of dry inert gas have no essential effect on the process of dissociation of hydroxide ions.

The run of the pyrohydrolysis of the chloride melts containing admixtures of the Lux bases, at least, is negligible. The first conclusion mentioned above makes it possible to use alkali metal hydroxides as strong bases for the studies of such a kind in high-temperature ionic melts. Nevertheless, with molten KCl–LiCl eutectic at 400 °C, bubbling dry argon does not result in complete removal of water, which is held strongly by the melt. However, heating of this melt with water traces up to 500 °C in the dry atmosphere, with subsequent return to 400 °C, leads to an e.m.f. value equal to that obtained after addition of the initial Lux base. Taking into account the fact that NaOH serves as a source of water in the melts, owing to its dissociation, the investigations of oxoacidity at temperatures near 400 °C should be performed with the use of Na_2O_2.

Since the experimental results hardly correspond to the stoichiometry of reaction (1.2.4) (they are characterized by a considerable shift of the value of the equilibrium constant calculated according to the equation mentioned). Besides the calculations according to the generally accepted equilibrium (1.2.4), the thermodynamic parameters were calculated for equilibrium (2.5.68), in which dimerized hydroxide ions take part. The results of the investigations in the molten KCl–LiCl eutectic at 400 and 500 °C are collectively displayed in Table 2.5.6.

The data presented in this table show that the calculated values of the equilibrium constant (1.2.4) are subject to considerable shifts according to an increase in the initial concentration of the added Lux base. It is interesting that the $E-pO^*$ dependences obtained in the KCl–LiCl eutectic for the cell with the metal-oxide indicator oxygen electrode, both in dry and the wet atmospheres, are linear and their slopes (see Table 2.5.6) are practically coincident. Nevertheless, the authors of Ref. [180] noted that in the wet atmosphere the slope of the $E-pO^*$ plot was twice as large as the, "theoretical" one corresponding to $1.15RT/F$ observed in the dry atmosphere. Such a deviation was explained by incomplete dissociation of OH^- in the molten KCl–NaCl. However, a simple calculation of the equilibrium concentrations performed based on the values of the constants obtained by the authors of Ref. [180] for the stoichiometry of reaction (1.2.4) at low partial pressures of water vapour shows that, as a rule, the degree of transformation of hydroxide ions in molten chlorides exceeds 95%. The slope of the $E-pO^*$ dependence obtained by the authors of the mentioned paper [180] in the wet atmosphere is in good agreement with the data on the behaviour of $Pt(O_2)|YSZ$ membrane oxygen electrode at low pO values, i.e. with excess O^{2-} in the ionic halide melts (see Part 4). In contrast, there are no similar deviations from the theoretical slope in the experiments whose results are presented in Table 2.5.6. Moreover, irrespective of the kind of oxygen

TABLE 2.5.6

Results of potentiometric investigations of hydroxide ion dissociation in the KCl–LiCl eutectic melt

Ar		Ar + H_2O		$pK_{(1.2.4)}$	$pK_{(5.68)}$
E (V)	pO	E (V)	pO		
400 °C, electrode Ni|NiO, $p(H_2O) = 0.0236$ atm.					
−0.153	2.69	−0.144	4.26	1.08	3.18
−0.267	2.27	−0.194	3.45	1.08	2.78
−0.287	1.99	−0.214	3.13	1.31	2.73
−0.295	1.81	−0.220	3.03	1.59	2.82
−0.299	1.59	−0.226	2.94	1.94	2.95
0.062 ± 0.009 V		0.058 ± 0.013 V		1.48 ± 0.4	2.82 ± 0.1
500 °C, electrode Ni|NiO, $p(H_2O) = 0.0236$ atm.					
−0.301	2.53	−0.214	3.49	0.55	2.53
−0.336	2.10	−0.244	3.15	1.11	2.64
−0.375	1.64	−0.288	2.66	1.52	2.60
−0.402	1.42	−0.315	2.36	1.63	2.60
−0.411	1.27	−0.333	2.15	1.72	2.45
0.089 ± 0.003 V		0.095 ± 0.006 V		1.31 ± 0.5	2.56 ± 0.1
500 °C, electrode Ni|NiO, $p(H_2O) = 0.017$ atm.					
0.837	2.26	0.984	3.18	0.94	2.63
0.755	1.69	0.906	2.69	1.49	2.71
0.717	1.49	0.877	2.50	1.81	2.73
0.680	1.27	0.841	2.27	2.05	2.72
0.643	1.01	0.806	2.05	2.30	2.76
0.157 ± 0.004 V		0.143 ± 0.003 V		1.72 ± 0.5	2.71 ± 0.1

electrode used as indicator (the metal-oxide electrode characterized by the slope of $1.15RT/F$, or the gas membrane electrode with the slope close to $2.3RT/F$ in basic media) the slopes of the E–pO* plots in dry and in wet atmospheres are practically coincident.

Similarly, the spread of the calculated values of the equilibrium constant of the reaction described by the stoichiometry of equation (2.5.68) is very small at all initial molalities of the Lux base and all the partial pressures of water over the melt studied. Hence, the equilibrium (2.5.68) is more correct for the description of the dissociation process for hydroxide ions, at least, in the melts based on alkali metal halides than the generally accepted reaction (1.2.4).

Investigations of the kinetics of the decomposition of hydroxide ions in a dry atmosphere may provide another reason in favour of the proposed stoichiometry of decomposition. Such an investigation was performed by measurements of the e.m.f. for the potentiometric cell with the indicator oxygen electrode during removal of the moisture from the ionic melt in the flow of dried inert gas. The calculations of the order of the decomposition reaction of hydroxide ions to O^{2-} in molten chloride mixture were made using the Van't Hoff method [322]:

$$
n = \left| \frac{\Delta\log\left(\frac{\Delta m}{\Delta \tau}\right)}{\Delta\log\bar{m}} \right|,
\tag{2.5.69}
$$

(where τ denotes the time counted from the start of bubbling the dry inert gas). The results shown in Table 2.5.7 give evidence that the order of the hydroxide-ion decomposition reaction with respect to OH^- (or $(OH)_2^{2-}$) is equal to 1. Although this result seemingly gives undoubted evidence of the correctness of equation (2.5.68), it cannot be considered ultima ratio in favour of the existence of $(OH)_2^{2-}$ ions as the predominant form in hydroxide solutions in ionic melts based on alkali metal halides. The reason is that the rate of a set of consecutive chemical processes is limited by the rate of the slowest stage, and so the reaction of hydroxide-ion dissociation in the melts may be divided into two stages. The first stage is the dissociation process itself, resulting in the formation of a solution of oxide ion and water in the melt studied:

$$
(OH)_2^{2-} \rightarrow O_{melt}^{2-} + H_2O_{melt}.
\tag{2.5.70}
$$

TABLE 2.5.7
Results of investigations of decomposition kinetics of hydroxide ion in the molten KCl–LiCl eutectic at 500 °C

τ (min)	E (V)	$\Delta m/\Delta \tau$	$\bar{m}$	n
0	0.877	–	–	0.83
7	0.735	0.00320	0.0183	
10	0.726	0.00110	0.0052	
0	0.887	–	–	1.16
8	0.757	0.00190	0.0098	
13	0.752	0.00022	0.0015	
0	0.973	–	–	0.94
4	0.860	0.00082	0.0031	
6	0.850	0.00032	0.0011	
Average: $n = 0.98 \pm 0.16$				

The second step is the evaporation of water from the melt to the gas phase (in a flow of dry argon):

$$H_2O_{melt} \rightarrow H_2O_{gas}. \tag{2.5.71}$$

It seems more probable that the latter stage, undoubtedly referred to as a reaction of the first order, is the limiting one. To obtain more powerful evidence to decide which of the two equilibria, (2.5.68) or (1.2.4), actually occurs in the ionic melt, one should analyze the dependence of the equilibrium molality of oxide ion upon its initial molality under a constant pressure of water vapour. For this purpose, we shall consider the dependence when the equilibrium in hydroxide solutions is described by equation (2.5.68). The corresponding equilibrium constant is expressed by the formula

$$K_{(5.51)} = \frac{m_{O^{2-}} p(H_2O)}{m_{(OH)_2^{2-}}}, \tag{2.5.72}$$

which can be rewritten as follows by using the value of the initial molality of oxide ion in the melt,

$$K_{(5.51)} = \frac{m_{O^{2-}} p(H_2O)}{m_{O^{2-}}^0 - m_{O^{2-}}}. \tag{2.5.73}$$

The sought dependence of pO under a constant partial pressure of water vapour (pO$_w$) against the corresponding pO* in a dry atmosphere is immediately obtained from this equation as

$$pO_w = pO^* - \log \frac{K_{(5.51)}}{K_{(5.51)} + p(H_2O)}. \tag{2.5.74}$$

Since the denominator in this equation is constant, the ratio of the slopes of plots of E–pO* in dry and wet atmospheres should approach 1, as actually takes place (see Table 2.5.6).

We shall now consider the similar dependence for the process (1.2.4), whose equilibrium constant can be expressed by the following equation:

$$K_{(1.2.4)} = \frac{m_{O^{2-}} p(H_2O)}{m_{OH^-}^2}. \tag{2.5.75}$$

After making some transformations in this equation one can obtain the dependence of $m_{O^{2-}}$ as a function of $m_{O^{2-}}^0$ as the positive root of the following square equation:

$$4K_{(1.2.4)}(m^0_{O^{2-}} - m_{O^{2-}})^2 - p(H_2O)m_{O^{2-}} = 0. \tag{2.5.76}$$

If the absolute value of the e.m.f. difference in dry and wet atmospheres corresponds to 1 pO unit, we can neglect the equilibrium-molality of oxide ion as compared with the initial molality of O^{2-}, i.e. assuming $m_{O^{2-}} \ll m^0_{O^{2-}}$ one can obtain the resulting dependence as follows:

$$m_{O^{2-}} = \frac{4K_{(1.2.4)}(m^0_{O^{2-}})^2}{p(H_2O)} = \text{const}(m^0_{O^{2-}})^2, \tag{2.5.77}$$

and, finally

$$pO_w = 2pO^* + \text{const}', \tag{2.5.78}$$

that is, the slope of the dependence $E-pO^*$ in the wet atmosphere should be twice as large as that obtained in a dry atmosphere. However, the results obtained in the molten KCl–LiCl eutectic, for which the assumption made is correct, there is no such doubling of the slope.

If the equilibrium concentration of oxide ions in the melt is comparable with the initial one (it is within one power of ten), the solution to equation (2.5.76) may be represented by the formula:

$$m_{O^{2-}} = m^0_{O^{2-}} - \frac{p(H_2O)}{8K_{(1.2.4)}}$$

$$- \sqrt{\left(m^0_{O^{2-}} + \frac{p(H_2O)}{8K_{(1.2.4)}}\right)^2 - (m^0_{O^{2-}})^2}. \tag{2.5.79}$$

This form of equation (2.5.79) does not allow us to make any conclusions about the form of the pO_w–pO^* dependence and its slope, since the transformation of this equation into logarithmic values does not result in explicit solutions. Therefore, the calculations of magnitudes were performed for specific values of oxide-ion concentrations, partial pressures of water vapour and equilibrium constants. The dependences obtained in such a manner are presented in Fig. 2.5.11. It should be noted that not all of these plots are strictly linear, and have an appreciable bending toward the "abscissa" axis.

Treatment by the least-squares method of the obtained data shows that, when equilibrium (1.2.4) is correct, the slope of the $E-pO^*$ dependence in the

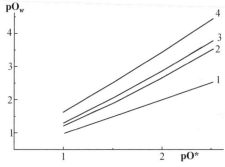

Fig. 2.5.11. Dependences pO_w against pO^*: 1—$p(H_2O) = 0$; 2—$p(H_2O) = 10^{-2}$, $K_{(1.2.4)} = 10^{-1}$; 3—$p(H_2O) = 2 \times 10^{-2}$, $K_{(1.2.4)} = 10^{-1}$; 4—$p(H_2O) = 10^{-2}$, $K_{(1.2.4)} = 10^{-2}$.

wet atmosphere should be appreciably higher than that of the corresponding value in the flow of dry inert gas. So, dependence 2 in Fig. 2.5.11 is characterized by a slope that is 1.52 times as high as that of the $E-pO$ calibration plot with additions of a strong Lux base. The corresponding values for dependences 3 and 4 are equal to 1.64 and 1.86, respectively. In contrast, the experiments performed demonstrate clear linear dependences for all the $E-pO^*$ plots, which are characterized by practically the same e.m.f. shifts for passing wet gas and the same slope values for the plots. This calculation shows undoubtedly that the dissociation of hydroxide ions in high-temperature ionic melts occurs according to equation (2.5.68) and not according to the stoichiometry in equation (1.2.4), as was assumed before. It should be noted that the experimental routine used in Ref. [180] should give reasons to revise the stoichiometry of the process of OH^- dissociation, since there is no appreciable directed shift of the calculated value of the equilibrium constant of equation (1.2.4) with small changes of the partial pressure of water vapour over the melt. Such a small directed shift was attributed by the authors of Ref. [180] to the scatter of the experimental data.

The use of the Ni|NiO metal oxide electrode for potentiometric investigations of hydroxide-ion dissociation makes it possible to neglect the so-called, "hydroxide" function of the conventional oxygen electrodes. Indeed, since the slope of the $E-pO$ calibration plots obtained with the use of the metal-oxide electrode is close to $1.15RT/F$, both in dry and in wet atmospheres, it is obvious that hydroxide ions do not participate in the electrode process. If this were not the case, the slope should be equal to

2.3RT/F owing to the following potential-defining reaction

$$NiO + H_2O + 2\bar{e} \rightleftharpoons Ni^0 + 2OH^-,$$

(2.5.80)

and considerable difficulties arise with attributing this slope to the peroxide or hydroxide function. For the Pt(O_2)|YSZ membrane oxygen electrode, the slope 2.3RT/F is not bound up with the hydroxide function, since the solid electrolyte membrane has no conductivity by hydroxide ions.

The pK value of the equilibrium (5.68) in the molten equimolar KCl–NaCl mixture at 700 °C is equal to 2.22 ± 0.06 [232]. If water vapour is absent in the atmosphere over the melt, then the equilibrium of the reaction (2.5.68) is shifted completely to the right. As for the forms of existence of hydroxide ions in molten salts, it should be emphasized that the dimers of OH^-, i.e. $(OH)_2^{2-}$ are the main form in solutions which only contain traces of OH^- and they have no effect on the behaviour of the equilibrium studied. The most obvious reason for dimerization of hydroxide ion to $(OH)_2^{2-}$ in molten salts is the formation of hydrogen bonds between the monomeric OH^- ions. Greenberg and Hallgren [323], Nonozhylov and Ptchelina [324] found such bonds in a NaCl melt containing OH^- by the method of IR spectroscopy. Gladyshev and Shakhetova analyzed the behaviour of different admixtures in molten alkali metal halides and came to the conclusion that these impurities were prone to aggregation in polynuclear associates and that this process starts just by the formation of the dimers [325]. It may be assumed that interaction of H_2O with O^{2-} initially leads to the formation of oxide hydrate $O^{2-} \cdot H_2O$, having the total chemical composition coinciding with OH_2^{2-}. The results of our investigations demonstrate that in molten alkali metal halides this ion is stable enough under the pressure of water vapour, even at temperatures close to 700 °C and above [232].

In summarizing the considered studies it should be emphasized that owing to the simplicity of the preparation and purification from water traces and carbonate admixtures, hydroxides of alkali metals can be used as strong Lux bases for the calibration of cells, including all oxygen electrodes, and for different potentiometric measurements at temperatures near and above 500 °C. Titration of the strongest Lux acids may give rise to certain deviations from the stoichiometry, owing to the basic properties of water (pyro-hydrolysis). This, however, does not result in appreciable distortion of the obtained results.

As to the problem of completeness of hydroxide ion dissociation in the atmosphere of inert gas, which is not purified from water traces, the situation is as follows. By taking into account the value of the equilibrium constant of reaction (2.5.68), which is equal to 0.006 at 700 °C in the molten KCl–NaCl equimolar mixture, one can estimate the degree of dissociation of hydroxide ion as 0.995 ($\sim$ 1) at the partial pressure of water vapour near 10^{-5} atm. (this is the usual concentration for gaseous extra-purity A), 0.857 at 0.001 atm. (for pure Ar) and 0.376 at 0.01 atm. (which is appropriate for nitrogen of technical quality). This means that only extra-pure argon can be used without preliminary purification for the creation of an inert atmosphere over the melts for oxoacidity studies, where the degree of hydroxide-ion dissociation approaches 1 (unity).

A more detailed description of the dissociation of BaO and other alkaline earth oxides in halide melts will be given in the next part. Nevertheless, the disadvantages of these substances as strong Lux bases are obvious enough. The first, and most important of them, is the limited (and relatively small) solubility of these oxides in ionic melts possessing rather weak oxoacidic properties. Another essential drawback of the use of alkaline-earth-metal oxides as strong Lux bases in molten alkali metal halides is the fact that their addition to the melts studied leads to the appearance of foreign multi-charged ions, such as Ba^{2+}, Sr^{2+}, Ca^{2+} in the melt. This may result in essential distortions of the experimental results because of the co-precipitation and formation of solid solutions with other oxides in the process of investigation, changing the oxoacidic properties of the melt. The use of the said bases as titrants when studying the acidic properties of anion acids such as $Cr_2O_7^{2-}$, or PO_3^- leads to the formation of slightly soluble neutral chromates or phosphates in the melt, which results in the overestimation of the acidic properties of the studied Lux acids.

On the basis of the facts given above, one may consider the following reagents to be strong (completely dissociated) Lux bases for oxoacidity studies in ionic melts:

- Na_2O and NaOH (KOH) are suitable for all halide- and oxygen-containing melts at temperatures near and above 500 °C;
- Na_2O_2 can be used for studies of ionic melts and reagents possessing weak reduction ability at temperatures above 450 °C;
- Li_2O can be used in acidic melts, e.g. KCl–LiCl, or Ca^{2+}-containing mixtures possessing oxobasicity indices of 3.5 and higher.

Traces of water or oxygen in the inert atmosphere over the melts result in a reduction of the equilibrium concentration of oxide ions in ionic liquids, and favour the formation of other weaker bases such as $(OH)_2^{2-}$ or O_2^{2-}. However, this mainly concerns the investigations with the use of the gas oxygen electrode, where the partial pressure of O_2 over the melts is 1 atm. Even the use of simple drying routines for the preliminary treatment of the commercial inert gases allows one to avoid this error.

Chapter 3

Equilibria in "Solid Oxide–Ionic Melt" Systems

Part 6. Characteristics of Oxide Solubilities and Methods of Their Determination

3.6.1 PARAMETERS DESCRIBING SOLUBILITIES OF SOLID SUBSTANCES IN IONIC SOLVENTS

Dissolution of solid oxides in ionic melts is accompanied by the simultaneous interaction of the dissolving material with the melt ions. Metal cations are surrounded by the melt anions and oxide ions are fixed by constituent cations of the ionic medium. This results in the formation of different ionic complexes. Besides, a certain part of the powder added is dissolved in the melt without dissociation. The sum of the changes taking place in these processes is called "solvation". So, any process of oxide dissolution in an ionic medium may be described by a system of equations. For the case of the formation of two-charged metal cation these are

$$MeO_s \rightleftharpoons MeO_1, \tag{3.6.1}$$

$$MeO_1 \rightleftharpoons Me^{2+} + O^{2-}, \tag{3.6.2}$$

$$Me^{2+} + nX^- \rightleftharpoons MeX_n^{2-n}, \tag{3.6.3}$$

$$O^{2-} + mKt^{n+} \rightleftharpoons Kt_mO^{mn-2}, \tag{3.6.4}$$

where Kt^{n+} denotes the most acidic constituent cation of the melt. We will consider below the ionic melts based on alkali-metal cations such as Cs^+, K^+, Na^+ and Li^+.

Since the process of complexation between the melt anion and different cations of close radii proceeds to practically equal degrees, one may assume that for a set of oxides dissolved in the same ionic solvent the solubility value should be dependent on the degree of "acidic cation–anion" interaction which is the Lux–Flood acid–base equilibrium. In this case, the constant of the acid–base equilibrium (0.1) (see p. XII) is the quantitative measure of the acidity of chemical substances. If all the participants of the acid–base reaction are in the dissolved state then the equilibrium achieved in the homogeneous system may be considered an acid–base one. However, most oxides possess limited solubilities in the ionic melts heated to temperatures lower than the melting points of the oxides. This is the main reason for the formation of two-phase systems "solid oxide–saturated solution of the oxide in the melt". This allows us to conclude that the reactions of cations with oxide ions lead to the removal of the oxide ions from the melt, not only owing to the formation of the dissolved acid–base complex, but also because of the formation of the solid phase of the oxide. Therefore, equilibrium (3.6.1) should be taken into account both in investigations by potentiometric titration techniques and in studies performed by the method of isothermal saturation. The scheme of the equilibria (3.6.1)–(3.6.4) leads to the conclusion that the process of fixation of oxide ions by metal cations, in the common case in heterogeneous reactions, results in the precipitation of a new phase (solid metal oxide). Hence, the acidic properties of Me^{2+} cation cannot be connected unambiguously with the completeness of the interaction $Me^{2+}–O^{2-}$ in ionic melts.

Taking into account the incomplete dissociation of metal oxides in ionic melts, the complete oxide solubility ($\sum s_{MeO}$) can be represented by the sum of the activities (concentrations) of metal cations and non-dissociated oxide:

$$\sum s_{MeO} = a_{s,Me^{2+}} + a_{s,MeO}, \tag{3.6.5}$$

where $a_{s,Me^{2+}}$ and $a_{s,MeO}$ are the activities of metal cation and non-dissociated oxide in the saturated solutions of the oxide. Since most metal oxides are slightly soluble in ionic melts, the concentration of non-dissociated oxide in saturated solutions is hardly detectable. This fact forces investigators to use solubility products for different practical purposes. The solubility product of oxide, P_{MeO}, may be expressed via the activities, by the following equation:

$$P_{MeO} = a_{s,Me^{2+}} a_{ss,O^{2-}}. \tag{3.6.6}$$

The saturated solution region implies that the metal cations and oxide ions are in equilibrium with both non-dissociated and the non-dissolved oxide and according to the law of mass action the dissociation constant of the oxide can be represented in the following manner:

$$K_{MeO} = a_{s,Me^{2+}} a_{s,O^{2-}} / a_{s,MeO}. \tag{3.6.7}$$

The comparison of equilibria (3.6.6) and (3.6.7) results in the following equation:

$$K_{MeO} = P_{MeO}/a_{s,MeO}. \tag{3.6.8}$$

The complete set of constants describing the equilibria taking place in the saturated solutions of oxides can be obtained if we know at least two parameters of the three presented in equation (3.6.8).

Since there are no well-developed and generally accepted methods for the determination of the activity coefficients and of their dependence on the concentration of the corresponding particles in molten salts, practical investigations of solubility usually result in the corresponding parameters being obtained based on the concentration, but not on the ion activities.

3.6.2 METHODS OF OXIDE SOLUBILITY DETERMINATION

Investigations of the solubilities of metal oxides in oxygen-containing and oxygen-free ionic melts are performed with saturated solutions obtained in different ways. During the past few decades two methods for the solubility studies have been examined and introduced into practice. They are the methods of isothermal saturation and of potentiometric titration.

3.6.2.1 Isothermal saturation method

The method of isothermal saturation consists of the addition of an excess quantity of oxide to the melt studied. Such an addition results in the formation of a saturated solution which is in equilibrium with the oxide precipitate. The fact that the equilibrium conditions have been achieved is detected in different ways; among them we should mention potentiometric measurements using different oxygen electrodes [238, 326] and titrimetric determination of the concentration of metal ions in a sample of the melt [327, 328]. The sum of the concentrations of ionic and non-dissociated forms of the oxide according to equation (3.6.5) is the main result of these determinations. Taking into account the thermodynamic parameters which describe equilibria in the saturated solutions, equation (3.6.5) may be rewritten as

$$s_{MeO} = \sqrt{P_{MeO}} + P_{MeO}/K_{MeO}. \tag{3.6.9}$$

Naturally, if the potentiometric titration routine is used for controlling the saturation of the solution, the preliminary calibration of such a cell with known quantities of a strong Lux base allows us to divide the total magnitudes of oxide solubility into those connected with the ionic and molecular forms of the oxide. However, such determinations are not described in the literature.

The isothermal saturation method implies the carrying out of a number of experiments to obtain a set of data suitable for statistical treatment. This makes the consumption of reagents and time higher than needed with the method of potentiometric titration. Some information about the duration of the experiments and the frequency of sample testing may be obtained from Fig. 3.6.1, which presents the results of investigations of SrO and BaO solubilities in the molten KCl–NaCl eutectic at different temperatures. It is seen that after keeping the solid oxide in the melt for 2–3 h the dependences achieve a plateau which corresponds to the formation of the saturated solution of oxide.

The analysis of the melt samples with dissolved oxides is often used [327, 328]; although it may give rise to considerable distortions of the data obtained, owing to inclusions of particles of the suspended oxide in the analysed sample. This fact must be especially taken into account, since the densities of the oxides studied are often 1.5–2 times higher than that of the ionic melt–solvent. In particular, such inclusions might be present in the

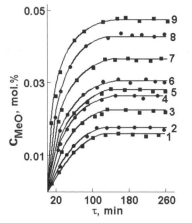

Fig. 3.6.1. Dependences of concentration of SrO ((1) 973 K, (2) 993 K, (3) 1023 K, (4) 1073 K) and BaO ((5) 973 K, (6) 993 K, (7) 1023 K, (8) 1043 K, (9) 1073 K) in the molten KCl–NaCl eutectic vs time of the phase contact. The data are cited according to Volkovitch [328, Fig. 1].

samples obtained by Woskressenskaja and Kaschtschejev [327] who poured the saturated melt out of the crucible, whereas the oxide precipitate remained on the bottom. In principle, the visual control of transparency of the melt cannot help to state the existence of fine suspended particles in the sample, especially if a melt–solvent is heated to temperatures above 700 °C. Another approach consists of preliminary sintering of the studied oxide at temperatures higher than that of the experiment [328]. The pellet obtained is assumed to remain in one piece (unbroken) under the experimental conditions without escape of fine oxide particles. Generally speaking, the determination of the solubility of different materials in ionic melts by the isothermal saturation method requires reliable prevention of the formation of suspensions in the molten salts, which usually complicates the experimental technique.

One of the obvious advantages of the isothermal saturation technique consists of the fact that it is very convenient for performing polythermal studies. However, it can hardly be appropriate for the determination of solubilities of slightly soluble oxides such as MgO, NiO or Al_2O_3. Since the experiment results in the determination of oxide solubility, but not of its logarithm, this method possesses a lower limit of solubility estimation. As a rule, the data obtained in studies of such a kind cannot be processed to give simultaneous estimates of the solubility product and the value of the dissociation constant apart from some particular cases where the oxide is completely dissociated or, on the contrary, is dissolved without appreciable dissociation. Since the "pure" ionic melts are usually polluted by various admixtures of oxygen-containing compounds the said investigations may give errors. These errors become too great when the oxide solubility is comparable with the oxide–ion concentration in the "pure" melt–solvent and the dissolution process is depressed to a considerable degree.

Naumann and Reinhardt [329] described the use of the radioactive marker method to determine the oxide solubilities. The advantages of this method concern only the procedure for the subsequent treatment of the obtained sample of the melt with the metal oxide, whereas its drawbacks are similar to those of the "classic" isothermal saturation methods. This radiochemical method has not been used extensively for solubility determinations in molten salts.

3.6.2.2 Potentiometric titration method

The potentiometric titration routine is now considered to be the easiest way to obtain a lot of information on the behaviour of metal oxides in various ionic melts. It allows determination of the solubility product of oxides over a wide range of solubilities. In some cases (when the region of non-saturation can be detected in the potentiometric curve), it permits the estimation of the dissociation constants of the oxide studied. As a rule, potentiometric investigations are performed in cells with the liquid junction; however, this does not give rise to appreciable errors, since the potentials of the liquid junction are extremely low [81].

The potentiometric titration method possesses some advantages character-istic of potentiometry. The measured e.m.f. values are dependent on the logarithm of the activity (concentration) of the potential-defining ion and this considerably widens the range of concentrations which can be detected. The experimental techniques and routine are rather simple and the obtained results are well reproducible. Owing to the stable activity coefficients of metal cations the measurements can be performed in sufficiently concentrated solutions of the corresponding metal halides. Performing the measurement does not imply any interference in the acid–base processes in the melt studied and the experiment is faster than the isothermal saturation technique discussed above.

Owing to these advantages, the potentiometric titration method is used widely for the investigation of various kinds of acid–base equilibria in molten media. Usually the concentration parameters of the said equilibria are obtained.

Carrying out the titration procedure results in the obtaining of the potentiometric curves constructed in the e.m.f. $(pO)-m^0_{O^{2-}}$ coordinates. The position of the curve in this diagram is defined by the solubility product and by the dissociation constant of the oxide.

In the general case, the potentiometric titration curves consist of four sections, as shown in Fig. 3.6.2 (curve 3).

1. The first section corresponds to the non-saturated solution characterized by a sharp pO decrease at small initial concentrations of a base-titrant; the behaviour of the potentiometric titration curves under these conditions is dependent on the magnitude of the dissociation constant of the formed oxide.

2. The second section is characteristic of the saturated solutions before the equivalence point and continues up to complete acid–base neutralization of the cation studied. A flat form of the potentiometric curve in this section is defined by the formation of a buffer system of composition "oxide precipitate/solution of the cation" possessing a higher buffer number than

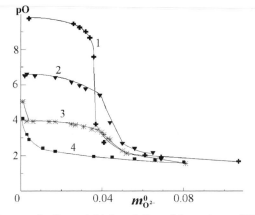

Fig. 3.6.2. Dependences of pO on initial molality of base-titrant 2NaOH ($m_{O^{2-}}^0$) at the potentiometric titration of metal cations in CsBr–KBr (2:1) at 700 °C: (1,2) practically insoluble oxides (NiO, 0.035 mol kg^{-1} and CdO, 0.040 mol kg^{-1}, respectively); (3) merely soluble oxide (SrO, 0.050 mol kg^{-1}); (4) soluble oxide (BaO, 0.050 mol kg^{-1}).

the solutions, whereas "solution of the cation/completely dissolved but non-dissociated oxide" belonging to the first section of the curve possesses a lower buffer number that causes sharp pO decrease (see item1).

3. The third section corresponds to the pO drop at the equivalence point and is characterized by a sharp pO decrease in the solutions possessing low buffer numbers. The solutions of this section are saturated with respect to the studied oxide and this section is of no interest for the calculation of the thermodynamic parameters describing the solubility.

4. The fourth section is characterized by a relatively flat decrease in pO in the region of excess titrant; it allows one to detect the acidic properties of the precipitated oxide if it has them.

The presence or absence of some of the mentioned sections in a potentiometric titration curve is the criterion which allows one to assign the formed oxide to the class of insoluble ones, slightly soluble or soluble in the given molten medium.

For example, the potentiometric titration curve obtained in the case formation of a practically insoluble oxide does not contain the first section (Fig. 3.6.2, curves 1 and 2) since the equilibrium concentration of O^{2-} in the pure melt, which arises as a result of the acid–base dissociation of oxygen-containing impurities, is sufficient to achieve the value of solubility product of the corresponding oxide and the latter precipitates even after introducing the metal cations in the "pure" melt. Under the experimental conditions, the pO value of a pure melt is usually within 4–4.5 apart from hygroscopic melts (Li^+ or Ca^{2+}-based halide mixtures) or those possessing strong reductive properties (iodides) where the pO value ranges from 3 to 4. As the molality of the metal halide added to the melt for the study is close to 0.05 mol kg^{-1}, it is possible to estimate the boundary between insoluble and slightly soluble oxides as

$$P_{MeO} \approx (1 \times 10^{-4})(5 \times 10^{-2}) = 5 \times 10^{-6},$$

$$pP_{MeO} = 5.3.$$

(3.6.10)

If the solubility product of oxide is less than $(2–5) \times 10^{-6}$ then the section corresponding to the non-saturated solution will be absent in the potentiometric titration curve.

The formation of a slightly soluble oxide can be recognized from the potentiometric curve with a short section of non-saturated solution (Fig. 3.6.2, curve 3), i.e. in this case the concentration of oxygen-containing impurities in the pure melt is insufficient for precipitation of the oxide from the melt. However, the precipitation of oxide starts when even the first small weight of Lux base-titrant is added to the melt.

The solubility product considerably exceeds the boundary value of 5×10^{-6}. Since the precipitation occurs at relatively high pO values, the corresponding potentiometric titration curves contain a sharp pO drop at the equivalence point.

In the case of formation of moderately soluble or well-soluble oxides the potentiometric curves are characterized by a wide non-saturated solution section and, as a rule, by a blurred pO drop at the equivalence point (sometimes this drop is absent) as is shown in Fig. 3.6.2 (curve 4). Precipitation of a soluble oxide is initiated in the solutions containing relatively high initial concentrations of oxide ions. In this case, the interaction can be detected from the dependence of the ligand number $\bar{n}$ on the initial concentration of the titrant: it deviates considerably from zero. To divide the sections of non-saturated and saturated solutions using only the titration curve is a very difficult problem, especially when precipitation takes place from the concentrated titrant solution.

All that is written above allows us to consider the solubility product value of $P_{MeO} = 5 \times 10^{-6}$ to be the boundary between practically insoluble and slightly soluble oxides [175, 330]. In other words, if the addition of the metal cation to the pure melt results in the precipitation of oxide then the oxide formed is an insoluble one. In the opposite case, it should be classified among moderately soluble or soluble oxides, but this division can only be made on the basis of quantitative characteristics obtained by processing the titration data (pP_{MeO}). After the equivalence point the formed deposit of oxide can, in turn, show acidic properties owing to fixation of excessive oxide ions which arise in the solution with the excess of titrant. This process is detected on the basis of the value of the ligand number, $\bar{n}$, exceeding 1, and the interaction leads to the formation of anion oxocomplexes, whose composition, for example, for Pb^{2+} cations in the KCl–NaCl equimolar mixture corresponds to PbO_2^{2-} for the dissolved part and to $(Na_xK_{1-x})PbO_2$ in the precipitated solid phase.

In concluding our consideration of the principal features of the potentiometric titration curves of metal cations it should be mentioned that a thorough analysis of the curves has not been presented before in the publications known to us. As a rule, the points coming out from the flat titration section (and this especially concerns some first points belonging to the unsaturated solution section) are not taken into account. Nevertheless, just these data were shown in Refs. [175, 330] to be the only suitable ones for calculation of the dissociation constants of oxides. This statement in the paper [175] is formulated as a guess, which is, nevertheless, in perfect agreement with the experimental data. Therefore, it is necessary to provide an accurate quantitative substantiation of the method of calculating the correct equilibrium parameters. We shall now consider such a substantiation.

From what we have said above, it follows that the acid–base equilibrium in the solutions containing metal cations and oxide ions in different sections of the titration curve is described either by the dissociation constant (in unsaturated solutions) or by the values of solubility product (in saturated solutions). In Refs. [175, 330] we proposed a method based on the analysis of the scatter in the calculated equilibrium parameters corresponding to the titration process. Indeed, in the unsaturated solution section there is no oxide precipitation and the calculated value of the solubility product increases monotonously (the directed shift) whereas the calculated value of the dissociation constant fluctuates about a certain value, which is the concentration-based dissociation constant of the studied oxide.

In the saturated solution, the behaviour of the thermodynamic parameters describing the process becomes the opposite: the calculated pK_{MeO} value increases monotonously, whereas the calculated pP_{MeO} values undergo fluctuation about a certain average value. In this case, the directed shift of the calculated pK value is caused by the following factors. In the saturated solution section the thermodynamic dissociation constant of the metal oxide is expressed via the equilibrium molalities as

$$K_{MeO} = \frac{m_{Me^{2+}} m_{O^{2-}}}{s_{MeO}}. \qquad (3.6.11)$$

The usual computational calculations are connected with the substitution of s_{MeO} by m_{MeO} in the denominator of equation (3.6.11), the latter actually being the sum of oxide concentrations in the different phases:

$$m_{MeO} = s_{MeO} + m_{MeO,s}, \qquad (3.6.12)$$

where $m_{MeO,s}$ is the quantity of the formed oxide precipitate, recalculated with respect to 1 kg of the ionic melt studied.

The substitution of equation (3.6.12) in equation (3.6.11) gives

$$K'_{MeO} = \frac{m_{Me^{2+}} m_{O^{2-}}}{m_{MeO,s} + s_{MeO}} \frac{s_{MeO}}{s_{MeO}} = K_{MeO} \frac{s_{MeO}}{m_{MeO,s} + s_{MeO}}. \qquad (3.6.13)$$

This expression shows that according to the addition of the titrant in the saturated solution section the denominator of equation (3.6.13) should increase, since s_{MeO} remains unchanged and the quantity of the precipitated oxide ($m_{MeO,s}$) increases, whereas K_{MeO} is obviously a constant. This finally causes a monotonous decrease in K'_{MeO} (or increase in pK'_{MeO}).

The experimental results and those derived from the potentiometric titration of lead cations with additions of NaOH in the molten KCl–NaCl equimolar mixture at 700 °C are typical examples of the division of the potentiometric titration curves into sections for unsaturated and saturated solutions. We shall now consider the data in Table 3.6.1 and Fig. 3.6.3.

The sequence of some points from the initial section of the potentiometric curve (namely, points 1–4) is characterized by a monotonous pP'_{MeO} decrease from 6.06 to 5.12, whereas the pK'_{MeO} values remain practically constant. This section of the titration curve corresponds to a sharp e.m.f. decrease at low initial concentrations of the titrant (see Fig. 3.6.3, the titration and differential titration curves). From the above-said it is seen that the PbO solution is unsaturated and the concentration constant of PbO can be estimated from the data of this section (points 2–4) as p$K_{PbO} = 3.29 \pm 0.04$.

Points 5–20 of the specified titration curve belong to the saturated solution section, which is flatter in Fig. 3.6.3. Here, the pP'_{MeO} values are practically constant apart from some experimental points (points 14–16) located near the

TABLE 3.6.1

Experimental and derived results of potentiometric titration of $PbCl_2$ (0.050 mol kg^{-1}) in the molten KCl–NaCl equimolar mixture at 700 °C

No.	$m^0_{O^{2-}}$	E (V)	pO	$\bar{n}$	pK'_{MeO}	pP'_{MeO}
$Pb^{2+} + O^{2-} \rightleftharpoons PbO$						
1	0.001	0.542	4.75	0.02	3.05	6.06
2	0.005	0.485	4.26	0.10	3.30	5.51
3	0.009	0.450	3.95	0.18	3.29	5.34
4	0.015	0.418	3.67	0.30	3.29	5.12
5	0.020	0.410	3.60	0.40	3.42	5.12
6	0.023	0.404	3.55	0.45	3.47	5.11
7	0.026	0.401	3.52	0.51	3.54	5.13
8	0.028	0.396	3.48	0.55	3.57	5.13
9	0.033	0.386	3.39	0.65	3.66	5.15
10	0.036	0.376	3.30	0.71	3.69	5.14
11	0.038	0.362	3.18	0.75	3.65	5.08
12	0.043	0.349	3.07	0.84	3.80	5.18
13	0.045	0.344	3.02	0.88	3.89	5.25
14	0.048	0.343	3.01	0.94	4.21	5.54
15	0.052	0.196	2.33	0.95	3.58	4.90
16	0.055	0.153	2.13	0.97	3.43	4.75
$PbO + O^{2-} \rightleftharpoons PbO_2^{2-}$						
17	0.065	0.092	1.85	1.02	–	
18	0.081	0.032	1.57	1.08	1.01	
19	0.096	−0.004	1.40	1.12	0.93	
20	0.108	−0.026	1.30	1.16	1.01	

pO drop at the equivalence point. This section is easily detected both from the titration data and the data of the differential titration curve. From points 4–13 the concentration pP_{PbO} value can be calculated as 5.12 ± 0.05. On the contrary, the pK'_{MeO} values are subjected to the directed shift (increase) for the reason described above. After reaching the equivalence point, addition of the excess of the titrant results in the formation of plumbate both in the deposit and in the solution. The constant of formation of PbO_2^{2-} ion is estimated in Ref. [175] as $pK = -0.98 \pm 0.20$. The value of this constant shows that the interaction of lead oxide with the strongest bases in molten salts having weak acidic properties is thermodynamically favourable.

Besides its relative simplicity, this method cannot be considered as universal, since it is applicable only to the case of formation of slightly soluble or practically insoluble oxides. Sometimes, the treatment of the experimental

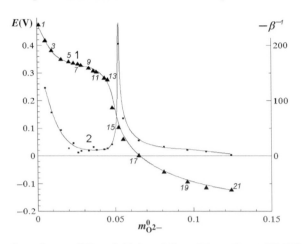

Fig. 3.6.3. The dependence of E vs initial molality of base-titrant 2NaOH ($m^0_{O^{2-}}$) at the potentiometric titration of Pb^{2+} (0.051 mol kg^{-1}) in the molten KCl – NaCl equimolar mixture at 700 °C (1), and the corresponding differential titration curve (2). The points in the titration curve are numbered according to the data in Table 3.6.1.

results leads to cases where the calculated values of both thermodynamic parameters (pK'_{MeO} and pP'_{MeO}) are subjected to a small directed shift. In this case, the question about the saturation of the solutions studied cannot be answered using the above-described method. The phenomenon mentioned of the change of oxide solubility accompanying the change of the concentration of the cation solution from which the oxide precipitates results in an appreciable shift of the solubility product value (reduction in pP) even in the saturated solutions, which complicates the use of the method for detecting the formation of the saturated solution.

Therefore, an essential development of the potentiometric method for treatment of titration results was required to mark the division of solutions formed after addition of a strong Lux base to metal cation solutions between unsaturated and saturated ones. The following method is based on the use only of potentiometric results [214] and is applicable for all cases taking place in the potentiometric titration of metal cations in molten salts.

(i) Treatment of experimental data for slightly soluble oxides

This method is suitable for metal oxides possessing negligible solubility in molten salts. If the titration process yields values of pO exceeding three in the cation excess section that is characteristic of slightly soluble or insoluble oxides, then the value of the equilibrium molality of oxide ion can be neglected as compared with the initial Lux base molalities in the calculation formulas. Such a neglect makes it possible to construct the dependence $E-p\theta$, where

$$p\theta = -\log|m^0_{Me^{2+}} - m^0_{O^{2-}}| \qquad (3.6.14)$$

for division of the potentiometric titration curve into the sections of unsaturated and saturated solutions [214]. The parameter $p\theta$ is practically coincident with the negative logarithm of the equilibrium molality of metal cation in the titrated solution. An example of such dependences is presented in Fig. 3.6.4.

On the basis of the formulas used for the calculation of the concentration parameters of the studied equilibria, let us consider, the behaviour of the $E-p\theta$ dependence in unsaturated and saturated solutions of metal oxides.

(a) Unsaturated solution

The expression for the equilibrium constant from the experimentally determined values can be written in the following form:

$$K_{MeO} = \frac{(m^0_{Me^{2+}} - m^0_{O^{2-}} + 10^{-pO}) \times 10^{-pO}}{m^0_{O^{2-}} - 10^{-pO}}. \qquad (3.6.15)$$

High values of pO (i.e. pO $\gg$ 3) characteristic of the titration of metal cations which form slightly soluble oxides in molten salts allow us to neglect the equilibrium molality of oxide ion as compared both with the

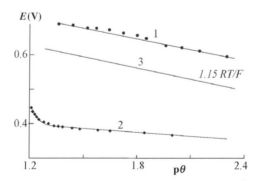

Fig. 3.6.4. $E-p\theta$ dependence for slightly soluble and practically insoluble oxides in the CsCl–KCl–NaCl eutectic melt at 700 °C: (1) NiO, (2) PbO, (3) the theoretical dependence.

initial molality of oxide-ion donor and the initial molality of the metal cation in the solution. In brief, this relationship can be expressed by the following inequality:

$$m_{O^{2-}}^0 \sim m_{Me^{2+}}^0 \gg 10^{-pO}. \tag{3.6.16}$$

In view of this inequality, equation (3.6.15) can be rewritten as

$$K_{MeO} = \frac{(m_{Me^{2+}}^0 - m_{O^{2-}}^0) \times 10^{-pO}}{m_{O^{2-}}^0} \tag{3.6.17}$$

and the concentration pK_{MeO} value is expressed as follows:

$$pK_{MeO} = -\log|m_{Me^{2+}}^0 - m_{O^{2-}}^0| + pO + pm_{O^{2-}}^0. \tag{3.6.18}$$

This equation makes it possible to present the pO value as a function of the thermodynamic and calculated parameters of the studied equilibrium:

$$pO = pK_{MeO} - p\theta - pm_{O^{2-}}^0. \tag{3.6.19}$$

It is obvious that the slope of the E–$p\theta$ dependence is to be calculated as the derivative of the complex function:

$$\left(\frac{\partial E}{\partial p\theta}\right)_T = \left(\frac{\partial E}{\partial pO}\right)_T \left(\frac{\partial pO}{\partial p\theta}\right)_T. \tag{3.6.20}$$

Substitution of the pO value from equation (3.6.19) into the derivative leads to the following expression:

$$\left(\frac{\partial E}{\partial p\theta}\right)_T = -\left(\frac{\partial E}{\partial pO}\right)_T \left(1 - \frac{\partial pm_{O^{2-}}^0}{\partial p\theta}\right)_T. \tag{3.6.21}$$

The fraction in parentheses

$$\left(1 - \frac{\partial pm_{O^{2-}}^0}{\partial p\theta}\right)_T$$

can be transformed in the following way (taking into account the fact that the numerator and the denominator are decimal logarithms, the multipliers 0.4343 are reciprocally cancelled):

$$\frac{\partial pm_{O^{2-}}^{0}}{\partial p\theta} = \frac{\dfrac{dm_{O^{2-}}^{0}}{m_{O^{2-}}^{0}}}{\dfrac{d(m_{Me^{2+}}^{0} - m_{O^{2-}}^{0})}{m_{Me^{2+}}^{0} - m_{O^{2-}}^{0}}} = \frac{dm_{O^{2-}}^{0}(m_{Me^{2+}}^{0} - m_{O^{2-}}^{0})}{dm_{O^{2-}}^{0} m_{O^{2-}}^{0}}$$

$$= 1 - \frac{m_{Me^{2+}}^{0}}{m_{O^{2-}}^{0}}. \tag{3.6.22}$$

Substitution of the above expression into equation (3.6.21) results in the following simple solution:

$$\left(\frac{\partial E}{\partial p\theta}\right)_{T} = -\left(\frac{\partial E}{\partial pO}\right)_{T} \frac{m_{Me^{2+}}^{0}}{m_{O^{2-}}^{0}}, \tag{3.6.23}$$

which allows us to make some important conclusions. First, the $E-p\theta$ dependence in an unsaturated solution should be non-linear. The slopes of the tangent lines exceed the corresponding slope of the calibration $E-pO$ plot and the increase in the initial molality of oxide ion leads to a gradual reduction in the slope of the tangent line (Fig. 3.6.4, dependence 2). Equation (3.6.23) shows that the decrease in the e.m.f. or pO at small initial concentrations of the titrant should be sharp enough. All we have said is observed for the titration of cations, which form slightly soluble or insoluble oxides. The construction of the $E-p\theta$ plot exhaustively confirms the saturation of the studied solution with respect to the oxide forming as a result of the acid–base reaction.

Therefore, the values of pK_{MeO} determined by averaging the titration data from the initial section of the titration curve (characterized by a sharp e.m.f. reduction at small titrant additions) are just the dissociation constant of the studied oxide, to an accuracy determined by the natural spread of the experimental data. It should be emphasized, however, that the values of the pK_{MeO} concentration constant calculated in such a way, contain an appreciable error, caused by the fact that the initial concentration of the titrant in the halide melt is comparable in magnitude with that of oxygen-containing admixtures in the "pure" melt.

Taking into account the concentration of oxygen-containing admixtures in equation (3.6.17) leads to the following formula;

$$K'_{MeO} = \frac{(m^0_{Me^{2+}} - m^0_{O^{2-}} - m'_{O^{2-}}) \times 10^{-pO}}{m^0_{O^{2-}} + m'_{O^{2-}}}. \tag{3.6.24}$$

The effect of such a substitution on the magnitude of the numerator is negligible since

$$m^0_{Me^{2+}} - m^0_{O^{2-}} - m'_{O^{2-}} \approx m^0_{Me^{2+}}. \tag{3.6.25}$$

After converting to logarithms, this effect becomes vanishingly small. Besides, with small titrant additions ($m^0_{O^{2-}} \sim m'_{O^{2-}}$) the magnitude of the denominator increases. So, the calculated dissociation constant value decreases as compared with the concentration (thermodynamic) one:

$$K'_{MeO} = \frac{1}{1 + \dfrac{m'_{O^{2-}}}{m^0_{O^{2-}}}} K_{MeO}. \tag{3.6.26}$$

If the initial concentration of the titrant is equal to that of oxygen-containing admixtures, the K_{MeO} value obtained is half as high as the true concentration constant. A rise of the initial molality of the titrant reduces this deviation, but the dissociation constant values are somewhat underestimated. The results in Table 3.6.1 show that only the data corresponding to the initial titrant molalities of 5×10^{-3} mol kg^{-1} and higher can be used for averaging pK_{MeO} without appreciable error.

(b) Saturated solution

Let us consider the behaviour of the E–$p\theta$ dependence in the section of the saturated solution of the studied oxide. At first, it should be noted that the acid–base equilibrium (3.6.1) in this section is described by the solubility product of the oxide (2.4.14), which can be expressed via the experimentally determined values in such a manner

$$P_{MeO} = (m^0_{Me^{2+}} - m^0_{O^{2-}} + 10^{-pO}) \times 10^{-pO}. \tag{3.6.27}$$

This formula can be transformed into the following one:

$$P_{MeO} = (m^0_{Me^{2+}} - m^0_{O^{2-}}) \times 10^{-pO} \tag{3.6.28}$$

using the assumption (3.6.16). The corresponding equation for the estimation of pP_{MeO} value will be as follows:

$$pP_{MeO} = -\log|m^0_{Me^{2+}} - m^0_{O^{2-}}| + pO. \qquad (3.6.29)$$

The dependence $E-p\theta$ for the saturated solution of metal oxide taking into account equation (3.6.29) can be rewritten as

$$\left(\frac{\partial E}{\partial p\theta}\right)_T = \left(\frac{\partial E}{\partial pO}\right)_T \frac{\partial(pP_{MeO} - p\theta)}{\partial p\theta} = -\left(\frac{\partial E}{\partial pO}\right)_T. \qquad (3.6.30)$$

The sequence of equations presented above leads to the conclusion that the dependence $E-p\theta$ in the saturated solution of oxide should be presented by straight line as shown in Fig. 3.6.4 (dependence 1) with the slope equal to that of the $E-pO$ calibration plot (Fig. 3.6.4, dependence 3) in absolute magnitude. The treatment of the titration curves obtained practically shows, however, that the slopes of the $E-p\theta$ dependences differ somewhat from the calibration slope, since the oxide powder deposited from the more dilute solution of the corresponding cation possesses a higher solubility in the melt, because of the smaller sizes of the precipitated oxide particles. This causes some increase in the value of the solubility product, which becomes slightly variable and, in its turn, an increase in the $E-p\theta$ slope. Therefore, if the $E-p\theta$ dependence for the cation studied is a straight line and has no initial non-linear section with a sharp pO decrease, then the oxide solution is saturated at all the titrant weights. In this case, the precipitate of the oxide is formed even after the addition of the initial weight of cation to the pure melt (as an anhydrous halide salt) owing to the interaction with traces of oxygen-containing compounds dissolved in the melt. As a rule, the accuracy of the calculated average values of pP_{MeO} is higher than that of the corresponding dissociation constants. However, to achieve this, it is necessary to choose the experimental points lying far from the equivalence point section, within the ligand numbers from 0.1 to 0.7. In this section the error caused by oxide-ion traces is negligible, whereas near the equivalence point the first multiplier in equation (3.6.27) approaches zero, and neglect of the admixture concentration distorts the derived results essentially. Moreover, as was said in the Part 5, the use of alkali-metal hydroxides results in partial hydrolysis of the melt containing the metal cations, by water formed by the dissociation of the hydroxides according to reaction (1.2.4):

$$Me^{2+} + H_2O + 2Cl^- \rightleftharpoons MeO\downarrow + 2HCl\uparrow . \tag{3.6.31}$$

This causes additional errors, which become appreciable just at the equivalence point. The third factor consists of changes of the solubility of the oxide precipitate formed from the solutions of the cation of different concentrations as was discussed above.

The mentioned features of potentiometric and derived dependences are observed with excess of the studied cation in the melt, i.e. in acidic solution. Nevertheless, the construction of similar plots after reaching the equivalence point is very helpful in the cases where doubts arise about the correctness of the reversible work of the electrode pair used. If the oxide precipitate obtained as a result of titration does not show acidic properties, then the E–$p\theta$ plot constructed on the basis of the data after the equivalence point become equivalent to the E–pO calibration plot, since the value of $-\log|m^0_{Me^{2+}} - m^0_{O^{2-}}|$ at the excess of titrant is equal to pO. The reaction of the precipitate with oxide ions results in an upward shift of the E–$p\theta$ plot. Changes of the potential of the reference electrode can cause deviations of e.m.f. in both directions (upwards or downwards). In this case, one should either perform additional calibration (this is the longer way), or calculate a correction, which is equal to the magnitude of the e.m.f. shift. Then the calibration parameters used for the calculations are corrected using the value estimated.

All the above speculations refer only to the case where pO values at excess of cation exceed the value of three. In the opposite case, the condition (3.6.16) is disturbed and the described method cannot be applied for analysis of the obtained potentiometric data. This requires another approach.

(ii) Treatment of results for moderately soluble oxides

If, under the titration conditions, there is the formation of a moderately soluble oxide, either in the dissolved state or as a precipitate then the treatment of the results has been shown in Section 3.7.3 to become very complicated, mainly because the potentiometric titration curve does not include any characteristic sections, apart from the initial sharp decrease in e.m.f. (pO), which is gradually transformed to a flatter dependence. The pO drop at the equivalence point is often absent and both pP'_{MeO} and pK'_{MeO} are subject to a small directed shift. A question arises of how to determine whether the solution is saturated so that we can calculate pP_{MeO} or whether it is unsaturated and the behaviour of the potentiometric curve is defined by the pK'_{MeO} magnitude and the latter can be estimated from the corresponding data. This question was exhaustively solved in our recent work [331] using the method of division of the titration curves into sections belonging to the saturated and unsaturated solution of the oxide. This method used the $E = f(-\log m^0_{O^{2-}})$ dependences. In order to unify the designations in equations and graphs, $-\log m^0_{O^{2-}}$ will hereafter be denoted as pθ).

We shall now consider the characteristic features of E-pθ dependences, which can be obtained for unsaturated and saturated solutions of the metal oxide possessing appreciable solubility in the studied melt.

(a) Unsaturated solution

Since the equilibrium concentrations of the oxide ion during the studied acid–base process are higher than those in the studies of slightly soluble oxides, they cannot be neglected and the expression for the dissociation constant equation (3.6.11) should be written in the full form

$$K_{MeO} \times 10^{p\text{O}}(m^0_{O^{2-}} - 10^{-p\text{O}}) = m^0_{Me^{2+}} - m^0_{O^{2-}} + 10^{-p\text{O}} \qquad (3.6.32)$$

transformed in such a manner

$$K_{MeO}m^0_{O^{2-}} 10^{p\text{O}} - K_{MeO} = m^0_{Me^{2+}} - m^0_{O^{2-}} + 10^{-p\text{O}}, \qquad (3.6.33)$$

$$m^0_{O^{2-}} = \frac{K_{MeO} + m^0_{Me^{2+}} + 10^{-p\text{O}}}{K_{MeO} + 10^{-p\text{O}}} \times 10^{-p\text{O}}, \qquad (3.6.34)$$

$$p\theta = p\text{O} - \log\frac{K_{MeO} + m^0_{Me^{2+}} + 10^{-p\text{O}}}{K_{MeO} + 10^{-p\text{O}}}$$

$$= p\text{O} - \log\left(\frac{m^0_{Me^{2+}}}{K_{MeO} + 10^{-p\text{O}}} + 1\right). \qquad (3.6.35)$$

The derivative of $p\theta$ with respect to pO is equal to

$$\left(\frac{\partial p\theta}{\partial pO}\right)_T = 1 - \frac{10^{-pO}}{\partial(10^{-pO})}\frac{m^0_{Me^{2+}}\partial(10^{-pO})}{K_{MeO} + 10^{-pO}}$$

$$\times \frac{1}{1 + \dfrac{m^0_{Me^{2+}}}{K_{MeO} + 10^{-pO}}} \qquad (3.6.36)$$

or

$$\left(\frac{\partial p\theta}{\partial pO}\right)_T = 1 - \frac{m^0_{Me^{2+}} \times 10^{-pO}}{(K_{MeO} + m^0_{Me^{2+}} + 10^{-pO})(K_{MeO} + 10^{-pO})}.$$

$$(3.6.37)$$

The value of the second summand is considerably lower than 1 (unity) at $K_{MeO} \sim 10^{-pO} \sim m^0_{O^{2-}}$. Therefore, the derivative of the inverse function can be calculated using the following well-known transformation:

$$\frac{1}{1 - x} \approx 1 + x, \text{ at } x \ll 1 \qquad (3.6.38)$$

that, finally, results in the expression

$$\left(\frac{\partial p\theta}{\partial pO}\right)_T = 1 + \frac{m^0_{Me^{2+}} \times 10^{-pO}}{(K_{MeO} + m^0_{Me^{2+}} + 10^{-pO})(K_{MeO} + 10^{-pO})}.$$

$$(3.6.39)$$

In particular, it follows from equation (3.6.39) that, in the case of a strongly dissociated and moderately soluble oxide in the section of unsaturated solution, the slope of the dependence E–$p\theta$ is approximately equal to that of the E–pO calibration plot. If the degree of dissociation is appreciably lower than 1 (unity), then the slope of the E–$p\theta$ plot (or the line tangent to this dependence) should be slightly higher than the slope of the E–pO plot. Both cases are realized in the titration of Ba^{2+} and Sr^{2+} in the CsBr–KBr (2:1) melt at 700 °C (see Fig. 3.6.5). The slopes of the corresponding E–$p\theta$ plots are seen from Fig. 3.6.5 to be practically coincident with $2.3RT/F$, which is the slope of the E–pO calibration plot at low pO values.

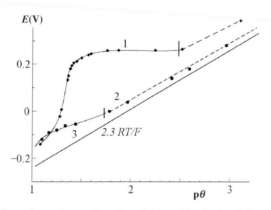

Fig. 3.6.5. E–$p\theta$ dependence for moderately soluble oxides in the CsBr–KBr (2:1) melt at 700 °C: (1) BaO, (2) SrO, (3) the E–pO calibration dependence. The unsaturated solution sections are denoted by dashed line.

(b) *Saturated solution*

We now consider the behaviour of the E–$p\theta$ dependence in saturated solutions of moderately soluble oxides. Similar to what has been said in Section (b), there are no assumptions which allow the neglect of any parameters or to simplify the calculations.

The solubility product of the oxide can be expressed using the values determined experimentally using equation (3.6.27) and then calculating the formula for $m_{O^{2-}}^0$ from this equation gives

$$m_{O^{2-}}^0 = m_{Me^{2+}}^0 - \frac{P_{MeO}}{10^{-pO}} + 10^{-pO}. \qquad (3.6.40)$$

Some transformation allows us to obtain the equation for $p\theta$ as

$$p\theta = -\log\left(m_{Me^{2+}}^0 - \frac{P_{MeO}}{10^{-pO}} + 10^{-pO}\right). \qquad (3.6.41)$$

The derivative

$$\left(\frac{\partial p\theta}{\partial pO}\right)_T$$

is equal to

$$\left(\frac{\partial p\theta}{\partial pO}\right)_T = \frac{10^{-pO} + \dfrac{P_{MeO}}{10^{-pO}}}{m^0_{Me^{2+}} - \dfrac{P_{MeO}}{10^{-pO}} + 10^{-pO}}.$$
(3.6.42)

The expression for the derivative

$$\left(\frac{\partial E}{\partial p\theta}\right)_T$$

is as follows:

$$\left(\frac{\partial E}{\partial p\theta}\right)_T = \left(\frac{\partial E}{\partial pO}\right)_T \frac{m^0_{Me^{2+}} - \dfrac{P_{MeO}}{10^{-pO}} + 10^{-pO}}{10^{-pO} + \dfrac{P_{MeO}}{10^{-pO}}}.$$
(3.6.43)

We shall now analyse some consequences of the obtained formula, which describes the behaviour of the E–$p\theta$ dependence in different sections of the potentiometric titration curve. The first consequence concerns the onset of the metal-oxide precipitation: the condition

$$\frac{P_{MeO}}{10^{-pO}} = m^0_{Me^{2+}}$$

corresponds to this situation

$$\left(\frac{\partial E}{\partial p\theta}\right)_T = \left(\frac{\partial E}{\partial pO}\right)_T \frac{10^{-pO}}{m^0_{Me^{2+}} + 10^{-pO}} < \left(\frac{\partial E}{\partial pO}\right)_T,$$
(3.6.44)

which means that at the beginning of metal-oxide precipitation the dependence E–$p\theta$ should possess a slope value which is less than that of the E–pO calibration plot in the corresponding e.m.f. region. It is obvious that the lower the solubility product of the metal oxide, the lower is 10^{-pO}, and hence, the lower is the slope of the dependence E–$p\theta$ at the beginning of the precipitation. Therefore, the bend is less pronounced for the oxides possessing higher values of solubility product.

The second consequence is bound up with the precipitation halfway, which is described by the condition

$$2\frac{P_{MeO}}{10^{-pO}} = m^0_{Me^{2+}}.$$

We can obtain the derivative of the e.m.f. with respect to $p\theta$, as

$$\left(\frac{\partial E}{\partial p\theta}\right)_T = \left(\frac{\partial E}{\partial pO}\right)_T \frac{\dfrac{P_{MeO}}{10^{-pO}} + 10^{-pO}}{10^{-pO} + \dfrac{P_{MeO}}{10^{-pO}}} = \left(\frac{\partial E}{\partial pO}\right)_T. \qquad (3.6.45)$$

In this case, the dependence E–$p\theta$ should possess a slope of the tangent line equal to that of the E–pO calibration dependence. This means that, in this case, the tangent line must be parallel to the calibration dependence.

The third consequence is related to the termination of the oxide's precipitation, which is described by the condition

$$\frac{P_{MeO}}{10^{-pO}} = 0.1m^0_{Me^{2+}}$$

the slope of the tangent line to the E–$p\theta$ dependence is expressed as

$$\left(\frac{\partial E}{\partial p\theta}\right)_T = \left(\frac{\partial E}{\partial pO}\right)_T \frac{0.9m^0_{Me^{2+}} + 10^{-pO}}{0.1m^0_{Me^{2+}} + 10^{-pO}} > \left(\frac{\partial E}{\partial pO}\right)_T. \qquad (3.6.46)$$

Equations (3.6.44)–(3.6.46) demonstrate that in the process of oxide precipitation from the solution the slope of the line tangent to the E–$p\theta$ dependence gradually increases and finally exceeds the slope of the E–pO calibration-dependence in the corresponding e.m.f. range.

The experimentally obtained E–$p\theta$ dependences for moderately soluble oxides are characterized by a considerable decrease in the e.m.f. in the unsaturated solution section. The formation of the saturated solution is accompanied by a reduction in the slope of the tangent, since under these conditions the weights of the titrant are consumed, mainly, for precipitation of the solid phase. Naturally, this process does not increase the oxide ion concentration in the melt. After complete precipitation of the oxide the slope of the tangent to the E–$p\theta$ dependence increases, and the dependence itself asymptotically approaches the E–pO calibration line, as can be seen from Fig. 3.6.5. This is because, at high initial concentrations of the titrant, the following approximation is correct:

$$-\log(m^0_{O^{2-}} - m^0_{Me^{2+}}) = -\log m^0_{O^{2-}} - \log\left(1 - \frac{m^0_{Me^{2+}}}{m^0_{O^{2-}}}\right)$$

$$\approx -\log m^0_{O^{2-}}.$$ (3.6.47)

The above-mentioned features give rise to the appearance of the loops in the dependences of E–$p\theta$, bent toward the ordinate axis. The size and coordinates of the start of the loop are dependent on the oxide's solubility: its increase causes a diminution of the loop and shifts the loop's start to lower pO values. So, Fig. 3.6.5 shows that BaO is more soluble in the CsBr–KBr melt at 700 °C than is SrO. Besides, the pO values at the half-way precipitation allow us to estimate (to a sufficient accuracy) the solubility product of the precipitated oxide. Indeed, it follows from equation (3.6.45) that

$$pP_{MeO} = pO - \log\frac{m^0_{Me^{2+}}}{2}.$$ (3.6.48)

Transformation of this equation, taking into account that $m^0_{Me^{2+}} = 0.05$ (the usual initial molality of the studied oxide), gives

$$pP_{MeO} = pO + 1.6,$$ (3.6.49)

which for the case of BaO and SrO solutions in the CsBr–KBr melt yields the solubility product values as $pP_{BaO} = 3.52$ (3.67 ± 0.3) and $pP_{SrO} = 5.25$ (5.33 ± 0.1). It is seen that the estimates are in excellent agreement with the average values obtained with the use of the standard routine using the potentiometric results (these values are presented in parentheses). It should be noted that the described method is used as a criterion for the choice of experimental points for the calculation of concentration values of pP_{MeO}.

The conclusions made are of general character and can be used in some other cases. So, the calibration E–pO dependence in the $CaCl_2$–KCl eutectic melt obtained with weights of KOH (Fig. 1.3.2) includes the said loop, which points to the onset of CaO precipitation from the saturated solution. For that, the estimation of the solubility product of calcium oxide performed according to equation (3.6.48) gives pO = 0.6, $m^0_{Ca^{2+}} = 2.83$ and $pP_{CaO} = 0.75(0.72 \pm 0.2)$.

(iii) Correction of solubility products using two indicator electrodes

All the routines described for the determination of the thermodynamic (concentration) parameters in metal oxide solutions include some indirectly obtained values. For example, the equilibrium concentration of metal cations is calculated proceeding from the quantity of the oxide-ion donor consumed for titration (precipitation). Direct determination of the concentration of metal cations in the melt (if it is possible) allows one to obtain more correctly the obtained solubility product values. Our paper [332] reports a method for correction of the solubility product values for oxides on the basis of the potentiometric titration data. The modification of the standard routine consists of the simultaneous use of two indicator electrodes, one of which is the membrane oxygen electrode and the other is a metal electrode, reversible to the cations the oxide consists of. This routine was used to estimate the solubility products of copper(I) and nickel(II) oxides in the molten KCl–NaCl equimolar mixture at 700 °C. Investigation of Cu_2O by the proposed method is of considerable importance since, as will be shown further, the process of dissociation/dissolution of copper(I) oxide in molten alkali-metal halides differs from the generally accepted one which was considered, e.g. in Ref. [119].

The scheme of the cell for carrying out the potentiometric study by the proposed method can be presented as follows (the cell structure provides simultaneous independent measurements of e.m.f. of cells $1-2$ and $1-3$ at two pairs of potentiometer contacts):

(1) Reference half-element: $KCl-NaCl, Me^{2+}, O^{2-}$

(2) Metal electrode: $Me^{2+}|Me$ (3.6.50)

(3) Oxygen electrode: $YSZ|Pt(O_2)$

The silver electrode, which is a silver wire with a platinum top immersed in a 0.2 mol kg^{-1} solution of AgCl in the molten KCl–NaCl equimolar mixture is used as reference electrode [332]. The metal electrode is made as a plate of the metal $(2-3 \times 1$ cm), whose oxide solubility is determined with the platinum top immersed in the solution studied.

The electrochemical cells $1-2$ and $1-3$ are preliminarily calibrated with known quantities of Me^{2+} and O^{2-}, respectively. The calibration of the metal electrode is performed in the chloride melt, thoroughly purified from the oxide-ion traces by the addition of ammonium chloride. This excludes distortions, which may be caused by the precipitation of the corresponding oxides from the solutions before the titration. The results of such calibrations are collected in Table 3.6.2.

The potentiometric titration routine for the case considered was quite similar to the standard titration routine. However, in the process of the titration, two e.m.f. values—namely, E_{1-2} and E_{1-3}—are registered simultaneously. After obtaining the equilibrium e.m.f. values, the equilibrium molalities of the metal cation and oxide ion were calculated on the basis of the calibration data (see Table 3.6.2):

$$pMe = \frac{E_{0,1-3} - E_{1-3}}{k_{1-3}}, \tag{3.6.51}$$

$$pO = \frac{E_{0,1-2} - E_{1-2}}{k_{1-2}}, \tag{3.6.52}$$

$$pP_{MeO} = \frac{E_{0,1-2} - E_{1-2}}{k_{1-2}} + \frac{E_{0,1-3} - E_{1-3}}{k_{1-3}}. \tag{3.6.53}$$

Since the equilibrium concentrations of the reagents are determined independently from each other, additional errors bound up with the use of the indirect concentrations of the reagents do not arise. The results of the treatment of the obtained potentiometric data are presented in Tables 3.6.3 and 3.6.4. To show the advantages of the proposed method as compared with the standard titration routine, the values of pP_{MeO} obtained according to the conventional method are collected in the same tables, the values calculated using the proposed method being denoted by an asterisk.

Titration of Ni^{2+} was performed with quantities of BaO, containing the carbonate and hydroxide admixtures, since the use of NaOH (KOH) for the titrimetric determination allows us to calculate the value of the solubility product of NiO to the required degree of accuracy.

TABLE 3.6.2

Coefficients of the $E = E_0 + kpm$ calibration dependences obtained against silver reference electrode ($Ag|Ag^+$, 0.2 mol kg^{-1}) in the molten KCl–NaCl equimolar mixture at 700 °C

Electrode	pm, parameter	E_0 (V)	k (V)
Pt(O$_2$)|YSZ	pO < 3	-0.382 ± 0.030	0.211 ± 0.01
	pO > 3	-0.069 ± 0.001	0.113 ± 0.001
Ni|Ni^{2+}	pNi^{2+}	0.104 ± 0.001	-0.095 ± 0.001
Cu|Cu$^+$	pCu$^+$	-0.060 ± 0.008	-0.184 ± 0.004
Cu|Cu$_2^{2+}$	pCu$_2^{2+}$	-0.117 ± 0.008	-0.184 ± 0.004

TABLE 3.6.3
Experimental and derived results of determination of Cu_2O solubility in the molten $KCl-NaCl$ equimolar mixture at 700 °C

$m_{O^{2-}}^0$	E_{1-2} (V)	pO	$-E_{1-3}$ (V)	pCu_2	pP^*	pP
0.0165	0.201	2.76	0.296	1.28	4.04	4.10
0.0317	0.201	2.76	0.298	1.29	4.05	4.08
0.0620	0.196	2.74	0.303	1.32	4.06	4.15
0.0742	0.190	2.71	0.310	1.36	4.07	4.17
0.1040	0.174	2.64	0.327	1.45	4.09	4.22
0.1300	0.155	2.55	0.347	1.55	4.10	4.30
0.1570	0.120	2.38	0.382	1.75	4.13	4.33
0.2050	0.034	1.97	0.466	2.20	4.17	4.53

$pP^* = 4.09 \pm 0.10$, $pP = 4.22 \pm 0.40$.

The comparison of the results obtained according to both the routines described above shows that the method with the use of two indicator electrodes allows us to narrow the confidence range of the determined parameter (0.1 against 0.4 for Cu_2O and 0.12 against 0.25 for NiO) by minimizing the number of indirectly obtained data in the solubility product calculations. However, this method can only be applied for the determination of oxide solubilities with such metal electrodes and metals, which are not prone to the formation of compounds (oxides or ions) in intermediate

TABLE 3.6.4
Experimental and derived results of determination of NiO solubility in the molten $KCl-NaCl$ equimolar mixture at 700 °C

$m_{O^{2-}}^0$	E_{1-2} (V)	pO	$-E_{1-3}$ (V)	pNi	pP^*	pP
0.0039	0.757	7.51	0.025	1.36	8.87	8.85
0.0145	0.746	7.41	0.037	1.48	8.89	8.86
0.0251	0.730	7.26	0.055	1.67	8.93	8.86
0.0337	0.709	7.07	0.077	1.91	8.98	8.86
0.0355	0.702	7.01	0.084	1.99	9.00	8.85
0.0375	0.700	6.99	0.087	2.01	9.00	8.89
0.0443	0.634	6.39	0.128	2.45	8.84	8.63
0.0450	0.625	6.31	0.138	2.56	8.87	8.61
0.0469	0.618	6.25	0.145	2.63	8.88	8.76
0.0486	0.590	5.99	0.177	2.97	8.96	8.84
0.0499	0.500	5.17	0.292	4.19	9.36	9.17

$pP^* = 8.92 \pm 0.12$, $pP = 8.80 \pm 0.25$.

oxidation states. For example, alkali-metal oxides cannot be studied by such methods owing to the relatively low melting points of the metals that provide a high solubility of free metal in the ionic melt.

The data of Table 3.6.4 lead to the conclusion that the use of BaO instead of alkali-metal hydroxides for the titration results in an overestimation of the value of the solubility product. This seems to be caused by Ba^{2+} cations possessing enhanced acidic properties which arise in the melt, thus leading to the fixation of some quantity of oxide ions from the melt and, consequently, to the increase in the solubility of the studied metal oxide. A similar conclusion may be drawn if we attribute the composition $BaCl_2$–KCl–NaCl to the melt studied. As seen in Part 3, such a melt possesses enhanced acidity that causes an increase in metal-oxide solubilities. We can estimate the oxobasicity index of the melt KCl–NaCl $+ 0.05$ mol kg^{-1} $BaCl_2$ as $+0.23$, although this value is comparable with the accuracy of the solubility calculations.

Another practically useful result of the application of the method described for the solubility studies is the confirmation of the stoichiometry of the dissociation process of Cu_2O, at least in chloride melts. This process results in the formation of one Cu_2^{2+} ion instead of two Cu^+ ions:

$$Cu_2O \downarrow \rightleftharpoons Cu_2^{2+} + O^{2-}. \tag{3.6.54}$$

The emergence of just this interaction is confirmed by the data of Table 3.6.3 since the pP' values (which are not presented in the table) calculated for the generally accepted reaction:

$$Cu_2O \downarrow \rightleftharpoons 2Cu^+ + O^{2-} \tag{3.6.55}$$

are subject to a considerable directed shift in the process of titration. In our opinion, these results (described in detail in Ref. [332]) undoubtedly point to the running of reaction (3.6.54) in the molten KCl–NaCl equimolar mixture. Consequently, the proposed method can be used to find or to confirm the stoichiometry of some Lux acid–base interactions in molten salts and, if necessary, to obtain more accurate parameters of the acid–base equilibria.

However, wide systematic investigations in molten salts require unified experimental techniques which provide a reduction in the data-spreads which arise from the systematic errors characteristic of different methods. This makes the obtained regularities better substantiated.

Therefore, the considerable advantages of the method with the use of two indicator electrodes can only be gained in some specific cases. Moreover, most oxides possess low-enough solubilities in halide melts and errors of the solubility product estimations are, as a rule, small enough. All the regularities which will be discussed in Part 7 were obtained using the simple method of solubility determination.

3.6.2.3 Sequential addition method

The sequential addition method (SAM) recently developed in our studies makes it possible to estimate the solubility of oxides having different values of specific (molar) surface and, probably, of surface energy at the "oxide—ionic melt" interface boundary [333].

The initial preconditions for the development of this method originated in the E–pO calibration dependences in some chloride melts, in particular, KCl–NaCl + BaO at 727 °C [239] obtained by Combes *et al.* and NaCl–ZnCl$_2$ + ZnO at 450–500 °C [173] obtained by Picard *et al.* In both the mentioned papers the E–pO* calibration plots are uniform: they end with "plateaus" arising because of the formation of saturated solutions of the oxide studied (Fig. 3.6.6). The beginning of the plateau is characterized by a bend toward the pO axis, which is not explained in the works mentioned. Maybe, it was considered to be caused by a natural scatter of the experimental data. Nevertheless, this scatter can hardly be called natural, since the reduction and subsequent growth of the e.m.f. are regular and well-reproducible. Of course, such a strange scatter of the calibration data has no relationship to the experimental errors. The melt's supersaturation cannot be considered among possible reasons of these deviations since the solutions undoubtedly contact the excess of solid oxide, either at the moment of addition, or when lying at the bottom of the experimental container-crucible. The origin of this fact should be found from considerations of another kind. In my earlier review [59], it is shown that the equilibrium solubility of metal oxides is appreciably dependent on the size of the solid particles of oxides or, at least, on the manner of the oxide's preparation. In principle, the dependences presented in Fig. 3.6.6 can be explained by partial dissolution of a weight of the oxide in the melt which immediately results in the formation of a saturated solution of the oxide. After such a partial dissolution, the average size (radius) of this weight becomes smaller and the powder consisting of such particles possesses higher solubility. If we create a routine which allows us to estimate quantitatively

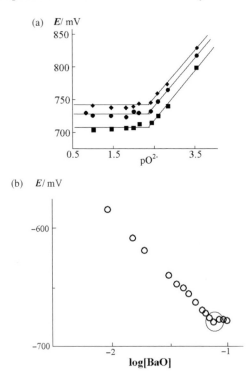

Fig. 3.6.6. E–pO^* calibration dependences: (a) KCl–NaCl + BaO at 727 °C by Combes *et al.* [239]; (b) NaCl–ZnCl$_2$ + ZnO, 450 (triangles), 470 (rings) и 500 °C (quadrates) by Picard and coworkers [173].

the changes of the particle size, we can find the values of surface energy at "solid oxide–ionic melt" interface boundary.

The essence of the method proposed consists of the following. A sequence of weights of the oxide studied is added to the melt, up to the formation of the saturated solution (similar to the conventional weight calibration of electrochemical cells with oxygen electrodes) and some weights are added after the saturation. The set of data obtained consists of two kinds of points: some of them lie on the inclined calibration-like section (unsaturated solution, Fig. 3.6.7, section 1), other points belong to the plateau section (saturated solution, Fig. 3.6.7, section 2). The inclined section obtained in the case of incompletely dissociated oxides lies above the conventional E–pO calibration dependence, although the entire weight is completely dissolved in the melt. This part of the experiment gives an opportunity to obtain certain information

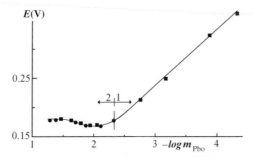

Fig. 3.6.7. SAM dependences for lead oxide in the molten KCl–NaCl equimolar mixture at 700 °C: (1) unsaturated solution section, (2) saturated solution section (results of different experiments are denoted by rings and quadrates, respectively).

on the behaviour of the oxide in its unsaturated solution and to calculate the dissociation constant of the oxide. Indeed, pO values can be obtained from the calibration data; after the pP_{MeO} calculation the dissociation constant is estimated according to the following equation:

$$K_{MeO} = \frac{10^{-2pO}}{m^0_{MeO} - 10^{-pO}} \qquad (3.6.56)$$

taking into account the fact that $m_{Me^{2+}} = m_{O^{2-}} = 10^{-pO}$.

The weight of the oxide which immediately causes the saturation of the melt (onset of a plateau) is not dissolved completely. Its partial dissolution leads to an appreciable reduction in the oxide particle sizes and, which is the same, the oxide powder formed possesses a considerably higher molar surface area than the initial metal oxide. After the partial oxide dissolution, the saturated solution is formed and this is in equilibrium conditions with the powder consisting of the reduced particles. Subsequent weights of the oxide added to the saturated solution are not dissolved and the same concerns the partially dissolved oxide. Addition of the initial oxide results in a reduction in the average molar surface area (increase in the average particle radius) and, consequently, the concentration of the saturated solution of the oxide decreases according to the predictions obtained using the Ostwald–Freunlich equation [334]:

$$\log s_2 - \log s_1 = \frac{2\sigma M}{2.3RTd}\left(\frac{1}{r_2} - \frac{1}{r_1}\right). \qquad (3.6.57)$$

The equation presented can be used for quantitative estimations only in the case of oxide powders consisting of particles of the same size and form. However, in commercial oxides the oxide particles are non-uniform, and therefore, the oxide solubility changes should be connected rather with the changes of the molar surface area of the powder than with their ionic radii. To verify this assumption let us consider the following equation, which gives the dependence of a solid substance's solubility in a liquid medium against its molar surface area:

$$\Delta G_x^f + \sigma S = \mu_x^0 + RT \ln m_x. \tag{3.6.58}$$

As follows from this equation, an increase in the molar surface area of the solid substance results in a rise in its solubility, according to equation (3.6.58). This dependence of solubility can be described by the following simplified equation:

$$\sigma S = \text{const} + 2.3\,RT \log m_x, \tag{3.6.59}$$

which has the form of a straight line in S vs $\log m_x$ coordinates. The slope of this line gives the possibility of estimating the magnitudes of the "effective" surface energy at the "solid oxide–ionic melt" interface boundary. For this purpose there should be developed a method for estimating the value of the molar surface area of the powder *in situ*, i.e. without separation of the precipitate from the solution in the ionic melt. Moreover, it is necessary to estimate the effect of the partial oxide dissolution (the dependence of the surface area of the partially dissolved powder against the fraction of the solid oxide remaining after the dissolution). And, finally, one should know how to calculate the average molar surface area of the mixture of the initial oxide with the partially dissolved one that is formed on the plateau section.

To calculate the surface energies at the "oxide–ionic melt" interface boundary by means of equation (3.6.57) it is necessary to know the radii of the oxide particles in contact with the saturated solution in the melt–solvent. In this connection there arises a question whether one can correctly estimate the values of the oxide particles' radii from the data on the molar surface area obtained using any experimental method of estimating the specific surface area, e.g. the BET treatment. In order to answer this question we first consider the relationship between the volume and surface area of a spherical particle.

So, the particle of radius r_0 occupies the following volume:

$$V' = \frac{4}{3}\pi r_0^3 \qquad (3.6.60)$$

and the corresponding surface area is equal to

$$S' = 4\pi r_0^2. \qquad (3.6.61)$$

It is obvious that if we know the particle's surface area and volume then we can calculate its radius according to the formula:

$$r_0 = \frac{3V'}{S'}. \qquad (3.6.62)$$

By taking n particles containing a quantity of the solid substance equivalent to 1 mol we shall obtain a similar relationship between the molar volume and the molar surface area of the solid, since in this case we should multiply both the numerator and the denominator of the ratios in equation (3.6.62) by n:

$$r_0 = \frac{3M}{dS}. \qquad (3.6.63)$$

To elucidate how the particle size obtained from equation (3.6.63) agrees with the actual particle sizes we determined the specific surface area of CaO, PbO, ZnO and CdO powders by the BET method and estimated the values of their molar surface areas (in $m^2\,mol^{-1}$) as 594 (CaO), 705 (PbO), 560 (ZnO) and 193 (CdO). On the basis of the obtained values we estimated the effective radii of the oxide particles (μm) according to formula (3.6.63) as 0.28 (CaO), 0.33 (PbO), 0.26 (ZnO) and 0.82 (CdO). Taking into account the fact that, as a rule, the average radius of oxide particles obtained commercially is of the order of 1 μm, it should be concluded that the investigated powders possess a developed surface and, in this case, the use of equation (3.6.59) seems to be more appropriate.

Now we consider how the treatment of the SAM data is performed. The experimental routine yields the dependence similar to that presented in Fig. 3.6.7. As explained above, the sloping section is for the unsaturated solution. Since the inclined section is located over the calibration's $E-pO$ one, the dissociation of PbO in the unsaturated solution is incomplete. We may calculate the dissociation constant of PbO, as follows. The pO value at each point is estimated using the calibration data. Then, taking into account the fact that the cation's molality is equal to that of oxide ion, the molality of the undissociated oxide is found as follows:

$$m_{MeO} = m^0_{MeO} - m_{O^{2-}}.$$ (3.6.64)

Finally, K_{MeO} is calculated as

$$K_{MeO} = \frac{(m_{O^{2-}})^2}{m^0_{MeO} - m_{O^{2-}}}.$$ (3.6.65)

It should be added that the $E - \log m_{MeO}$ dependence can be used for subsequent calculations, since the values of the e.m.f. in the saturated solution section permit the determination of the total concentration of the oxide dissolved in the melt and, consequently, the quantity of the precipitated oxide. The first point, which falls off the linear dependence (it is located above the approximating line), is the point of the partial oxide powder dissolution. The formed solution is saturated with respect to the oxide powder, which possesses a higher molar surface area as compared with the initial commercial powder. In this case, the quantity of oxide dissolved in the melt is calculated using the dependence $E - \log m_{MeO}$. The difference between the initial molality of the oxide and its equilibrium one yields the quantity of the precipitated oxide powder.

The magnitude of the molar surface area of the partially dissolved oxide can be estimated using the following considerations. If a solid particle of spherical or cubic shape undergoes the partial dissolution its proportions remain unchanged (i.e. the cube is transformed into a smaller cube and a sphere is transformed into a smaller sphere). Let the mass of the said particle be reduced by a factor of x. In such a case, the volume of this particle diminishes by a factor of x, as well and its linear dimensions are reduced by a factor of $x^{1/3}$ only. Since the surface area of the particle is proportional to the square of its dimensions (i.e. it is $6d^2$ for a cube of a size d, and πd^2, for a sphere of a diameter, d), a reduction in the linear dimensions by a factor of $x^{1/3}$ decreases the surface area ($S_0 \rightarrow S_1$) by $x^{2/3}$.

We shall now determine the change in the specific surface area of the particle. In order to bring the initial powder and the partially dissolved one to the same mass, we should take one particle of the initial powder and x partially dissolved particles. Then, the ratio of surface area of the initial powder to that of the partially dissolved one will be as follows:

$$\frac{S_0}{xS_1} = \frac{S_0}{xS_0x^{2/3}} = x^{-1/3}, \tag{3.6.66}$$

that is, the increase in the specific surface area at the partial dissolution is proportional to the cube root of x. All we have said remains correct when calculating the molar surface areas.

Subsequent addition of the solid phase to the saturated solution does not result in dissolution of the oxide and leads to a reduction in the saturated solution's concentration owing to decrease in the average molar surface area of the oxide. This change can easily be calculated if we know the quantities of both powders and their molar surface areas.

According to equation (3.6.59) the obtained data should be represented by a straight line in the $-\log s_{MeO} - f(S)$ coordinates. Such a dependence for the case of PbO is presented in Fig. 3.6.8. From equation (3.6.59) it follows that the surface energy at a "solid oxide–ionic melt" boundary can be calculated from the slope of the $-\log s_{MeO} - f(S)$ dependence according to the equation

$$\sigma = 2.3RT \frac{d \log m_{MeO}}{dS}. \tag{3.6.67}$$

The approximation of the dependence (3.6.59) by the least-squares method allows us to obtain an equation to be used for the calculation of the molar

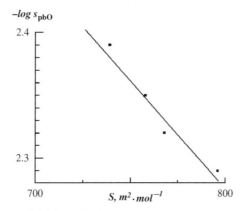

Fig. 3.6.8. Dependence of PbO solubility in the molten KCl–NaCl equimolar mixture at 700 °C on the magnitude of molar surface area (S) of the precipitate contacting with the saturated solution.

surface area of oxide which is in contact with the saturated solution of a known concentration.

Thus, in the present part we have considered the key features of the main methods used for the determination of oxide solubilities in molten salts. Such an analysis of their advantages and drawbacks will be very helpful for the consideration and understanding of the data on metal-oxide solubility in alkali-metal halide melts which will be discussed in Part 7.

Part 7. Regularities of oxide solubilities in melts based on alkali and alkaline-earth metal halides

Investigations of metal-oxide solubilities in the melts based on alkali-metal halides now represent one of the most developed branches of oxoacidity studies, which is of great importance for a variety of practical applications [59, 181]. These applications include processes for the production of various functional materials with the use of different halide melts as media for synthesis (melts−solvents) or as starting materials (this concerns the single-crystal growth).

For use of ionic melts as electrolytes in high-temperature chemical power sources it is necessary to know the metal-oxide's solubilities in these media, since the oxides may be formed as a result of reactions of electrochemically active substances with oxide ions which exist in the pure halide melts. This interaction is especially undesirable for the case of rechargeable high-temperature chemical power sources, since it results in irreversible removal of electrochemically active particles from the reaction sphere.

In view of the above-said it is necessary to investigate oxide solubilities in the melts based on alkali-metal halides and to establish regularities in "metal-oxide−ionic melt systems". Such data are required for predicting the stability of melts doped with additions of metal cations, and of electrochemical cells containing chlorides of transition metals, under the action of oxide ions formed in the melt owing to oxidation or pyrohydrolysis of its constituent parts.

As a rule, the investigations of oxide solubility are performed within the temperature range 500−800 °C. At these temperatures the choice of substances suitable for this purpose is essentially limited by their chemical and thermal stability. All the above-described methods of solubility determination imply the existence of the following equilibrium in the melt:

$$MeX_2 + R_2O = MeO + 2RX. \tag{3.7.1}$$

where "R" denotes an alkali metal.

Since oxides and halides of alkali metals possess sufficient stability under the experimental conditions, the possibility of determining the parameters of equilibria with the participation of metal-oxide is dependent on its stability and on the stability of the corresponding halide.

The substance is considered suitable for the investigations if its oxide and halide do not decompose at temperatures below 1100 K and both compounds exist in condensed phases (liquid or solid). The temperatures of the phase transitions of some of mentioned substances are collected in Table 3.7.1. It should be emphasized that some substances are difficult to obtain and their stability is extremely low ($FeCl_2$, $CrCl_2$, VCl_2). Therefore, there are no literature data on the behaviour of the corresponding oxides in ionic melts. The data presented in Table 3.7.1 allow one to choose the range of oxides which are among the most available for investigations. They are $MgO-CaO-SrO-BaO$, $ZnO-CdO$, $SnO-PbO$, $MnO-CoO-NiO-ZnO$ and Cu_2O. Examination of oxides from the three former sequences allows us to find

TABLE 3.7.1

Temperatures (°C) of melting, boiling and decomposition of some metal oxides and halides and crystallographic radii of the corresponding cations from data of Refs. [114, 121]

$Me^{(n)}$	Oxide		Chloride		Bromide		Iodide		$r(Me^{(n)})$ (nm)
	t_{mp}	t_{bp}	t_{mp}	t_{bp}	t_{mp}	t_{bp}	t_{mp}	t_{bp}	
Mg^{II}	2800	–	714	1418	711	1250	633	1014	0.074
Ca^{II}	2600	–	782	1627	730	1830	783	860	0.104
Sr^{II}	2415	–	875	2027	657	1970	538	1900	0.120
Ba^{II}	1917	–	962	1189	857	1980	711	1900	0.138
Mn^{II}	1780	–	650	1190	698	–	638	–	0.091
V^{II}	2077	3127	1000	1387	–	–	800^s	–	0.072
Cr^{II}	–	–	815	1330	842	–	867^d	–	0.083
Fe^{II}	1369	–	677	1026	688	968	594	935	0.080
Co^{II}	1805	–	724	1050	678	927	515	570^d	0.078
Ni^{II}	1950	–	1001	987^s	963	919^d	797	–	0.074
Cu^{II}	800^d	–	596	$-^d$	498^d	–	–	–	0.080
Cu^{I}	1229	$-^d$	430	1367	489	1355	600	1320	0.098
Zn^{II}	1975	–	283	732	394	670	446	624^d	0.083
Cd^{II}	1500	–	568	980	568	865	388	708	0.099
Sn^{II}	1042	1527	247	623	232	641	320	718	0.102
Pb^{II}	886	1472	498	954	370	893	412	872	0.126
Pd^{II}	$^d877$	–	678	$^d581$	717^d	–	350^d	–	–
Pt^{II}	–	–	$^d581$	–	180^d	–	300^d	–	–

The designation "d" before the temperature means that decomposition of the substance begins up to achieving this temperature, the designation "d" after the temperature means the phase transition with the decomposition, designation "s" denotes sublimation of the substance.

the regularities of oxide solubilities for metals with different atomic numbers, which belong to the same subgroup of the Periodic Table. The fourth sequence allows us to trace the same effect in the same period. Unfortunately, the necessary sets of experimental data on the solubility in the given melt are often absent. This requires us to use the correlations found for estimating the oxide solubility. We now proceed to a consideration of the existing literature data.

3.7.1 MOLTEN ALKALI-METAL HALIDES AND THEIR MIXTURES

3.7.1.1 KCl–LiCl (0.41:0.59) eutectic mixture

This melt was studied extensively as a solvent for oxide materials at different temperatures. Picard, Seon and Tremillon investigated the solubility of iron(II) and iron(III) in the KCl–LiCl eutectic melt at 470 °C [335, 336]. Potentiometric and X-ray examinations show that precipitation of iron cations with carbonate ion result in the formation of FeO precipitate, whereas Fe^{3+} cations are deposited as a solid solution of $LiFeO_2$ ($Li_yFe_{1-y}O$) composition. The solubility products of FeO, Fe_3O_4 and Fe_2O_3 have the following values: $10^{-5.4}$, $10^{-36.3}$ and $10^{-29.16}$, respectively. The authors of Refs. [335, 336] showed that Fe^{3+} cations oxidized the chloride melt with the evolution of chlorine. An obvious disadvantage of the said investigations is the use of a weak base (Na_2CO_3) as a titrant, which undoubtedly affects the values calculated for the solubility products. Besides, it is not clear how the deposit of Fe_2O_3 was obtained, if the corresponding cation oxidized chloride ions in the melt transforming into the Fe^{2+} ion.

Delarue reported the qualitative study of metal-oxide solubility in the molten KCl–LiCl eutectic at a temperature "near 500 °C" [319, 320]. This examination used visual solubility control. To obtain the metal-oxide, a small quantity of KOH was added to a solution of the studied metal chloride in the molten KCl–LiCl. If such an addition did not result in the precipitation of the oxide, then the oxide was referred to as a soluble one:

$$MeCl_2 + 2KOH \rightarrow MeO \downarrow + 2KCl + H_2O \uparrow \qquad (3.7.2)$$

examples were CdO, PbO, BaO, SrO, CaO, MnO, Ag_2O, Sb_2O_3, Bi_2O_3. Concurrent polarographic measurements showed the oxoacidic properties of alkaline-earth cations in Li^+-containing chloride melts to be levelled to those of Li^+. No distinctions were observed between the polarograms of CaO and $CaCl_2$ solutions in the molten KCl–LiCl eutectic. If the KOH addition did cause the oxide's precipitation, it was classified among the slightly soluble (e.g. CoO, NiO, ZnO) or insoluble (e.g. MgO, BeO, Al_2O_3) oxides in the KCl–LiCl eutectic. ZnO was found to possess appreciable acidic properties since it dissolved easily in the excess of the Lux base. Copper(II) oxide was shown to be extremely unstable in the chloride melt; its dissolution was accompanied with chlorine evolution and with the simultaneous reduction in Cu^{2+} to Cu^+ ions. Similar behaviour was characteristic of other chlorides formed by metals in their highest degrees of oxidation, in particular, $TlCl_3$ (Tl^{III}), $FeCl_3$ (Fe^{III}) and $AuCl_3$ (Au^{III}). Addition of these chlorides to the molten KCl–LiCl eutectic containing deposits of other oxides resulted in dissolution of the latter owing to interactions similar to the reaction presented here:

$$2CuCl_2 + ZnO \downarrow \rightarrow ZnCl_2 + 2CuCl + \tfrac{1}{2}O_2 \uparrow . \qquad (3.7.3)$$

Delarue also noted that there was a considerable difference between the solubilities of the oxide added to the melt as a powder and the one obtained *in situ*, i.e. by addition of KOH to the solution of the corresponding chloride in the studied melt. The solubilities of CaO and CdO formed as a result of the exchange reaction (3.7.2) in the studied melt were considerably higher: the added oxides were incompletely dissolved in the melt, whereas the addition of KOH to the corresponding chloride solution did not cause oxide precipitation [319, 320]. It is of interest to add that, apart from the behaviour of oxides in the molten KCl–LiCl eutectic, Delarue reported the study of the solubility of sulfides, sulfites and sulfates in the said melt [337]. The interactions of metal cations with sulfide ions were performed by addition of $Na_2S \cdot 9H_2O$ to the melt containing the studied cation (metal chloride). The running of the following reaction was as follows:

$$MeCl_2 + Na_2S \rightarrow MeS \downarrow +2NaCl. \qquad (3.7.4)$$

Sulfides of alkali- and alkaline-earth metals including magnesium, manganese(II) and thallium(III) were found to be soluble in the melt and their solutions were colourless. On the contrary, the addition of Na_2S to solutions of other cations caused precipitation of solid sulfides which were black, apart from ZnS (white), CdS (orange) and Ce_2S_3 (grey). It should be noted that the interactions of Cu^{2+}, Fe^{3+} are accompanied by the formation of metal sulfides in lower degrees of oxidation (Cu_2S and FeS); Au^+ and Hg^{2+} cations are reduced to free metals, and the simultaneously formed sulfur is dissolved in the KCl–LiCl eutectic melt. The latter was coloured blue and the formed sulfur monochloride (S_2Cl_2) was evaporated from the melt. Sulfides of iron and copper were shown to be destroyed by the addition of strong complex-forming anions such as cyanides. For example, for the case of iron:

$$FeS + 6CN^- \rightarrow [Fe(CN)_6]^{4-} + S^{2-}.$$
(3.7.5)

The corresponding oxides did not undergo such a destruction.

The addition of sodium sulfite to solutions of metal chloride in the molten KCl–LiCl eutectic causes chemical interactions whose directions are dependent on the cation acidity and the solubility of the corresponding oxides and sulfides. So, interaction with acidic cations proceeds according to the following equation:

$$Me^{2+}SO_3^{2-} \rightarrow MeO \downarrow + SO_2 \uparrow.$$
(3.7.6)

If the metal sulfide is insoluble in the KCl–LiCl melt then disproportionation of sulfite ions is observed according to the equation:

$$4SO_3^{2-} \rightarrow S^{2-} + 3SO_4^{2-}$$
(3.7.7)

and the sulfide ions formed precipitate the metal cations from the melt. It is interesting that the presence of cations of alkali-, alkaline-earth metals, magnesium, manganese and thallium does not result in visible destruction of sulfite ions, even if the process of decomposition according to reaction (3.7.7), would also be insignificant.

As for the behaviour of sulfate compounds in the molten KCl–LiCl eutectic, it should be noted that practically all sulfates are soluble in the melt and the addition of Na_2SO_4 does not cause any visible changes. Only two cations possessing the strongest acidic properties, namely, Al^{3+} and Ti^{4+} destroy sulfate ions with subsequent precipitation of the corresponding oxides, e.g.

$$Al_2(SO_4)_3 \rightarrow Al_2O_3 \downarrow + 3SO_3 \uparrow.$$
(3.7.8)

Shapoval *et al.* investigated saturated solutions of some oxides in the above-mentioned melt at 450 °C [176, 177, 338] by a polarographic method to estimate the dissociation constants of the oxides. The interaction of the cations with oxide ions in the melt was detected by changes in the potentials, a shift of the polarogram, and change in its shape as compared with the theoretical dependences. The stage of the acid–base dissociation of oxides was shown by polarographic studies to be slower than the reduction in the cation at the electrode. Therefore, the dissociation process was the limiting stage, which substantiated the estimations performed of the equilibrium constants. The stability constants of the oxides were as follows (mol%): CoO, 99($\pm$ 19); NiO, 6.8($\pm$0.8) $\times$ 10^3; PbO, 3.4($\pm$ 1.3) $\times$ 10^3; Bi$_2$O$_3$, 8.7($\pm$3.5) $\times$ 10^3. Lead oxide possesses acidic properties and the stability constant of PbO$_2^{2-}$ was equal to 2($\pm$1.6) $\times$ 10^2. CdO, Cu$_2$O and Ag$_2$O are unstable and decompose completely under the experimental conditions. These results agree with the data of Delarue [319, 320]. The solubility of Cu$_2$O in the chloride melt was found in Ref. [176] to increase according to addition of acids (PO$_3^-$) and decrease in the melts containing strong Lux bases.

Laitinen and Bhatia studied the solubilities of some metal-oxides in the molten KCl–LiCl eutectic at 450 °C to determine the possibility of the use of the corresponding metal-oxide electrodes as reversible to oxide ions [230]. The solubility values obtained by measurements at 450 °C are the following (mol kg^{-1}): NiO, 3.3 $\times$ 10^{-4}; BiOCl, 6.8 $\times$ 10^{-4}; PdO, 9.4 $\times$ 10^{-3}; PtO, 3.32 $\times$ 10^{-2}; Cu$_2$O, 3.8 $\times$ 10^{-2}. These results allow one to confirm by the thermodynamic data some conclusions made in Part 3 concerning the upper limit of reversible work for the Pt(O$_2$) gas–oxygen electrode. As follows from the obtained data, PtO possesses a higher solubility in molten salts in comparison with NiO, and therefore, the oxide film on the surface of platinum should be destroyed in strongly acidic solutions of an acidity close to that of Ni^{2+}/NiO solutions containing considerable quantities of nickel cations.

Duyckaerts, Landresse and Lysy reported investigations of the solubility of various uranium and neptunium oxides in the molten KCl–LiCl eutectic at 660 °C [231, 252]. The solubility product of NpO$_2$ in the chloride melt was estimated to be pP_{NpO_2} = 20.3 $\pm$ 0.5 [252].

The former metal was found to exist in the melt in the following oxidation states of 0, +3, +4, +5 and +6. As for the behaviour of uranium oxides in molten salts, Vorobey et al. investigated the chlorination of uranium oxides (UO$_2$, U$_3$O$_8$ and UO$_3$) in molten individual alkali-metal chlorides, and the eutectic KCl–LiCl and KCl–NaCl mixtures at 600–800 °C [339]. In the case of UO$_2$, the chlorination process is none other than addition of chlorine to the uranium oxide and it proceeds according to the following equation:

$$UO_2 \downarrow + Cl_2 \uparrow + 2Cl^- \rightarrow UO_2Cl_4^{2-}. \tag{3.7.9}$$

Complete transformation of the oxide into the complex anion was detected by the disappearance of the solid phase. However, one of the products of reaction (3.7.9), namely, UO_2Cl_2, is subject to partial decomposition under the experimental conditions:

$$UO_2^{2+} + Cl^- \rightarrow UO_2^+ + \tfrac{1}{2}Cl_2 \uparrow \tag{3.7.10}$$

and the values of the corresponding equilibrium constants are $(1.5 \pm 0.5) \times 10^{-2}$ for LiCl at $700\,°C$, $(1.8 \pm 0.6) \times 10^{-4}$ for NaCl at $800\,°C$, $(2 \pm 0.6) \times 10^{-6}$ for the KCl–NaCl melt at $700\,°C$ and $(2.7 \pm 0.8) \times 10^{-5}$ for the KCl–LiCl eutectic at the same temperature. These results show that the acidity of the melts, expressed by their ability to fix chloride ions, considerably affects the stability of the complexes of U^{VI}: an increase in the melt-acidity makes the uranium complex more prone to transformation into the lower oxidation degrees of U. The duration of the chlorination process decreases with the decrease in the melt acidity, as well.

The chlorination of U_3O_8 and UO_3 is a process of substitution of O^{2-} by Cl^-, by which oxide ion is oxidized to oxygen which, in turn, oxidizes the whole of the uranium to the upper oxidation state (U^{VI}):

$$U_3O_8 \downarrow + 4Cl_2 \uparrow + 4Cl^- \rightarrow 3UO_2Cl_4^{2-} + O_2 \uparrow. \tag{3.7.11}$$

This reaction runs through the stage of intermediate formation of uranates such as UO_4^{2-} or $U_2O_7^{2-}$, which are subject to the following substitution of oxide ions by chloride ions. It should be noted that this reaction is reversible, i.e. under the action of traces of oxygen, $UO_2Cl_4^{2-}$ ions are transformed into a mixture of non-stoichiometric oxides of uranium.

Potentiometric studies of the solubility of some metal-oxides (MgO, CoO and NiO) at $700\,°C$ and of magnesium oxide at 600, 700 and $800\,°C$, are reported in our works [162, 189, 190, 315]. In these works, the solubility product indices of the oxides are determined to be lower than the corresponding parameters in the molten KCl–NaCl and CsCl–KCl–NaCl mixtures at the same temperatures, by approximately 3.5. The obtained potentiometric curves are shown in Fig. 3.7.1 (titration at $700\,°C$) and

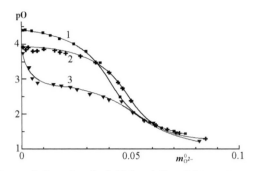

Fig. 3.7.1. Dependence of pO against the initial molality of titrant at potentiometric titration of Mg^{2+} (1, $m = 0.038$), Ni^{2+} (2, $m = 0.040$), Co^{2+} (3, $m = 0.045$) with NaOH weights in the molten KCl–LiCl eutectic at 700 °C.

Fig. 3.7.2 (polythermal investigations of MgO solubility). The solubility parameters at 700 °C are collected in Table 3.7.3.

As is seen from Fig. 3.7.1 the pO values at the excess of the studied cation are sufficiently low, and owing to the enhanced acidic properties of the KCl–LiCl melt, the increase in the melt acidity results in a considerable increase in the oxide solubility, so that CoO, which is practically insoluble in the molten KCl–NaCl equimolar mixture, becomes appreciably soluble with a sharp pronounced section of the unsaturated solution (see Fig. 3.7.1, curve 3). The existence of the said section allows us to calculate the dissociation constant of CoO in the molten KCl–LiCl eutectic at 700 °C using the potentiometric data for three initial points of the calibration curve (the corresponding treatment results are collected in Table 3.7.2), and its average value is presented in Table 3.7.3. The fourth point of the titration curve is the boundary one between the saturated and unsaturated solution, and therefore, it is available for calculations of the values of both the dissociation constant and the solubility product.

Comparison of the dissociation constant value with that obtained in the same melt at 450 °C by Shapoval [176] shows that the stability constant of oxides in molten salts decreases the rise in temperature. The pK_{CoO} value obtained in Ref. [176] is $pK_{CoO} = 2.7$ in molality scale, the agreement of this value with the one presented in Table 3.7.3 is good enough, taking into account the fact that this value was obtained by polarographic measurements.

As is shown in Part 3 (Fig. 1.3.3) the acidic properties of cadmium and lead cations in the molten KCl–LiCl eutectic are completely levelled to those of

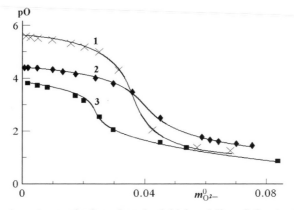

Fig. 3.7.2. The dependence of pO against the initial molality of titrant at potentiometric titration of Mg^{2+} ions with NaOH in the molten KCl–LiCl eutectic at 600 (1), 700 (2) and 800 °C (3) in the molten KCl–LiCl eutectic at 700 °C.

Li^+ cation, which possesses an appreciably stronger acidity. Owing to this fact, the degree of dissociation of CdO and PbO in the said melt is close to 1, at least, it concerns unsaturated solutions of the oxides.

Returning to the polythermal studies of the solubility of magnesium oxide in the molten KCl–LiCl eutectic mixture, we note that it has been found to increase with the temperature [316, 340]. This is easily seen from the potentiometric titration (PT) curves obtained at different temperatures (see Fig. 3.7.2): the calculations yield the following solubility products: $pP_{MgO} = 6.99 \pm 0.08$ (at 600 °C) → 5.87 ± 0.05 (at 700 °C) → 5.41 ± 0.04 (at 800 °C) [316]. The increase in the solubility product (pP_{MgO}) is irregular, i.e. the elevation of the melt temperature from 600 to 700 °C results in the increase in pP_{MgO} value by approximately 1, whereas in the 700–800 °C temperature range this change is half as much. Nevertheless, such an increase in the temperature does not result in the appearance of the unsaturated solution section in the potentiometric titration curve. This means that MgO remains practically insoluble in the molten KCl–LiCl eutectic at all the temperatures, and even the enhanced acidity of the molten background does not provide a sufficient solubility rise.

Generally speaking, the molten KCl–LiCl eutectic (owing to its enhanced oxoacidic properties) can dissolve sufficiently larger quantities of various oxide materials as compared with chloride melts containing the constituent cations of lower acidity such as Na^+, K^+ and Cs^+. The data

TABLE 3.7.2
Experimental and derived results of potentiometric titration of Co^{2+} ($m^0 = 0.040$ mol kg^{-1}) with NaOH weights in the molten KCl–LiCl at 700 °C

$m_{O^{2-}}^0$	E (V)	pO	$\bar{n}$	pK_{CoO}	pP_{CoO}
0.0006	0.596	3.72	0.010	1.73*	5.12
0.0035	0.552	3.31	0.075	2.22*	4.74
0.0049	0.518	3.00	0.097	2.03*	4.44
0.0074	0.504	2.87	0.151	2.12*	4.34*
0.0146	0.499	2.83	0.328	2.51	4.40*
0.0182	0.495	2.79	0.414	2.64	4.42*
0.0223	0.491	2.75	0.513	2.78	4.46*
0.0254	0.483	2.68	0.583	2.82	4.46*
0.0313	0.469	2.55	0.712	2.94	4.49*
0.0374	0.461	2.48	0.852	3.24	4.70
0.0400	0.452	2.39	0.899	3.34	4.79
0.0458	0.446	2.34	1.031		
0.0510	0.411	2.02	1.035		
0.0604	0.378	1.72	1.029		
0.0709	0.349	1.45			
0.0817	0.306	1.21			

p$P_{CoO} = 4.43 \pm 0.11$, p$K_{CoO} = 2.03 \pm 0.50$; points selected for the averaging are denoted with $*$.

on the solubility products of oxides in this melt point to a relatively low probability of the formation of precipitates owing to the interactions of cations of alkaline-earth and transition metals with the traces of oxide ion impurities in the said melt. Nevertheless, the conclusion made is completely correct only for the solubility product values. However, besides the ionized form, some quantity of the oxide is dissolved without dissociation (the Shreder's component of solubility) which is not subject to the action of the acidic cations of the melt.

TABLE 3.7.3
The equilibrium parameters of solubility and dissociation of some oxides in the molten KCl–LiCl eutectic at 700 °C

Oxide	pK_m	pP_m	pP_N	p$P_{N,KCl-NaCl}$	ΔpP
MgO	–	5.87 ± 0.06	8.38	11.62	3.24
NiO	–	5.34 ± 0.2	7.85	11.38	3.53
CoO	2.03 ± 0.2	4.43 ± 0.11	6.94	10.24	3.30

3.7.1.2 KCl–NaCl (0.50:0.50) equimolar mixture

There are many data on oxide solubility in the melts of the KCl–NaCl system, obtained by all the above-described methods. We shall begin our consideration with the isothermal saturation method.

Esin and Lumkis [341] investigated the behaviour of NiO, CoO, Cu_2O and FeO in the molten KCl–NaCl equimolar mixture at 800 °C. Addition of any oxide to the melt caused the formation of two immiscible layers containing 20–25% (in the bottom layer) and 0.05–0.15% (in the top layer) of a heavy metal-oxide. The interface boundary was sharp in all the experiments. The bottom layers of the melt had the colour characteristic of the corresponding oxide: light green for NiO, dark green (close to blue) CoO, brownish Cu_2O and black FeO. The upper layers possess the same colour but are considerably less intense. Nevertheless, it was found that the distribution of the metal-oxides between two phases did not obey the distribution law and, therefore, serious doubts arose regarding the obtained layers' homogeneity.

Passing electric current through the melt saturated by the studied oxides results in the deposition of the corresponding metals on the cathode surface. This means that the formation of some ionized oxide in the solutions, but not only the solid oxide suspensions, takes place in the process of dissolution. The authors of Ref. [341] calculated the activity coefficients of metal-oxides contained in the bottom and top layers, and found that the activity coefficients in the bottom layer were lower than those in the top one. Besides the KCl–NaCl equimolar mixture, the behaviour of the said oxides was studied in $BaCl_2$ and $CaCl_2$ melts at 1000 °C. Addition of the oxides to the former melts led to the formation of two layers. This can be explained by a relatively small increase in the oxoacidic properties of the said melt. It should be mentioned that Ba^{2+}-based chloride melts possess oxobasicity-index values close to $pI_L = 2$. The use of $CaCl_2$ as a solvent for oxides results in the disappearance of the two-layered structure of the melt, and the solutions of NiO and CoO are quite homogeneous: for the latter oxide, the formation of the homogeneous solution is accompanied by its change in colour from dark green to black, caused by the change in the complex stability in the melts. It is interesting to note that the addition of strong Lux bases led to a decrease in MeO concentration in the top layer and its increase in the bottom one. This fact might be considered as an argument confirming the formation of the oxide precipitate under the experimental conditions.

Naumann and Reinhardt reported the polythermal study of the solubility of alkaline-earth oxides (CaO, SrO and BaO) in the molten KCl–NaCl equimolar mixture and individual KCl or NaCl melts in the temperature ranges from the melting point of the corresponding melt to 900 °C. The examinations were fulfilled using the isotope method [329]. The oxide solubility increased in the sequence CaO → SrO → BaO, as was expected. The solubility products of oxides in the KCl and KCl–NaCl melts were found to be close to one another, both values being lower than the corresponding solubility parameter in molten NaCl. This was explained by higher oxoacidic properties of Na^+ ion as compared with K^+. A comparison of the experimental data with the results of thermodynamic calculations, assuming the existence of only-molecular or only-ionized forms of the oxides demonstrated that both the non-dissociated oxide and the ions (Me^{2+} and O^{2-}) existed simultaneously in the saturated solutions, since the experimental solubility values were intermediate between the two polar cases.

A similar investigation of alkaline-earth metal solubility in the molten KCl–NaCl equimolar mixture was reported by Volkovitch [328], the studies were performed in the temperature range 700–810 °C. The values of the oxide's solubility product was determined by complexometric titration of the aqueous solution obtained by dissolving a solidified sample of the melt containing the studied oxide, which allowed the sum of Me^{2+} and MeO concentrations in the melt to be obtained. Comparison of the calculated solubility products with the theoretical estimates gave an opportunity to determine the values of the activity coefficients of the oxide ions and the metal cations in the studied saturated solutions. Nevertheless, the activity coefficients obtained can hardly be considered correct, because of the incomplete dissociation of the oxide (and the presence of MeO in the melt, in addition to the mentioned ions) demonstrated in Ref. [329]). This means that the deviation of the solution properties from ideality is caused by the presence of three kinds of foreign impurities dissolved in the melt: doubly charged Me^{2+} and O^{2-} and uncharged MeO, whose action on the properties of the melt differ appreciably. The opportunity of a specific action for each of the said particles was not allowed for in the interpretation of this work.

Amirova et al. [342] reported an investigation of V_2O_3 solubility in molten KCl, NaCl and the equimolar KCl–NaCl mixture at 700 °C (only for the last melt), 850 and 950 °C. The solubility of vanadium(III) oxide was determined by the isothermal saturation method. The concentration of V(III) in the melt was determined by two methods. The first consisted in the dissolution in water of a cooled sample of the saturated solution obtained in the melt, followed by titrimetric determination of V(III). The second method comprised the smelting of the cooled sample with sodium peroxide to transform V(III) into V(V), with the subsequent determination of vanadate concentration. The results obtained by both methods were in good agreement. The solubilities of V_2O_3 in these molten systems are presented in Table 3.7.4. The data show that a change in the cation composition of the melt does not result in appreciable changes of V_2O_3 solubilities in it. Moreover, the solubility decreases in the sequence KCl–(KCl–NaCl)–NaCl, although from general considerations the opposite changes could be expected. This gives us evidence that V_2O_3 dissolved in a molten salt is not subject to an appreciable degree to acid–base dissociation with the formation of oxide ions: the fraction of V^{3+} in the saturated solution of KCl–NaCl mixture at 700 °C is determined by the chronopotentiometric method to be in the order of several parts per thousand.

Shapoval *et al.* studied the oxoacidic properties of Ca^{2+}, Li^+ and Ba^{2+} cations in the chloride melt at 700 °C [343]. The authors succeeded in finding the equilibrium constants of the reaction inverse to equation (3.10) estimated as $K = 3.53 \times 10^2$, and that of the reaction of BaO formation, was calculated to be $K = 81$.

$$Ba^{2+} + O^{2-} \rightleftharpoons BaO, \tag{3.7.12}$$

These data demonstrate that the said cations possess an appreciable acidity which increases from Ba^{2+} to Li^+, which agrees qualitatively with the order of the melt-acidity in molten chlorides at 700 °C discussed in Part 3.

Barbin and Nekrasov studied the solubility of lithium oxide in the molten KCl–NaCl equimolar mixture in the temperature range 973–1073 K by the isothermal saturation method [344, 345]. The dependence of Li_2O solubility in

TABLE 3.7.4
Solubility of V_2O_3 in molten KCl, NaCl and the equimolar KCl–NaCl mixture

Melt	V_2O_3 solubility, molar fractions		
	700 °C	850 °C	950 °C
KCl		1.04×10^{-3}	1.32×10^{-3}
NaCl		7.42×10^{-4}	1.01×10^{-3}
KCl–NaCl	6.66×10^{-4}	9.10×10^{-4}	1.17×10^{-3}

the melt against the inverse temperature is described by the following equation:

$$\ln[\text{Li}_2\text{O}] = 0.107 - 5.221T^{-1}, \tag{3.7.13}$$

which corresponds to the solubility changes from 0.52 to 0.86 mol% in the said temperature range. The value of Li_2O solubility at 700 °C found by Barbin and Nekrasov is approximately twice as that obtained by Kaneko and Kojima by the same method (0.31 mol%) [346]. Such a considerable difference for the readily soluble oxide is caused by the effect of the grain-size of the oxide powder upon its solubility, which will be discussed further. There are also considerable positive deviations from ideality in the Li_2O solutions in the molten KCl–NaCl, caused by the interactions of Li^+ cations and oxide ions with the constituent components of the melt, which results in the formation of complex oxocompounds of lithium, sodium and potassium.

Ovsyannikova and Rybkin studied the oxoacidic properties of metal cations in the molten KCl–NaCl equimolar mixture at 700 °C, and for this purpose the authors constructed the acidity scale of cations [65]. The acidic properties of the cations of the main subgroups of the Periodic Table were shown to weaken with the increase in the atomic number of the element, but such a correlation was not observed in the case of elements of the secondary subgroups [65]. Nevertheless, a detailed analysis of the data presented in Ref. [65] shows that the oxoacidic properties of the cations weaken in the sequence Zn–Cd. Although the performed study can be classified among the systematic ones, it contains no quantitative data available for estimations of the oxide solubilities in the molten KCl–NaCl equimolar mixture. The point is that the main acid–base equilibrium with the participation of the cation (equation (3.6.2)) in the chloride melt is superimposed on an additional one (equation (1.2.2)) since all the studied cations were added to the melt as sulfates but not as chlorides. Unfortunately, quantitative estimation of the effect of the equilibrium (1.2.2) on the process of dissociation of the oxide derivatives of cations was not performed in the said work. The use of the sulfates instead of chlorides will obviously result in the evaporation of SO_3 from the solutions of the most acidic cations, with the simultaneous precipitation of the corresponding metal-oxides from the melt, as it takes place in the solutions of $\text{Al}_2(\text{SO}_4)_3$ or $\text{Ti}(\text{SO}_4)_2$ mentioned above (see equation (3.7.8)). However, this is not the case if the cations possess relatively weak acidic properties: under these conditions the sulfate ions are stable enough. So, there are two kinds of solutions studied in this work. The first is homogeneous solutions, whose acidities are mainly defined by the acidic properties of the studied cation, with some distortions owing to the formation of the complexes $Me^{2+}:SO_4^{2-}$. The second kind is the heterogeneous solutions, whose acidity is defined by the solubility (but not the acidity) of the corresponding oxide plus the acidity of SO_3. These systems are obviously non-equivalent.

Apart from what we have said above, the results of Ref. [65] are in direct contradiction with the phenomenon of levelling of acidic (basic) properties by the constituent ions of the melt. For example, according to the potentiometric data presented, the equilibrium concentration of O^{2-} in a 0.01 mol kg^{-1} solution of $BaSO_4$ (the melting point is $1580\,°C$) is 10 times as high as that in the solutions of K_2SO_4 and Na_2SO_4 of the same molality. For the solutions containing other alkali- and alkaline-earth metal cations similar deviations are observed, namely, in the solutions of Cs_2SO_4, $SrSO_4$ and Rb_2SO_4 the oxide ion concentration is higher by a factor of 3.2, 2 and 1.2, respectively. If we consider the reaction, which obviously takes place in the said solutions of the metal sulfates owing to their thermal dissociation, we will note the following.

$$MeO + 2Na^+ \rightleftharpoons Me^{2+} + Na_2O. \tag{3.7.14}$$

If the metal cation in the chloride melt consisting of sodium and potassium salts shows appreciably lower acidic properties than the Na^+ ion, then equilibrium (3.7.14) is shifted to the left and the solution's basicity (acidity) are limited by the basic properties of Na_2O in the said melt. Therefore, a small quantity of a salt whose cation possesses weaker acidic properties than Na^+ should cause changes equal to those arising after addition of the corresponding sodium salt. Taking into account the fact that the authors investigated solutions having concentrations of about 0.01 mol kg^{-1}, where the ratio Me^{n+}/Na^+ is equal to $0.0026/n$, the effect caused by the addition of such a weight of the studied substance on the melt's properties should also be negligible, since it does not result in appreciable changes of the cation composition of the melt–solvent. This gives the final evidence that all the results obtained in Ref. [65] are incorrect.

Delimarskii *et al.* reported investigations of the solubility of ZnO, MgO, NiO and SrO in the molten KCl–$NaCl$ equimolar mixture at $700\,°C$ [236]. The solubility was studied by the method of potentiometric titration of the solutions containing 0.01 mol kg^{-1} of the corresponding chloride with KOH, and of KOH solutions with quantities of the studied chloride (the reverse titration).

On the basis of the obtained experimental data, the equilibrium parameters corresponding to the solubility product and the dissociation constant were calculated. Unfortunately, there was no statistical treatment of the obtained calculated values, and Ref. [236] did not contain the average values of the equilibrium parameters; this averaging was made by Combes *et al.* in Ref. [239]). The values of K_{MeO} and P_{MeO} calculated for the same oxide at the direct and reverse titration were demonstrated in Ref. [239] to differ measurably. This discrepancy in the data shows that systematic investigations of oxide solubilities aimed at finding common regularities and correlations should be performed using the same investigational techniques. Among all methods described in Part 6, direct potentiometric titration is the most convenient way of obtaining such results. Returning to Ref. [236], it should be noted that the authors concluded with the following main result that the solubility of the studied oxides increased in the sequence MgO → NiO → ZnO → SrO.

The solubilities of some oxides in molten KCl–NaCl at 1000 K were reported by Combes *et al.* [239]. According to their data, the potentiometric titration curve of Mg^{2+} with BaO contains two consecutive titration stages in the cation excess section (Fig. 3.7.3). The first corresponds to the formation of Mg_2O^{2+} and the second one is connected with MgO precipitation. In this relation, the following fact should be mentioned. Although the potentiometric titrations of Mg^{2+} ions in the molten KCl–NaCl equimolar mixture, and in the related melts, were performed more than once by different groups of investigators, the two-step titration curves were not observed again. This makes the obtained shape of the titration curve doubtful; as in the other cases, the titration curves were similar to those observed for the practically insoluble oxides: they contained only one pO drop corresponding to reaction (1.3.11).

Combes and Koeller reported the investigation of SrO solubility in the molten KCl–NaCl equimolar mixture at 727 °C [347]. Potentiometric titration was performed with weights of BaO that allowed calculation of the solubility product of SrO as $pP_{SrO} = 4.2$ (on the molality scale). This is appreciably higher than the values obtained by other authors by the potentiometric titration method. Such a fact is seemingly caused by the formation of BaO–SrO solid solutions instead of pure SrO under the titration conditions. The oxide material formed obviously possesses an appreciably lower solubility.

The use of BaO as a titrant in melts possessing weakly acidic properties cannot be recognized as correct at all, apart from its use for the titration of other alkaline-earth metal cations (since, in this case, the solid solutions

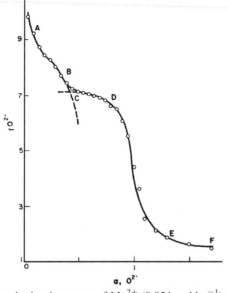

Fig. 3.7.3. Potentiometric titration curve of Mg^{2+} (0.054 mol kg^{-1}) with BaO weights in the molten KCl–NaCl equimolar mixture at 727 °C, by data of Combes et al. [239].

of composition $Ba_xMe_{1-x}O$ will be precipitated from the melts instead of the pure oxides). Of course, the solubilities of these solid solutions differ from those of the pure alkaline-earth metal-oxides. As mentioned in Part 6, the use of BaO for titration results in the accumulation of foreign acidic cations in melts–solvents, which in turn, results in an over-estimation of the equilibrium parameters. For example, the value of pP_{MgO} obtained in Ref. [239] is equal to 9.00 ± 0.15 against 9.27 ± 0.06 from the data of Ref. [330], where the MgO solubility is studied using NaOH as titrant. The main conclusion made in Ref. [239] is that the solubility products of alkali-metal oxides increase in the sequence $MgO \rightarrow CaO \rightarrow BaO$ (see Table 3.7.5). In that paper, the solubility of MgO was determined by the potentiometric titration method whereas the other oxides were studied by the isothermal saturation techniques.

Komarov and Krotov studied the solubility of ZrO_2 in the molten KCl–NaCl equimolar mixture at various temperatures, using the method of isothermal saturation [246]. According to their data, the thermal dependence of ZrO_2 solubility can be approximated by the following equation:

$$N_{ZrO_2} = (-5.7 \pm 3.1) \times 10^{-6} + (7 \pm 3) \times 10^{-9}T. \qquad (3.7.15)$$

Calculations using this equation show that the solubility of ZrO_2 in that melt at 700 °C is also negligible, even in a strongly acidic medium [246]. Therefore, the use of the membrane oxygen electrodes made of stabilized ZrO_2 should not cause appreciable pollution by Zr(IV) compounds of the studied halide melts under these conditions. As mentioned in Part 4, the upper limit of reversible work of a YSZ-based membrane oxygen electrode is located near pO = 11, where destruction of the membrane material is noticeable.

We reported the solubilities of 11 oxides of the MeO type (s^2- and 3d-metals) in the molten KCl–NaCl equimolar mixture at 700 °C [175, 330, 332, 348] (see Table 3.7.5), studied by the method of direct potentiometric titration. The titration curves of Mg^{2+}, Ni^{2+}, Co^{2+}, Mn^{2+} and Zn^{2+} do not contain an unsaturated solution section, and therefore, they are referred to the ions practically insoluble in the specified chloride melt. The potentiometric curve for Mg^{2+} does not contain two steps in the acidic section, and it is smooth enough (Fig. 3.7.4, curve 1): the calculated pP_{MgO} values undergo small fluctuations about the average value, 9.27. It follows that there is no formation of intermediate compounds similar to Mg_2O^{2+} in the chloride melt, at least, with the initial concentrations of Mg^{2+} ($MgCl_2$) close to 0.05 mol kg^{-1}.

During the processes of titration of Cd^{2+}, Ca^{2+} or Cu_2^{2+} the solutions remain unsaturated after the addition of the corresponding chloride into the pure melt, but even the first small quantity of the titrant immediately results in the formation and precipitation of the corresponding oxide from the melt. This is seen from a sharp drop in pO after the first addition of the titrant, and

TABLE 3.7.5
Solubility products of some oxides in the molten KCl–NaCl equimolar mixture at 700 °C ($-\log P_{MeO}$), molalities

Oxide	Refs. [239, 240]	Ref. [236]	Ref. [328]	Ref. [330]
MgO	9.00 ± 0.15	8.46		9.27 ± 0.06
CaO	~5		6.62	4.36 ± 0.06
SrO		3.00	5.84	3.08 ± 0.4
BaO	2.31 ± 0.05		4.22	2.30 ± 0.15
NiO	11.2	8.32		9.03 ± 0.06
ZnO		6.18		6.93 ± 0.2
Cu_2O	5.4			4.17 ± 0.3
MnO				6.78 ± 0.05
CoO				7.89 ± 0.03
PbO				5.12 ± 0.05
CdO				5.00 ± 0.03

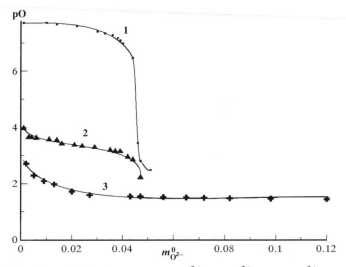

Fig. 3.7.4. Potentiometric titration curves of Mg^{2+} (1), Cd^{2+} (2) and Sr^{2+} (3) with NaOH weights in the molten KCl–NaCl equimolar mixture at 700 °C.

starting from this moment the titration curve becomes flat and smooth as shown in Fig. 3.7.4, curve 2. This feature of the titration curves allows us to classify CdO, CaO and Cu_2O as being among the oxides which are appreciably soluble in the KCl–NaCl melt at 700 °C.

A relatively wide section of unsaturated solution appears in the potentiometric curves for Pb^{2+}, Sr^{2+} and Ba^{2+}, and the last two cations are characterized by titration curves with the absence of a pO drop at the equivalence point (e.g. Fig. 3.7.4, curve 3). This gives evidence both of a relatively high solubility of BaO and SrO in the studied melt, and of a relatively weak acidity of the corresponding cation—at least, in diluted solutions. In this case, the division of the potentiometric titration curve into sections of saturated and unsaturated solution was performed according to the method for soluble oxides, as described in detail in Section 3.7.1.1. Examination of unsaturated solutions of BaO does not allow estimation of the corresponding dissociation constant, whereas the case of incomplete dissociation of SrO is detected clearly. It should be emphasized that the solubility product values in the chloride melts are well reproducible, as can be seen from the narrow confidence range (as a rule, it is less than 0.1), although the parameters of the homogeneous acid–base equilibria are estimated with considerably higher errors (usually ±0.3–0.5).

Analysis of the processes running in saturated oxide solutions in ionic melts shows that their sum of MeO and Me^{2+} concentrations usually exceeds the value obtained only from data on the solubility products owing to incomplete dissociation of the formed oxide in the solution. The sum of the particle concentrations can be calculated according to the formula (3.6.5). For example, for the moderately soluble lead oxide, the Σ^s_{MeO} value is equal to 1.75×10^{-2} mol kg^{-1}, which is six times as high as the concentration of Pb^{2+} ions in a saturated solution of PbO in the molten KCl–NaCl equimolar mixture.

A noticeable regularity of oxide solubilities in halide melts is a strict dependence of pP_{MeO} values on the inverse square of the cation radius (i.e. its square root) or other parameters

$$pP_{MeO} = pP_{r^{-2} \to 0} + \frac{\partial pP_{MeO}}{\partial(r_k^{-2})} r_k^{-2}. \qquad (3.7.16)$$

including such a multiplier. In particular, the dependence, pP_{MeO}, on the polarizing action by Goldshmidt (Zr^{-2}) of the cation introducing in the oxide composition is shown in Fig. 3.7.5. This dependence will be shown below to be characteristic of the alkaline-earth and transition metals (belonging to the first transition row). Only PbO drops out of this dependence, but lead is referred to the p-elements and has a relatively low melting point (886 °C). It is possibly owing to this fact that its behaviour in the melt differs from the other oxides studied.

Another characteristic which is dependent on r_k^{-2} is the intensity of the cation field, which is calculated as $ne/(4\pi r^2)$ [349] and manifests itself as the ratio of the cation charge to the area of a sphere having its radius equal to

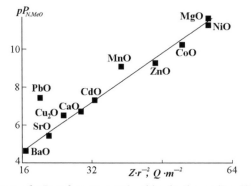

Fig. 3.7.5. Dependence of pP_N of some metal-oxides in the molten KCl–NaCl equimolar mixture at 700 °C against polarizing action of the cations by Goldschmidt (Zr_k^{-2}).

that of the cation. The said characteristics are distinguished only by a definite constant multiplier (Z in the case of the polarizing action of a cation, and $ne/(4\pi)$ in the case of the intensity of the electron-field one). Without any preference for any of the parameters of cations discussed above, we shall consider hereafter the dependence of pP_{MeO} against Zr_k^{-2}.

In certain cases, the results obtained by different authors are in good agreement, which can be explained by the similar experimental conditions. They are the method of solubility determination, the donor of oxide ion used and the initial molality of the cation forming the oxide being studied. However, the scatter of the solubility data is usually too large. Such a scatter of the data has not been explained. Naturally, the impossibility of systematising of the obtained results has not favoured the development of oxide-solubility investigations in molten salts.

The solubility data can be divided into two groups; depending on the methods of determination (Fig. 3.7.6), the solubilities of alkali-metal-oxides belonging to the same group are close. A surprising result from comparison of the available literature data is that the solubilities of oxides, determined by the isothermal saturation method, are several times lower than those estimated on the basis of the potentiometric titration data. Indeed, in the former case one determines the characteristic sum, $\sum s_{MeO}$, whereas in the latter case, only $s_{Me^{2+}}$ can be calculated. In terms of the solubility products and the dissociation constants, the difference in total solubility and the concentration of the metal cations in the saturated solution can be expressed by the following inequality:

$$\sum s_{MeO} - s_{Me^{2+}} = \left(\sqrt{P_{MeO}} + \frac{P_{MeO}}{K_{MeO}} \right) - \sqrt{P_{MeO}} = \frac{P_{MeO}}{K_{MeO}} \geq 0.$$

$$(3.7.17)$$

The fact that the data obtained by the isothermal saturation method are lower than those from the potentiometric titration leads us to the conclusion that

$$\frac{P_{MeO}}{K_{MeO}} < 0.$$

This observation has not been published before, although it is hard to believe that authors of the later papers have not been familiar with the results of the

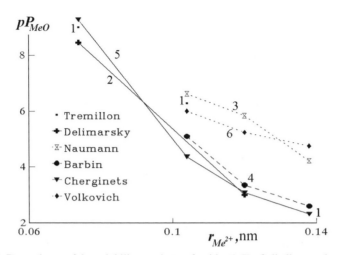

Fig. 3.7.6. Dependence of the solubility products of oxides (pP) of alkaline-earth metals in the molten KCl–NaCl equimolar mixture against the cation radius determined by the methods of potentiometric titration (1, 2, 5) and isothermal saturation (3, 4, 6).

investigations published. Probably, this discrepancy has been concealed owing to the absence of reasonable explanations.

As was said in Section 3.7.1.1, Delarue revealed the higher solubilities of CaO and CdO obtained *in situ* as a result of the acid–base reaction, in comparison with the solubilities of the powdered oxides [319, 320], although there was no explanation of the observed results.

The solubility of any solid substance in a liquid phase is well known to be dependent on the size of its particles (crystals) which are in contact with the saturated solution. For substances existing in solutions in the molecular form, the dependence of their solubility against the particle size is expressed by the Ostwald–Freundlich equation (3.6.57). From this equation it follows, in particular, that an increase in the crystal (particle) dimensions results in a decrease in solubility. The oxide particles precipitated from concentrated solutions should be larger than those formed from more dilute solutions. The increase in the particle sizes in concentrated solutions of cations and oxide ions may be explained by ageing oxide, i.e. by the re-crystallization processes, whose rate increases with the concentration of the dissolved substance.

Under the conditions of sedimentational titration, finely dispersed metal-oxide particles are formed in the melt, which immediately begins to age because of the re-crystallization processes, i.e. enlargement of the initially formed crystallites and decrease in the molar surface area S, and in turn, the oxide solubility (the solubility product). Keeping the oxide powder in the melt at high temperatures during a long period of time favours the process of re-crystallization, as is caused by high concentrations of the corresponding reagents in the solution. The former case is realized in the course of investigations by the isothermal saturation techniques, when the oxide studied is prepared in the form of sintered pills, which are subjected to the melt action for several hours. It is obvious that in the latter case the transfer of the substance from smaller crystals to larger ones is faster.

It should be noted that the solubility product values obtained in Ref. [236] are slightly higher than those reported in our own work. This may be explained by the fact that Delimarskii *et al.* performed their investigations in more diluted solutions of the cations (in the order of 0.01 mol kg^{-1}), where ageing is slower than in the concentrated solutions [236]. An essential disadvantage of solutions having low cation concentrations is the very considerable relative change in the cation concentration in the process of the investigation and, consequently, worse reproducibility of the solubility data. The changes of NiO solubility with the initial cation concentration in the molten KCl–NaCl equimolar mixture at 700 °C can be mentioned as an example. Titrimetric study of the cation solution of concentration 0.01 mol kg^{-1} results in the value of $pP_{NiO} = 7.95 \pm 0.20$ (e.g. in Ref. [236] this value is equal to 8.32), whereas titration of 0.05 mol kg^{-1} solution of Ni^{2+} yields $pP_{NiO} = 9.03 \pm 0.06$. Nevertheless, even in this case the solubility products differ by approximately an order of magnitude. In contrast, the results presented in Refs. [239, 348] for magnesium- and barium-oxide solubilities, obtained by titration of 0.05 mol kg^{-1} cation solutions, are in excellent agreement with the data obtained by our group. Near the equivalence point, the calculated values of pP_{MeO} often undergo a slight directed shift towards the higher oxide solubility. This may be caused by the difference in size of the oxide crystals precipitated from ionic melts. Despite this rather obvious reason for such deviations, the effect of the size of oxide crystals (particles) (or some derived parameters) on their solubilities in high-temperature ionic melts, has not been studied.

Our recent papers [333, 350] reported investigations of the effect of the grain-sizes of lead, cadmium, calcium and zinc oxides on their solubilities in the molten KCl–NaCl equimolar mixture at 700 °C. These studies were performed by the sequential addition method (SAM) described in detail in

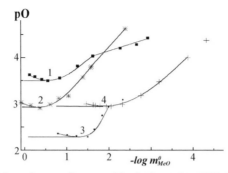

Fig. 3.7.7. The SAM dependences of some oxides in the molten KCl–NaCl equimolar mixture at 700 °C: (1) ZnO; (2) CdO; (3) CaO; (4) PbO.

Section 3.6.2.3, and resulted in the $pO = f(-\log m_{O^{2-}}^0)$ dependences. The latter are shown in Fig. 3.7.7.

All the presented dependences are practically of the same kind: they consist of an inclined section, whose position is exclusively defined by the value of the dissociation constant of the oxide, and a plateau beginning with the minimum (bending to the abscissa axis). This minimum occurs owing to partial dissolution of a portion of the oxide, immediately resulting in the saturation of the solution with respect to the oxide added. For zinc oxide there is an inflection at the first section of the SAM plot because, after small additions of ZnO, the equilibrium molality of oxide ions in the melt is comparable with that of the traces of oxygen-containing admixtures in the studied melt.

An example of the SAM data treatment for zinc oxide is presented in Table 3.7.6. The quantities of ZnO corresponding to points 1–6 do not result in the formation of a saturated solution: the e.m.f. (pO) data make it possible to estimate the dissociation constant of ZnO in the KCl–NaCl melt as $pK_{ZnO} = 6.37 \pm 0.08$. Point 7 corresponds to the weight of ZnO which provides saturation of the melt by zinc oxide, thereat the added powder of oxide is dissolved in the melt only partially. Subsequent additions of ZnO to the saturated melt result in an appreciable reduction in the saturated solution's concentration (decrease in the solubility product of ZnO) owing to the reduction in the averaged molar surface area of the deposit and, in turn, the oxide solubility. This reduction ends after the true plateau is achieved, when the changes of molar surface-area are negligible.

In another example studied, similar behaviour of the $pO - (-\log m_{O^{2-}}^0)$ plots is observed. The data which belong to the unsaturated solution section

TABLE 3.7.6
Results of the investigation of ZnO solubility in the molten KCl–NaCl equimolar mixture at 700 °C (SAM)

N	$-\log m_{ZnO}$	pO	pP_{ZnO}	pK_{ZnO}
1	2.92	4.43	8.86	5.92
2	2.66	4.29	8.57	5.90
3	2.27	4.21	8.43	6.15
4	1.62	4.04	8.08	6.46*
5	1.26	3.82	7.63	6.37*
6	0.83	3.56	7.12	6.29*
7	0.55	3.50	7.00**	–
8	0.41	3.53	7.06**	–
9	0.25	3.59	7.18**	–
10	0.12	3.63	7.27**	0.00

*Points chosen for pK_{ZnO} calculation.
**Points chosen for pP_{ZnO} calculation.

allow determination of the dissociation constant values, which are found in Ref. [350] to be $pK_{CdO} = 5.37 \pm 0.07$, $pK_{CaO} = 3.38 \pm 0.3$ and $pK_{PbO} = 3.50 \pm 0.07$ (PbO). It should be noted that the last value is in good agreement with the dissociation constant value obtained by potentiometric titration, which yields $pK_{PbO} = 3.29 \pm 0.04$ [175]. This means that the grain sizes have no effect on the dissociation constants, since this kind of Lux acid–base equilibria runs without the phase boundaries, and is not affected by the interface energy.

The results of the experimental SAM data treatment for the cases of ZnO and CaO are collected in Tables 3.7.7 and 3.7.8, respectively. As is seen, addition of the initial commercial powder to the saturated solution in contact with the partially dissolved oxide deposit leads to a reduction in the saturated solution's concentration by a factor of 1.5 for ZnO solutions, and approximately 1.25 for CaO solutions. Similar changes of the solubilities are observed if the CdO and PbO powders are examined by the SAM method. It should be noted especially that, in all the cases, the changes of e.m.f. (pO) are approximately 10 times as high as the experimental error of the determination of these parameters.

The results obtained allow us to construct the dependences of oxide solubilities ($-\log s_{MeO}$) against the molar surface area of the studied oxide. These dependences are presented in Fig. 3.7.8 for some of the oxides studied. All the dependences are close to linear plots. Their slopes allow estimation of

TABLE 3.7.7

Calculated parameters describing the change of the equilibrium ZnO concentration in the saturated solution in the molten KCl–NaCl equimolar mixture at 700 °C

$- \log m_{ZnO}$	$- \log s_{ZnO}$	$s_{ZnO}/s_{p,ZnO}$	S (m^2 mol^{-1})
−0.55	0.73	1.00	800
0.41	0.78	0.89	660
0.25	0.88	0.71	610
0.12	0.95	0.61	600

the surface-energy values at the "metal-oxide–alkali-metal halide melt" interface boundary, using equation (3.6.67). The corresponding data are contained in Table 3.7.9.

The values of the surface energy at the "oxide–chloride melt" interface boundary were calculated in Ref. [350] to be within the range 30–40 J m^{-2}. These values considerably exceed the corresponding parameters estimated for aqueous solutions of electrolytes, thus giving rise to essential changes of the solubility with the particle sizes (see Fig. 3.7.7).

From the known values of the plots of the dependence of solubility on the particle-sizes for the oxides studied we can calculate the molar surface area of the oxide particles which are precipitated from the saturated solutions obtained under the conventional potentiometric titration conditions. The results of these calculations are presented in Table 3.7.9. From these data, it follows that the oxide particles deposited under the conditions of potentiometric titration are characterized by considerably higher molar surface areas than with the commercial oxide powders. We now compare these values: the commercial calcium oxide used possesses a molar surface area of 594 m^2 mol^{-1} against 780 m^2 mol^{-1} characteristic of the

TABLE 3.7.8

The calculated parameters describing the change of equilibrium CaO concentration in the saturated solution of the molten KCl–NaCl equimolar mixture at 700 °C

$- \log m_{CaO}$	$- \log s_{CaO}$	$s_{ZnO}/s_{p,CaO}$	S, m^2 mol^{-1}
1.50	1.89	1.00	707
1.23	1.91	0.95	639
1.01	1.93	0.92	619
0.87	1.94	0.89	611
0.79	1.96	0.86	608
0.52	1.99	0.79	601

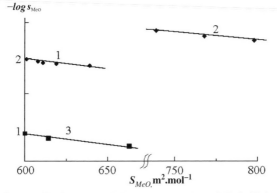

Fig. 3.7.8. Dependences of $-\log s_{MeO}$ of CaO (1), PbO (2) and ZnO (3) in the molten KCl–NaCl equimolar mixture at 700 °C against the molar surface area (S_{MeO}) of solid oxide particles contacting with the melt.

precipitated powder. For other oxides, these magnitudes are 1240 and 705 (PbO), 880 and 560 (ZnO), 719 and 193 (CdO) $m^2\,mol^{-1}$, respectively. Although the molar surface areas are usually in the order of $10^3\,m^2\,mol^{-1}$, the high values of the surface energies provide considerable differences in the powders' solubilities. The data mentioned can be used for estimating the effective molar surface area of sintered calcium oxides (in pills), which are placed in the melt to study their solubility by the isothermal saturation method. Combes *et al.* reported the solubility product value for this material as $pP_{CaO} \sim 5$ [238]. This allows one to calculate the molar surface area to be near $400\,m^2\,mol^{-1}$, this value being 1.5 times less than that of the commercially produced powder. So, pressing, and prolonged thermal treatment of oxide powders cause an appreciable diminution of its molar

TABLE 3.7.9

Parameters describing the effect of particle sizes on the oxide- solubility changes in saturated solutions in the molten equimolar KCl–NaCl mixture at 700 °C

	pP_{MeO} (SAM)	The coefficients		pP(PT) [175]	S(PT)
		$-\log s_0$	$-tg \times 10^3$		
CaO	4.58–4.71	3.09 ± 0.4	1.86 ± 0.6	4.36 ± 0.06	780
PbO	5.87–6.00	3.65 ± 0.2	1.72 ± 0.3	5.12 ± 0.05	1240
CdO	5.83–6.06	0.93 ± 0.2	1.45 ± 0.6	5.00 ± 0.03	719
ZnO	7.00–7.27	2.35 ± 0.3	2.37 ± 0.4	6.93 ± 0.20	880

surface area which, in turn, leads to a measurable decrease in the solubility of such an oxide in ionic melts.

We can add that the high values mentioned of the surface energies at the "metal-oxide–alkali-metal halide melt" interface boundary are not unexpected if the following considerations are taken into account. In low-temperature molecular or ionizing solvents, the "non-polar solid organic substance–solvent" interface boundary is characterized by surface energies in the order of $10^{-2}\,\mathrm{J\,m^{-2}}$ [121]. The ion–dipole interaction in the "solid ionic substance–water" system increases σ up to values in the order of $1.5\,\mathrm{J\,m^{-2}}$ [334]. Finally, the strongest electrostatic interaction in the "ionic solid substance–high-temperature ionic solvent" results in an increase in the surface energies by several decades of $\mathrm{J\,m^{-2}}$. Similarly, solid oxides are usually characterized by high melting points, exceeding $1500\,^\circ\mathrm{C}$, which means that the difference of the melting points of the oxide and the melt studied is in the order of $1000{-}1500$ K, in contrast with the "organic substance–molecular solvent" systems where this difference does not exceed $100{-}200$ K. Therefore, simplified equations such as the Antonov rule [351]:

$$\sigma_{\mathrm{MeO/KCl}} = \sigma_{\mathrm{MeO}} - \sigma_{\mathrm{KCl}}, \qquad (3.7.18)$$

where $\sigma_{\mathrm{MeO/KCl}}$ is the surface tension at the "oxide–chloride melt" interface boundary, σ_{MeO} and σ_{KCl} are the surface tensions at the "solid substance–smelted substance" boundaries (at their melting points) cannot give reasonable values for such surface energies.

Another question arises: why are the saturated solutions of fine dispersed oxides so stable, and the processes of the re-crystallization very slow? This is explained by cluster formation in the molten salts. The transfer of oxide from smaller particles to larger ones is considerably retarded since it consists not in the replacement of Me^{2+} and O^{2-} ions from one particle to another, but in the formation of MeX_n^{2-n} and Kt_mO_{m-2} complexes. Khokhryakov and Khokhlova reported the formation of mixed complexes of composition $MeOX_4^{4-}$ in the molten salts, such complexes preventing precipitation of the metal-oxide to the solid phase [352].

Over-estimation of the values of surface energies at the "oxide–chloride melt" interface boundary (if it actually takes place) can be caused by the action of factors other than surface effects, but it is not yet possible to find out these factors, even they exist at all.

3.7.1.3 CsCl–KCl–NaCl (0.455:0.245:0.30) eutectic

As was mentioned in previous sections, the oxoacidic properties of the specified melt are close to those of the KCl–NaCl equimolar mixture. An essential advantage of the ternary mixture is its considerably lower melting point (near 480 °C). This creates the possibility of performing investigations of the metal-oxide solubilities at relatively low and at relatively high temperatures, and to elucidate the effect of the melt temperature on the oxide solubilities in molten alkali-metal chlorides. Apart from the thermal dependence, such studies allow the obtention of solubility data, which serve as initial points for the estimations of relative acidic properties in other halide melts at the corresponding temperatures. It should be noted again that the molten KCl–NaCl equimolar mixture cannot be used for this purpose at temperatures below 658 °C. The data presented below were obtained in our own investigations: no similar data have been reported by other authors.

The potentiometric titration of cations was performed at 600 and 700 °C with the use of NaOH as a base-titrant [214, 353, 354]. The results of these studies are collected in Tables 3.7.10 and 3.7.11, respectively.

For all the metal cations studied, the potentiometric titration curves at different temperatures show no essential distinctions, and resemble the corresponding dependences for the KCl–NaCl melt. The solubility product calculations demonstrate that an increase in the melt temperature causes the increase in the oxide solubility (pP_{MeO}) by 0.3–0.4. The obtained data show

TABLE 3.7.10

Metal-oxide solubilities in the molten CsCl–KCl—NaCl eutectic at 600 °C, molality scale

Oxide	pP_{MeO}	pK_{MeO}	s_{MeO}	$s_{Me^{2+}}$	$\sum s$
BaO	2.85 ± 0.10	–	–	3.75×10^{-2}	
SrO	3.95 ± 0.40	1.76 ± 0.3	7.07×10^{-3}	1.11×10^{-2}	1.82×10^{-2}
CaO	5.83 ± 0.05	2.76 ± 0.2	8.51×10^{-4}	1.22×10^{-3}	2.07×10^{-3}
MgO	10.78 ± 0.11	–	–	4.07×10^{-6}	
MnO	7.83 ± 0.15	–	–	1.22×10^{-4}	
CoO	9.02 ± 0.06	–	–	3.09×10^{-5}	
NiO	10.70 ± 0.08	–	–	4.47×10^{-6}	
Cu$_2$O	5.05 ± 0.10	2.34 ± 0.3	1.95×10^{-3}	2.98×10^{-3}	4.93×10^{-3}
ZnO	6.90 ± 0.07	–	–	3.55×10^{-4}	
CdO	5.64 ± 0.48	–	–	1.51×10^{-3}	
SnO	9.38 ± 0.02	–	–	2.04×10^{-5}	
PbO	6.31 ± 0.05	4.13 ± 0.1	6.61×10^{-3}	7.00×10^{-4}	6.68×10^{-2}

TABLE 3.7.11

Metal-oxide solubilities in the molten CsCl–KCl–NaCl eutectic at 700 °C, molality scale

Oxide	pP_{MeO}	pK_{MeO}	s_{MeO}	$s_{Me^{2+}}$	$\sum s$
BaO	2.59	–	–	5.07×10^{-2}	
SrO	3.64 ± 0.13	1.19 ± 0.22	3.54×10^{-3}	1.51×10^{-2}	1.86×10^{-2}
CaO	4.90 ± 0.13	2.04 ± 0.25	1.38×10^{-3}	3.55×10^{-3}	4.94×10^{-3}
MgO	9.78 ± 0.04	–	–	1.29×10^{-5}	
MnO	7.59 ± 0.14	–	–	1.60×10^{-4}	
CoO	8.60 ± 0.03	–	–	5.01×10^{-5}	
NiO	9.52 ± 0.06	–	–	1.74×10^{-5}	
Cu_2O	4.45 ± 0.05	1.64 ± 0.22	1.55×10^{-3}	5.96×10^{-3}	7.51×10^{-3}
ZnO	6.25 ± 0.03	3.12 ± 0.20	7.40×10^{-4}	7.50×10^{-4}	1.49×10^{-3}
CdO	5.19 ± 0.05	–	–	2.54×10^{-3}	
PbO	5.14 ± 0.02	3.08 ± 0.20	8.70×10^{-3}	2.69×10^{-3}	1.14×10^{-2}

that the metal-oxide solubilities in chloride melts of different cation composi-tions but practically the same oxoacidic properties, at the same temperature, are approximately the same if the solubilities are expressed in molar fractions.

The data calculated even at two temperatures differing by 100 °C allow one to perform some estimations which connect the oxides' solubilities with their melting points and with the melt temperature.

The melting point of a substance in some approximations can be considered as a measure of the bonding energy between the neighbouring particles (ions) in the crystalline lattice. Therefore, it allows us to assume that the stronger the bond between these particles, the greater should be the energy applied to the solid substance to break it down to constituent particles and, hence, the higher should be the melting point of the said material's increase. Such strong bonding, as well as increasing the melting point, retards the processes of dissolution of the substance in solvents (naturally, the reduction in the dissolution rate is only a kinetic factor, and has no direct connection with the substance's solubility). From this fact, it may be concluded that the metal-oxide solubility should decrease with an increase in its melting point. The dependence of the solid substance's solubility in different solvents, against the solvent temperature and the melting point of the solid substance, is expressed by the Le Chatelier–Shreder equation which is valid for ideal solutions in the following notation [326, 355]:

$$\ln N_{MeO} = \frac{\Delta H_{m,MeO}(T - T_{mp,MeO})}{RTT_{mp,MeO}}. \tag{3.7.19}$$

We shall now recall the considerations lying at the basis of this equation. Thermodynamic analysis of the phase diagrams of binary systems provides the most suitable instrument for obtaining information about the interactions in the "solid dissolved substance–liquid" system. Let us consider the similar dependence for interactions in a binary system.

For the liquidus curve in the pre-eutectic region of the phase diagram with a simple eutectic, the phase-equilibrium condition means $\Delta G = 0$, which can be expressed by the following equation:

$$G_{A,s} = \bar{G}_{A,l}. \tag{3.7.20}$$

Here, $G_{A,s}$ is the Gibbs energy of the solid phase of composition "A" (solid solvent), and $\bar{G}_{A,l}$ is the partial molar Gibbs energy of component "A" in the liquid phase; the molar fraction of component "A" in the liquid phase is equal to x_A.

The well-known Clausius–Clapeyron equation for the process of evaporation of pure component "A" is represented as

$$\frac{d \ln p_A^0}{dT} = \frac{\Delta H_{evap}}{RT^2}, \tag{3.7.21}$$

where p_A^0 is the pressure of the saturated vapour of the solvent over the solution, and ΔH_{evap} is the heat of evaporation of the solvent. The corresponding expression for the process of sublimation of crystals of component "A" can be written as

$$\frac{d \ln p_{cryst}^0}{dT} = \frac{\Delta H_{subl}}{RT^2}, \tag{3.7.22}$$

where p_{cryst}^0 is the pressure of saturated vapour over solid component "A" under equilibrium conditions and is the heat of its sublimation. Subtracting equation (3.7.21) from equation (3.7.22) and taking into account the fact that

$$\Delta H_{subl} - \Delta H_{evap} = \Delta H_m, \tag{3.7.23}$$

one can obtain the expression

$$\frac{d \ln(p_{cryst}^0/p_A^0)}{dT} = \frac{\Delta H_m}{RT^2}. \tag{3.7.24}$$

According to equation (3.7.20), under the equilibrium conditions between the solution, solid, solvent, and vapour, the partial pressure over the solution is equal to the vapour pressure over the crystalline phase of composition "A", i.e.

$$p_A = p_{A,cryst}. \tag{3.7.25}$$

Substituting this equation into equation (3.7.24) we obtain

$$\frac{d \ln N_A}{dT} = \frac{\Delta H_m}{RT^2}. \tag{3.7.26}$$

Assuming that over a small temperature range, $\Delta H_m = const$ and dividing the variables, with subsequent integration with respect to T, from T to $T_{mp,A}$, and with respect to x from N_A to 1, one can obtain the final expression

$$-\ln N_A = \frac{-\Delta H_m}{R} \left(\frac{1}{T_{mp,A}} - \frac{1}{T} \right) = \frac{\Delta H_m (T_{mp,A} - T)}{RTT_{mp,A}}. \tag{3.7.27}$$

The Le Chatelier–Shreder equation presents the thermal dependence of the liquidus in the pre-eutectic region of the phase diagram with the simple eutectic. The same equation expresses the thermal dependence of the solubility of substance "B" in the solvent "A". This equation allows us to draw some conclusions concerned with the interactions in eutectic systems, and the behaviour of the substance's solubility during variation of the solution temperature. The most general of these conclusions are as follows:

1. The solubility of a solid substance increases with a rise in the temperature.
2. If two substances possess approximately the same melting points, then the substance with a lower heat of melting is more soluble than the other substance. The properties of the dissolved substances are not taken into account, as follows from equation (3.7.27).
3. If two substances possess approximately the same heats of melting, then the substance with the lower melting point is more soluble than the other substance.

The calculations show, however, that the solubilities of metal-oxides in the melts based on alkali-metal halides are considerably lower than the values predicted by the Le Chatelier–Shreder equation, i.e. the properties of the oxide solutions are characterized by negative deviations from ideality. This is caused by the fact that the sum of the interaction energies of the "dissolved substance–dissolved substance" system (non-dissociated oxide, $Me^{2+}-O^{2-}$)

and those of "solvent–solvent" system (cations and anions of the melt) exceeds the similar energy of the "solvent–dissolved substance" system (Me^{2+}–Cl^- and Na^+–O^{2-}). Nevertheless, by transforming equation (3.7.19) it is possible to derive some conclusions predicting the solubility changes with the increase in the melt temperature, and with the certain physico-chemical characteristics of the oxides, although these conclusions may be correct only in a relatively narrow temperature range. To estimate the oxide solubility changes with increase in the melt temperature, we transform equation (3.7.19) with the use of the following consequence from the expression for the solubility product of a 1:1 substance:

$$N_{Me^{2+},Hac} = \sqrt{P_{N,MeO}}. \tag{3.7.28}$$

Apart from the above-said it is well known that there exists a practically linear relationship between the enthalpy of melting of a substance and its melting point [356], which becomes especially clear if one considers only substances of closely similar compositions (ionic halides, oxides, etc.). According to Rabinovitch and Khavin, this dependence can be approximated by the equation [357]

$$\Delta H_{mp,MeO} = 25(\pm 4)T_{mp,MeO} \approx 3RT_{mp,MeO}. \tag{3.7.29}$$

As mentioned in the cited handbook [357], equation (3.7.29) remains correct for all inorganic substances, and deviations from this rule for given specific substances do not exceed 17%. Naturally, for metal-oxides MeO, as a kind of substances of similar composition and character of bonds, these deviations should be considerably smaller. Having performed some transformations of equation (3.7.19) taking into account the relationship (3.7.29) we finally obtain the following derivatives, which make it possible to connect the changes of oxide solubilities in molten salts with their melting point and the melt temperature:

$$\log P_{MeO} = 2.6 - 2.6\frac{T_{mp,MeO}}{T}, \tag{3.7.30}$$

$$\left(\frac{\partial \log P_{MeO}}{\partial T_{mp}}\right)_T = -\frac{2.6}{T}, \tag{3.7.31}$$

$$\left(\frac{\partial \log P_{MeO}}{\partial T}\right)_{T_{mp}} = \frac{2.6T_{mp}}{T^2}. \tag{3.7.32}$$

Equation (3.7.31) shows how the metal-oxide's solubility should be changed in the melt with the change in its melting point. This derivative possesses a negative value, and its calculated values are equal to 0.004 and 0.003 at temperatures of 600 and 700 °C, respectively.

The diagram which demonstrates the dependence of metal-oxide solubilities in the molten CsCl–KCl–NaCl eutectic at 700 °C against their melting points is shown in Fig. 3.7.9. For all the melts studied, these dependences are quite similar. As is seen, the solubilities of metal-oxides belonging to the same group (alkaline-earth, transition metals, and sequence Zn–Cd) decreases monotonically with the increase in their melting points. For the groups of analogous elements, such as Zn–Cd and Ba–Sr–Ca the slopes of $pP_{MeO}-T_{mp}$ plots coincide with the theoretical one (0.003) predicted by equation (3.7.31). The corresponding straight lines are practically parallel. From this it follows that the solubility of the chemically analogous metal-oxides decreases with the increase in the melting point, according to the thermodynamic calculations derived from the Le Chatelier–Shreder equation. A possible cause of the observed regularity is the fact that the deviations from the ideality in the solutions of the oxides formed by chemically analogous elements are practically the same. This means that the behaviour of the saturated solutions deviates from the Le Chatelier–Shreder equation in the same direction and in close proportions.

As for magnesium oxide, it should be mentioned that this drops out of the dependence for alkaline-earth oxides, because the chemical properties of magnesium are closer to those of aluminium than to the properties of other members of alkali-metal subgroup (it is the so-called diagonal periodicity). The dependence of oxide solubilities against their melting points exists for the

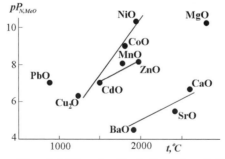

Fig. 3.7.9. Dependence of the solubilities of oxides in the molten CsCl–KCl–NaCl eutectic at 700 °C against their melting points.

elements of the first transition metal row, but its slope considerably exceeds the theoretical one.

In general, for all the studied oxides there is no uniform dependence of the oxide solubility products in the given melt and the melting point of the oxide. This seems to be connected with different crystalline structures of the said oxides, and with different degrees of covalence of the oxides formed by metals belonging to the s- and d-element groups.

The value of the derivative

$$\left(\frac{\partial \log P_{\text{MeO}}}{\partial T} \right)_{T_{\text{mp}}}$$

from equation (3.7.32) is bound up with the thermal coefficient of solubility predicted by the Le Chatelier–Shreder equation (the derivative contained in the right-hand part):

$$2.3 \left(\frac{\partial \log P_{\text{MeO}}}{\partial T} \right)_{T_{\text{mp}}} = \left(\frac{\partial P_{\text{MeO}}}{P_{\text{MeO}} \partial T} \right)_{T_{\text{mp}}}. \tag{3.7.33}$$

As seen from equation (3.7.32), this coefficient is dependent both on the melt temperature and on the melting point of the metal-oxide. In contrast with the derivative from equation (3.7.31), the value of the thermal coefficient of solubility is positive, and consequently the Le Chatelier–Shreder equation predicts an increase in the solubility of any substance together with elevation of the solvent temperature. The data on the oxide solubilities obtained for the molten $CsCl–KCl–NaCl$ eutectic at different temperatures allow us to verify the formulated conclusions. For this purpose, we write equation (3.7.32) in the form of finite differences, taking into account the fact that $pP_{\text{MeO}} = -\log P_{\text{MeO}}$:

$$\left(\frac{\partial P_{\text{MeO}}}{P_{\text{MeO}} \partial T} \right)_{T_{\text{mp}}} = 2.3 \times \frac{pP_{\text{MeO},T_1} - pP_{\text{MeO},T_2}}{T_2 - T_1}. \tag{3.7.34}$$

A question arises: which temperature should correspond to the derivative value from the left-hand side of equation (3.7.34)? This equation, as a matter of fact, is none other than a reformulation of the Lagrange theorem about the finite increment [358], which is written as follows: if the function $f(pP_{\text{MeO}})$ is uninterrupted in the $[a, b]$ segment (600–700 °C) and differentiated in this segment, then there can be found at least one point within this segment for which the following relationship

$$f'(x_0) = \frac{f(b) - f(a)}{b - a} \tag{3.7.35}$$

holds.

Since the dependence $pP_{MeO}-T$ agrees with these criteria, it is most expedient to attribute the derivative to a point from the central part of the 600–700 °C segment, i.e. 650 °C (923 K). The values of relative thermal coefficients of solubility obtained using the Le Chatelier–Shreder equation and the experimentally obtained coefficients are collected in Table 3.7.12.

A comparison of the experimental and the predicted data shows that the changes (increase) in solubilities of most of the metal-oxides studied with the elevation of the melt temperature is in good agreement with the Le Chatelier–Shreder equation. The estimation of the change in the oxide solubility in the vicinity of the experimental temperature can be made successfully using equation (3.7.32).

It is interesting to examine the relationship between the deviations of the relative thermal coefficient of solubility with the size of the oxide cations. We present this dependence against the parameter proportional to the inverse squared cation radius, i.e. Zr_k^{-2}, shown in Fig. 3.7.10. The results presented in this figure give evidence that the experimentally obtained thermal coefficients of solubility are, as a rule, in good agreement with those predicted by the

TABLE 3.7.12
Solubility products of oxides in the molten CsCl–KCl–NaCl eutectic and the thermal coefficients of the solubility

Oxide	$pP_{N,600}$	$pP_{N,700}$	T_{mp}	$-2,3(\Delta pP_{MeO}/\Delta T)$	$(\partial P_{MeO}/P_{MeO}\,\partial T)_T$	Col. 5/col. 6
1	2	3	4	5	6	7
MgO	12.68	11.68	3073	0.023	0.021	1.1
CaO	7.73	6.86	2873	0.020	0.020	1.0
SrO	5.81	5.52	2688	0.007	0.019	0.4
BaO	4.75	4.5	2190	0.006	0.015	0.3
MnO	9.73	9.49	2053	0.008	0.014	0.6
CoO	10.92	10.50	2028	0.011	0.014	0.8
NiO	12.60	11.42	2223	0.027	0.016	1.8
Cu_2O	6.94	6.34	1502	0.014	0.011	1.3
ZnO	8.80	8.19	2248	0.014	0.016	0.9
CdO	7.54	7.06	1773	0.011	0.012	0.9
PbO	8.21	7.06	1159	0.027	0.009	3

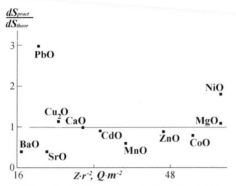

Fig. 3.7.10. Dependence of the relative thermal coefficient of solubility of oxides in the molten CsCl–KCl–NaCl eutectic on Zr_k^{-2}.

Le Chatelier–Shreder equation, which is valid for the case of ideal solutions, i.e. for the solutions, where the interactions between the dissolved particles and between the solvent constituents are close to that between the solvent and the dissolved substance.

The fact that the values of the relative thermal coefficients of solubility are close to the ideal ones shows that the deviations from ideality of the behaviour of the metal-oxides in their saturated solutions in molten salts are not so considerable, and that these deviations depend in the same proportion on the temperature changes. This conclusion may be substantiated by the concentration dependence of the activity coefficients of cations in molten chlorides, which is discussed in Ref. [81]: the activity coefficients of metal cations in diluted solutions are practically unchanged up to concentrations in the order of 0.1 mol kg^{-1}.

Essential deviations from the theoretical predictions are observed in the saturated solutions of oxides formed by large cations such as PbO, BaO, SrO, which possess appreciable solubilities in alkali-metal halide melts. The experimentally obtained relative thermal coefficient of solubility of lead oxide in the CsCl–KCl–NaCl eutectic melts is approximately three times as high as the theoretical one. This may be explained by the distinction of the electron structure of lead from other studied metals (lead belongs to p-elements) and because of this, lead drops out of all the found regularities. Another possible reason consists in the closeness of the melt temperature (600 or 700 °C) to the melting points of lead oxide (886 °C), and the complete solubility of PbO in all the studied chloride melts is very appreciable [359].

The deviations of the relative thermal coefficients of solubility from their theoretical values in the case of appreciably soluble oxides of type MeO can probably be connected with relatively the high concentrations of the oxides in their saturated solution. This concerns both the ionic ($Ba^{2+} + O^{2-}$, $Sr^{2+} + O^{2-}$) and non-dissociated (PbO) forms of oxides, which cause considerable distortions of the solution structure which, in turn, result in deviations from ideality.

It should be emphasized that the increase in the solubility of substances with rising temperature is not an undoubted regularity. It is known that, for aqueous solutions, there are numerous exclusions concerning slightly soluble substances with cations having an enhanced ability to solvation (Li^+, Ca^{2+}). For example, $Ca(OH)_2$, $CaSO_4$, Li_2CO_3 and $CaCO_3$ are characterized by anomalous changes of solubility with increasing solution temperature. Therefore, a slight solubility of a substance in the given solvent is the reason requiring thorough verification of the general rule of increase in solubility with the temperature of the solvent. In the case of the oxide solubility in ionic melts the above said remains true, since practically all metal-oxides are slightly (or moderately) soluble in these solvents.

Investigations in the molten CsCl–KCl–NaCl eutectic again confirm that the metal-oxide solubility in molten alkali-metal halides is dependent on the radius of the oxide's cation. In the ternary CsCl–KCl–NaCl eutectic the plot (3.7.16) remains correct: these dependences are presented in Fig. 3.7.11 for both experimental temperatures.

From Fig. 3.7.11 it is seen that the solubilities of all the studied oxides increase with temperature, but the slope of the dependence (3.7.16) decreases. This means that the solubilities of slightly soluble oxides increase faster than those of moderately soluble or readily soluble oxides.

The dependences of metal-oxide solubilities on the cation radius, inverse cation radius, and the sum of the radii of the cation and the oxide ion, are non-linear. There is no sharp correlation between the solubility and the common ionic moment of the cations (this parameter was introduced by Sementchenko) and the specific charge of the cation by Protsenko (and see Ref. [349]).

A correlation between the metal-oxide solubility in molten alkali-metal halides and the electronegativity of the metal cation is shown in our work [360] to consist of two regions: alkaline-earth metal-oxides belong to one of them and the second group consists of oxides of transition metals. The dependence of metal-oxide solubilities on the Pauling electronegativities

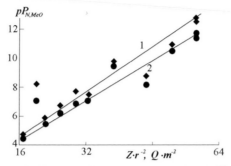

Fig. 3.7.11. Dependence of $pP_{N,MeO}$ in the molten CsCl–KCl–NaCl eutectic at 600 (1) and 700 °C (2) against Zr_k^{-2} [330].

in Ref. [361] that contains only one sharp is a practically linear plot for alkaline-earth metal-oxides, whereas there is no clear correlation for the transition metal-oxides (see Fig. 3.7.12).

Besides Pauling's electronegativity scale, estimated on the basis of the energies of bond rupture, there exist some other electronegativity scales based on various assumptions. Among these, the scales obtained by Mulliken, Allred and Rochow and Sanderson, should be mentioned. All these scales, together with the principles of their construction and the values of the electronegativities, can be found in handbooks.

The calculations of the attraction force between an atom (nucleus) and a valence electron remote from the nucleus by the distance of the covalent radius form the basis of the electronegativity scale constructed by Allred and

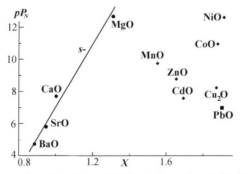

Fig. 3.7.12. Dependence of metal-oxide solubilities in the molten CsCl–KCl–NaCl eutectic at 600 °C against the electronegativity by Pauling.

Rochow, which is widely used by chemists:

$$F = \frac{Z^* e}{r^2}.$$ (3.7.36)

Using Slater's rules for the estimation of the effective charges of atoms (Z^*), but taking into account screening by all the electrons of the atom, which increases the Z^* values by 0.35 as compared with Slater's, Allred and Rochow obtained the following expression for the calculation of X:

$$X = \frac{0.00359Z^*}{r^2} + 0.744.$$ (3.7.37)

Thus, the electronegativity values are linearly dependent on a parameter similar to the polarizing action of the cation. The dependence of the solubility on the Allred–Rochow electronegativity is divided into two sharply bounded and practically linear plots with close slopes for $3d$-elements and for alkaline-earth metal-oxides (see Fig. 3.7.13). Since the slopes of these dependences are close, different positions of these plots may be explained by different relationships between the nuclear charge and Z^* for metals characterized by different electronic configurations. From the above-said it may be concluded that in high-temperature alkali-metal halide melts a correlation of metal-oxide solubilities with the crystal-lochemical radii of the cations is considerably simpler, i.e. it does not require the introduction of any corrections of the nucleus charge, such as Z^*.

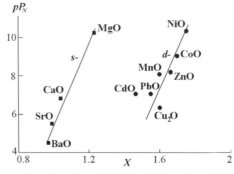

Fig. 3.7.13. Dependence of metal-oxide solubilities in the molten CsCl–KCl–NaCl eutectic at 700 °C against electronegativity by Allred and Rochow.

The said dependences of metal-oxide solubilities on the electronegativities are qualitatively similar for all the studied alkali-metal halide melts, there are no essential distinctions in their behaviour.

3.7.1.4 CsBr–KBr (0.66:0.34) melt

The metal-oxide solubilities have been shown in our papers [362, 363] to decrease considerably in the sequence "chloride–bromide". For example, strontium oxide, which is referred to as moderately soluble in low-acidic chloride melts becomes slightly soluble in the bromide melt, and the corresponding potentiometric titration curve has a sharp pO drop at the equivalence point (see Fig. 3.7.14, curve 3). Calcium and cadmium oxides, which are slightly soluble in chloride melts, become practically insoluble (Fig. 3.7.14, curve 2). This can be detected from the absence of the unsaturated-solution section in the corresponding titration curves. The basic properties of BaO in the bromide melt essentially weaken (Fig. 3.7.14, curve 4), and its degree of dissociation in the saturated solution diminishes considerably. It should be noted, however, that the bromide melt is a very capricious ionic solvent, especially as concerns its behaviour at high pO values occurring in the titration resulting in the formation of insoluble oxides. An obvious cause of this consists in the stronger reductive properties of bromide ion as compared with chloride ion. The presence in the melt of a

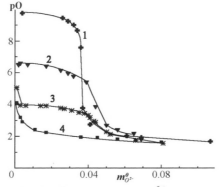

Fig. 3.7.14. Potentiometric titration curves for Ni^{2+} (1, $m = 0.035$ mol kg^{-1}), Cd^{2+} (2, $m = 0.040$ mol kg^{-1}), Sr^{2+} (3, $m = 0.050$ mol kg^{-1}), Ba^{2+} (4, $m = 0.050$ mol kg^{-1}) in the CsBr–KBr (2:1) melt at 700 °C.

strong oxide-ion acceptor favours the running of the following redox reaction:

$$\tfrac{1}{2}O_2 \uparrow + 2Br^- \rightarrow Br_2 \uparrow + O^{2-}. \tag{3.7.38}$$

Although the concentration of free oxygen in the extra-pure argon used for the creation of an inert atmosphere over the melt is low (approximately 0.002 vol%), the rate of reaction (3.7.38) is too high because of fixation of the formed oxide ions by metal cations. Passing the inert gas through the melt results in practically complete removal of the cations, forming slightly soluble oxides from the solutions owing to their "titration" by the oxygen admixture in the inert gas. The results of the potentiometric titration of the cations which form oxides moderately soluble in the bromide melt show that the corresponding solutions are not influenced by such an oxidation process, since the addition of these cations to the melt does not cause considerable pO increase.

Some progress in the studies of strongly acidic cations was achieved by passing the initial argon through a titanium sponge. Owing to fixation of gaseous oxygen by the sponge, its concentration reduces to approximately 10^{-8} vol%. This purification makes it possible to determine the solubility products of oxides such as MgO, NiO, CoO, MnO and ZnO. The effect of the process (3.7.38) on the studied acid–base reactions is negligible.

The oxide-solubility parameters obtained in the bromide melt by the method of potentiometric titration are collected in Table 3.7.13. It should be noted that their reproducibilities (confidence ranges) are appreciably worse than those in the molten alkali-metal chlorides. Lower values of oxide

TABLE 3.7.13
Parameters of metal-oxide solubilities in the molten CsBr–KBr (2:1) mixture at 700 °C, molality scale

Oxide	pP_{MeO}	pK_{MeO}	s_{MeO}	$s_{Me^{2+}}$	$\sum s$
BaO	3.67 ± 0.30	1.70 ± 0.43	1.07×10^{-2}	1.46×10^{-2}	2.53×10^{-2}
SrO	5.33 ± 0.12	3.01 ± 0.50	4.79×10^{-3}	2.16×10^{-3}	6.95×10^{-3}
CaO	8.44 ± 0.06	–	–	6.02×10^{-5}	
MgO	11.27 ± 0.06	–	–	2.32×10^{-6}	
MnO	9.21 ± 0.04	–	–	2.48×10^{-5}	
CoO	9.84 ± 0.20	–	–	1.20×10^{-5}	
NiO	11.22 ± 0.3	–	–	2.46×10^{-6}	
ZnO	9.54 ± 0.16	–	–	1.70×10^{-5}	
CdO	7.86 ± 0.16	–	–	1.17×10^{-4}	
PbO	7.03 ± 0.07	5.06 ± 0.36	1.07×10^{-2}	3.05×10^{-4}	1.10×10^{-2}

solubilities, together with an enhanced sensitivity of the bromide melt to oxidants (traces of free oxygen), give rise to additional sources of experimental errors.

As shown in Refs. [362, 363] all the correlations of metal-oxide solubilities with the physico-chemical characteristics of the oxides and the cations found in the chloride melts remain correct for the molten CsBr–KBr (2:1) mixture. Also, on the basis of the data presented in Table 3.7.13, the correlation dependence was obtained of the metal-oxide solubilities against the parameter a of the cubic crystalline lattice of the studied oxides. This is presented in Fig. 3.7.15. The same correlations were found later to be true for chloride melts. Moreover, there are no basic distinctions in the said dependences.

From the dependence shown in Fig. 3.7.15 it can be concluded that the growth of the parameter a of the cubic crystalline lattice of the oxide increases its solubility in ionic melts based on alkali- and alkaline-earth metal halides. This may be explained by the fact that the increase in the parameter a of the crystalline lattice results in the weakening of the interaction between ions and molecules in oxide crystals, which favours dissolution of these oxide crystals in molten salts. The solubilities of ZnO and PbO are not shown in Fig. 3.7.15, since at the temperature of the experiment both oxides possess hexagonal crystalline lattices.

The dependences constructed for the oxides belonging to alkaline metal-oxide and transition metal-oxide groups are practically linear and they can be

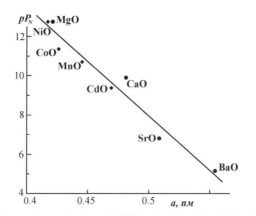

Fig. 3.7.15. Dependence of metal-oxide solubilities in the molten CsBr–KBr (2:1) mixture at 700 °C against the parameter a of cubic crystalline lattice of the oxides.

TABLE 3.7.14
The coefficients of $pP_{MeO}-a$ dependences for some alkali halide melts

Melt, t (°C)		$pP_{a=0}$	K
CsCl–KCl–NaCl (600)	All	43 ± 8	−74 ± 17
	AEM	46 ± 6	−80 ± 12
	TM	50 ± 16	−90 ± 36
CsCl–KCl–NaCl (700)	All	40 ± 6	−69 ± 12
	AEM	42 ± 15	−72 ± 32
	TM	45 ± 10	−80 ± 23
CsBr–KBr (700)	All	35 ± 5	−55 ± 10
	AEM	38 ± 13	−60 ± 26
	TM	37 ± 15	−58 ± 35

AEM: alkaline-earth metal oxides; TM: the transition metal oxides.

approximated by the following plots:

$$pP_{MeO} = pP_{MeO,a=0} + \left(\frac{\partial pP_{MeO}}{\partial a} \right)_T \times a = pP_{MeO,a=0} + ka. \quad (3.7.39)$$

The parameters of these plots for some investigated melts, obtained by the least-squares method, are presented in Table 3.7.14.

It may be concluded that the dependence of metal-oxide solubilities in molten alkali-metal halides upon the lattice parameters of the oxides is linear and well reproducible (with small confidence ranges at the confidence level of 0.95). This dependence is correct for all the metal-oxides, MeO studied and can be used for the prediction of their solubilities on the basis of their lattice parameters.

3.7.1.5 Molten CsI, 700 °C

The range of oxides suitable for investigations in this melt is appreciably narrowed as compared with the molten CsBr–KBr (2:1) mixture. The potentiometric titration method allows us to determine the solubilities of moderately soluble oxides in the iodide-melt, since the pO values in the section of the excess of the metal cations do not exceed 5–6. The solubility products of alkaline-earth oxides (excluding MgO), cadmium and lead oxides, were determined in Ref. [364]. The values of solubility products obtained, and the derived oxide solubilities in molten CsI at 700 °C are presented in Table 3.7.15.

TABLE 3.7.15
Solubility products of some oxides in molten CsI at 700 °C, molality scale

Oxide	pP	$s_{Me^{2+}}$
BaO	3.80 ± 0.15	1.26×10^{-2}
SrO	5.10 ± 0.30	2.82×10^{-3}
CaO	6.32 ± 0.14	6.29×10^{-4}
CdO	6.82 ± 0.06	3.89×10^{-4}
PbO	6.52 ± 0.10	5.50×10^{-4}

Studies of the acidic sections for the practically insoluble metal-oxides, such as MgO, NiO, CoO, could not use the normal experimental techniques. As a rule, the initial addition of the studied metal halide resulted in a considerable increase in pO accompanied by intense evaporation of iodine from the CsI melt, and by a slow but appreciable shift to lower values of the e.m.f. of the potentiometric cell. After stabilization of the e.m.f., the first small quantity of the titrant caused a sharp pO drop, and the solution exhibited basic properties as if the acidic substance had not been added to the iodide-melt. Nevertheless, after the termination of the experiment, the precipitated metal-oxide—whose quantity corresponded to the introduced quantity of the metal halide—was found at the bottom of the crucible-container. This means, that strongly acidic solutions in iodide-melts are extremely unstable, and are oxidized with the evolution of iodine and the formation of oxide ion in the melt, although the source of this oxygen has not been found yet.

Additional thorough purification from oxygen traces (by titanium sponge) of the inert gas used reduced its content below 10^{-8} vol% of O_2, and evacuation of the potentiometric cell in a tightly closed quartz shell did not result in any useful changes. Hence, the redox process in iodide-melts at high pO values runs without the participation of gaseous oxygen in the atmosphere over the melt, and it is not caused by the action of the electric current used for heating the tube furnace containing the potentiometric cell.

One of the most probable reasons for such a behaviour of strongly acidic solutions in iodide-melts, under the potentiometric titration conditions is transfer of oxygen through the solid electrolyte membrane to the iodide-melt. The following processes could be assumed to occur in the said system.

Gaseous oxygen on the inner surface of the membrane gas–oxygen electrode is transformed into oxide ions and holes, both of which charged objects are transferred to the external surface of the electrode that is in contact with the melt. Here, the holes withdraw electrons from iodide ions, which leads to the formation of free halogen evaporating from the studied melt. This process is favoured by the relatively low oxidation potential of iodine (in comparison with other halogens). If the said process actually takes place, then the decreasing partial pressure of oxygen in the inner electrode space should reduce the redox potential at the membrane oxygen electrode thus retarding the process of the melt oxidation. In this case, the problem should be solved by using membrane electrodes with "metal/metal-oxide" powders providing the constant partial pressure of gaseous oxygen.

It should be noted that iodine evaporation from the strongly acidic solutions was observed by Rybkin and Seredenko in molten caesium iodide at 650 °C (PO_3^- solutions) [63] and by Rybkin and Banik [64] who studied solutions of PO_3^-, $H_2PO_4^-$, VO_3^-, etc. in molten NaI at 700 °C.

However, let us return to the solubility data in caesium iodide-melt at 700 °C. It can be seen that the oxide solubilities in this melt are intermediate between those in KCl–NaCl (CsCl–KCl–NaCl) and CsBr–KBr melts. Since for the said iodide-melt the range of oxides suitable for the investigations is too small, compared with other halides, and the range of solubilities is too narrow, some of the correlations considered above have not been analysed for molten CsI. Nevertheless, for chloride, bromide and iodide-melts, linear correlations between pP_{MeO} and r_k^{-2} can be found, and the corresponding parameters of equation (3.7.16), with their confidence ranges, are presented in Table 3.7.16.

The investigations performed in the chloride melts demonstrate that, for melts of the same anion composition containing cations of close acidities (and recall that the oxobasicity indices of CaCl–KCl–NaCl and KCl–NaCl melts are practically the same), the metal-oxide solubilities have very close values. This means that, at the same anion composition, the oxide solubility products depend on the acidity of the constituent cation of the melt

TABLE 3.7.16
Coefficients of the dependences of pP_{MeO} against r_k^{-2} (equation (3.7.16)) for molten alkali-metal halides (confidence level, 0.95)

Melt	t (°C)	$pP_{r^{-2}\to 0}$	$\partial pP_{MeO}/\partial(r_k^{-2})$
KCl–NaCl	700	1.8 ± 0.9	0.053 ± 0.01
CsCl–KCl—NaCl	600	1.7 ± 2.0	0.057 ± 0.01
	700	1.7 ± 0.3	0.054 ± 0.02
CsBr–KBr	700	3.2 ± 1.8	0.053 ± 0.01
CsI	700	2.65 ± 0.6	0.052 ± 0.01

(the solubility products in the molten KCl–LiCl eutectic mixture are considerably higher) and on the complexation ability of the melt-anion rather than on the presence of holes in such a melt, which proves the dissolution of oxides according to the physical mechanism. The correlation dependence (3.7.16) found makes it possible to propose the obtained regularities for estimation of metal-oxide solubility in the melts of the corresponding anion composition at a given temperature. It should be noted that, while performing such estimations, one must take into account the method of preparation of the saturated oxide solution (precipitation from the metal chloride solutions, dissolution of the oxide powder, etc.).

The close slopes of the $pP_{MeO} - r_k^{-2}$ dependences found (equation (3.7.16)) allow us to assume that the main cause of the changes of metal-oxide solubilities in melts at the melt-anion substitution is the same for all the oxides, and consists of the changing properties of the halide melt (e.g. the ion association).

The interaction between charged particles, which we can observe in the solutions of metal-oxides (Me^{2+}, O^{2-}), is superimposed upon a number of factors complicating the description of such interactions. Ion solvation is to be mentioned among them. It should be mentioned that, at relatively low temperatures, the solvate shells have rather large sizes, and for these systems there is no direct correlation between the crystallochemical radius of the metal cation (r_k) and the radius of the solvated metal cation (r_s), since the latter depends on the configuration of the outer electron layer of the metal cation (s-, d-, or p-). Elevation of the melt temperature favours destruction of the solvate shell, and under such conditions r_s approaches r_k. This is confirmed, in particular, by the fact that the confidence range of the dependence of pP against r_k^{-2} narrows with the temperature rise (see Table 3.7.16). The dependences obtained for oxide solubility against the metal-cation radii lead to the conclusion that, at high temperatures, the properties of metal cations maximally approach the properties of a hard charged sphere, and the effects caused by cation solvation reduce.

It is of interest to determine the ratio of the concentrations of the ionized (Me^{2+}) and non-dissociated (MeO) forms of metal-oxides in their saturated solutions in different alkali-halide melts. Naturally, the necessary data can only be obtained for oxides with appreciable solubilities in the ionic melts, which can be related to the appearance of the unsaturated solution section in the corresponding titration curves. Equations (4.14) and (3.6.7) make it possible to calculate the said ratio of the ionized and non-dissociated forms of metal-oxides, and their sum, in the saturated solutions. These total characteristics give a complete picture about the behaviour of the metal-oxides in their solutions, since incomplete dissociation of the metal-oxides is taken into account. Moreover, these values allow us to elucidate the behaviour of the said ratio in molten alkali-metal halides, upon substitution of the melt-anion in the "chloride–bromide" sequence, since the determination of the dissociation constants of the oxides in molten CsI fails. As the average molecular masses of the alkali-metal chloride and bromide melts differ essentially, the magnitudes expressed on the molar-fraction scale are to be compared.

It is obvious that the molar fractions of all the studied oxides in their saturated solutions are negligible in comparison with the molar fractions of the melt–solvent, which approaches 1 (unity). This means that these solutions may be considered as the metal-oxide solubility in 1 mol of the melt–solvent, and in this case the said comparison will be adequately substantiated.

The concentrations of the ionized and non-dissociated forms of the metal-oxides in their saturated solutions in the studied alkali-metal halide melts (on the molar-fraction scale) and the ratios of these forms are presented in Table 3.7.17.

TABLE 3.7.17
Concentrations of ionized and undissociated forms of some metal oxides in their saturated solutions in alkali-metal halide melts (molar-fraction scale)

Oxide	$s_{Me^{2+}}$	s_{MeO}	Σs	$\Sigma s/s_{Me^{2+}}$
CsCl–KCl–NaCl, 600 °C				
CaO	1.36×10^{-4}	9.54×10^{-5}	2.31×10^{-4}	1.70
SrO	1.24×10^{-3}	7.95×10^{-4}	2.04×10^{-3}	1.65
Cu$_2$O	3.35×10^{-4}	2.18×10^{-4}	5.53×10^{-4}	1.65
PbO	7.85×10^{-5}	7.42×10^{-4}	8.21×10^{-4}	10.5
CsCl–KCl–NaCl, 700 °C				
CaO	3.71×10^{-4}	1.35×10^{-4}	5.06×10^{-4}	1.36
SrO	1.70×10^{-3}	3.89×10^{-4}	2.10×10^{-3}	1.23
Cu$_2$O	6.76×10^{-4}	1.78×10^{-4}	8.54×10^{-4}	1.26
PbO	2.95×10^{-4}	9.34×10^{-4}	1.23×10^{-3}	4.17
CsBr–KBr, 700 °C				
SrO	3.89×10^{-4}	8.68×10^{-4}	1.26×10^{-3}	3.24
BaO	2.63×10^{-3}	1.95×10^{-3}	4.58×10^{-3}	1.74
PbO	5.50×10^{-5}	1.95×10^{-3}	2.01×10^{-3}	36.45

These data show that when the total sum of solubility of metal-oxides is determined only from data on their solubility products, there arises an error whose value increases together with the strengthening of the oxoacidic properties of the metal cation. It should be emphasized that for oxides such as CaO, SrO and Cu_2O, whose cations possess close radii, the ratios of the total solubility to the concentration of the ionized form in the chloride melts are close—varying from 1.5 to 2. For lead oxide, this ratio is considerably higher, although the radii of Pb^{2+} and Sr^{2+} cations are practically coincident. This gives further evidence of the anomalous behaviour of Pb^{2+} ions in ionic melts, since lead belongs to the p-elements, whose properties differ appreciably from the s- and d-elements. As seen from Table 3.7.17, the elevation of the ionic melt temperature results in the increase of the dissociation degrees of metal-oxides in their saturated solutions and, consequently, the contribution of the concentration of ionized form of the metal-oxide to the total solubility rises as compared with the concentration of the undissociated form. This fact is confirmed by the reduction upon temperature rise of the ratio $\sum s/s_{Me^{2+}}$ for the oxides studied in the molten CsCl–KCl–NaCl eutectic at 600 and 700 °C. From what we have said, it follows that the temperature increase leads to a faster rise in the degree of dissociation as compared with the total oxide solubility in the melt, i.e. the acidic properties of the metal cations reduces.

The substitution of the melt-anion in a molten alkali-metal halide, i.e. proceeding from chloride melts to bromides and iodides, results in the appreciable reduction in the contribution of the ionized constituent (Me^{2+} concentration) in the total oxide solubilities in the said melts. The data allow us to trace the effect of the constituent halide ion on the solubilities of metal-oxides in the corresponding melt. It should be noted that the substitution of chloride ions with bromide ones results in an appreciable reduction in the solubilities of all the metal-oxides studied, i.e. the values of $pP_{MeO,N}$ become lower by 1–2 orders. At the same time, the oxide dissociation in the saturated solution of the oxide reduces. Thus, a considerable reduction in metal-oxide solubilities in bromide melts as compared with chloride ones can be explained in a similar way. Equilibrium (2.4.13) which takes place at the dissolution of a metal-oxide in an ionic halide melt is superimposed upon the interactions of the metal cation with the anions of the melt–solvent, which are denoted hereafter as X^- (equation (3.6.3)).

It is well known that the "absolute" equilibrium concentration of the metal cations in the solutions in molten halide melts cannot be measured. The term "equilibrium metal cation concentration" usually includes the sum of the molalities of both the "free" cation and its different halide complexes possessing the common composition MeX_n^{2-n}, which can be called the "molality of the solvated cation". The value of the solubility product obtained practically is $P_{MeO,exp}$, where

$$P_{MeO,\varepsilon\xi\pi} = [Me^{2+}]\left(1 + \sum_{i=1}^{n} K_i[X^-]^n[O^{2-}]\right)$$

$$= P_{MeO,true} + P_{MeO,true}\sum_{i=1}^{n} K_i[X^-]^n[O^{2-}]. \qquad (3.7.40)$$

This distinction from the "true" solubility product, p_{true}, becomes more essential with the increase in the complexation ability of the halide anion and its equilibrium molality in the melt. From equation (3.7.40), it is seen that there exist two possible reasons for the change in the solubility product of metal-oxides in the studied melts:

– the change in the complexation ability of the melt-anion, which is expressed quantitatively by the complexation constant;
– different degrees of association of the melt–solvent ions (the equilibrium molality of the "free" halide ions).

The stability of halide complexes is known to increase in the sequence $Cl^- \rightarrow Br^- \rightarrow I^-$ [81, 365–369]. This follows from the fact that the corresponding stability constants, estimated on the basis of molten alkali-metal nitrates, increase in the sequence given. However, alkali-metal nitrate melts are referred to media possessing the lowest ability of metal–cation solvation. Moreover, the molalities of the metal cations studied in the above-said works do not exceed 0.02 mol kg^{-1} and the corresponding molalities of halide ions are in the order of 0.3 mol kg^{-1}. It is obvious enough that there is no appreciable association of halide ions in complexes with alkali-metal halides in such diluted solutions.

For molten alkali-metal halides (KtX), the description of the equilibrium system is essentially complicated owing to the auto-complex structure of these melts. This kind of the melt structure can be presented as follows [370–373]:

$$KtX^{(n-1)-} + (n-1)Kt^+. \qquad (3.7.41)$$

The results of investigations performed in molten individual alkali-metal halides show that the complexation ability of bromide ions is minimal as compared with the other halides considered. This means that molten alkali-metal bromides, with constituent potassium and caesium cations, contain a relatively large quantity of "free" cations, and for this reason their oxoacidic properties are stronger. The increase in the melt acidity is the factor causing the increase in metal-oxide solubilities.

However, a reduction in the complexation ability of the melt-anion leads to a shift in the equilibrium (3.6.3) to the left, which finally results in a reduction in the metal-oxide's solubility. The substitution of bromide ions by iodide ions (transfer from bromide to iodide-melts) leads to the strengthening of the basic properties (which cause a decrease in the oxide solubility) and of the complexation ability (which results in the solubility increase). For molten CsI, the simultaneous action of both factors leads to an increase in the metal-oxide solubilities. However, iodide ion belongs to the group of soft bases, and this is the main reason for the solubilities of oxides in iodide-melts remaining lower than that in the molten alkali-metal chlorides.

A conclusion similar to that given above can be made on the basis of simpler considerations. We shall analyse the scheme of interparticle interactions in a solution of metal-oxide in molten alkali-metal halides:

$$
\begin{array}{ccc}
Kt^+ & \overset{I}{\leftrightarrow} & X^- \\[2mm]
II\updownarrow & & \updownarrow III \\[2mm]
O^{2-} & \overset{IV}{\leftrightarrow} & Me^{2+}
\end{array}
\tag{3.7.42}
$$

If the sum of interactions I and IV is close to that of II and III, then the saturated solution of the metal-oxide in the KtX melt will be ideal or close to ideality. However, interaction IV for the case of interaction of two doubly charged ions should be considerably stronger than the interactions between singly charged ions of the ionic solvent and single-double-charge interaction of the solvent particles (1–1 ions) with those of the dissolved metal-oxide (2–2 ions). Therefore, negative deviations should be observed from ideality in the "metal-oxide–alkali halide melt" system. The calculations performed on the basis of the Le Chatelier–Shreder equation show that this is the case. The substitution of the halide anion of the ionic melts does not result in an appreciable change in the energy of interaction IV. A reduction in the complexation ability of the melt-anion leads to a weakening of the interaction I and, (to a larger extent), of the interaction III, whereas the degree of interaction II increases. The above-mentioned considerations show that the oxide-solubility changes should be dependent on the competition of interactions of types I (the melt association), II (the melt acidity), with the strongest interaction III (the complexation ability) with changes of the ionic melt composition.

Consequently, while passing from the chloride melts to the CsBr–KBr one, the reduction in the complexation ability of the melt-anion turns out to be the main factor which causes the reduction in the metal-oxide solubilities, whereas in iodide-melts the solubilities strengthen. Nevertheless, in the case of iodide-melts, we should take into account an additional factor: the interactions between hard and soft acids and bases. So, a "Cl–Br" anion substitution reduces the oxide solubilities, but on the contrary, for the case of ("Br–I") substitution there is a certain increase in the oxide solubility, which can be partially caused by the difference in the cation composition of the studied halide melts.

3.7.1.6 Molten potassium halides

Papers devoted to the studies of metal-oxide solubilities in molten potassium halides are few. In particular, Guyseva and Khokhlov reported the determination of the solubility product of MgO in molten KCl in the 1067–1195 K temperature range [374]. The study was performed by the potentiometric method using the $Pt(O_2)$ gas-oxygen electrode. The experimental method consisted of the addition of a known quantity of $MgCl_2$ to the pure melt, and measurements of e.m.f. for the following cell:

$$O_{2(gas)}|Pt|MgO_{(s)} + MgCl_{2(melt)} + KCl_{(melt)}\|KCl_{(melt)}|C, Cl_{2(gas)}.$$
(3.7.43)

The relationship between the equilibrium molality of the oxide ion and the e.m.f. of the cell (equation (3.7.43)) can be expressed by the following equation:

$$[O^{2-}] = \frac{p_{O_2}}{p_{Cl_2}} \exp\left(\frac{2F}{RT}(E - E^*)\right),$$
(3.7.44)

where E^* is the conventional standard potential of the O_2/O^{2-} system vs the chlorine electrode, which varies with the temperature according to the equation

$$E^* = -1.70 + 3.16 \times 10^{-4}T \pm 0.010 \text{ V}, \tag{3.7.45}$$

at $p_{O_2} = P_{Cl_2} = p$. This equation allows us to estimate the solubility product of MgO by using the formula:

$$P_{MgO} + \frac{[MgCl_2]}{p} \exp\left(\frac{2F}{RT}(E - E^*)\right). \tag{3.7.46}$$

Although the obvious drawback of this work consists of passing O_2 through the acidic melt, which leads to oxidation of chloride ions in the excess of the Lux acid, nevertheless, the authors obtained the thermal dependence of MgO solubility in molten KCl as follows:

$$pP_{MgO} = 2.198 + 11,158.3T^{-1} \pm 0.086, \text{ molar fraction scale.} \tag{3.7.47}$$

According to this equation the solubility of MgO in the said melt at 800 °C can be estimated by a value of $pP_{MgO} = 12.59$, in molar fractions, which corresponds to 10.33 in molalities. This value corresponds to an equilibrium concentration of Mg^{2+} and O^{2-} in the melt of 2.1×10^{-6} mol kg^{-1}. A similar investigation performed by the isothermal saturation method [375] yields the total solubility of MgO at 800 °C. Taking into account the fact that, according to the data of Ref. [375] the concentration of the non-dissociated oxide is 4.9×10^{-2} mol kg^{-1}, the data of Ref. [374] give evidence of incomplete dissociation of magnesium oxide in molten KCl at 800 °C.

From the above-mentioned studies [374, 375] it can be concluded that, up to now, systematic solubility investigations in molten potassium halides have not been performed. This also concerns other alkali-metal halide melts, although these results would be very useful for the estimation of the effect of melt acidity and anion composition on metal-oxide solubilities at ~ 800 °C.

Recently we have studied the oxide solubilities in molten KCl at 800 °C [376]. An example of the potentiometric titration curve is depicted in Fig. 3.7.16 (curve 1), and the solubility data obtained by treatment of the titration results are presented in Table 3.7.18. The data collected in this table show that the order of arrangement of oxides according to increase in their

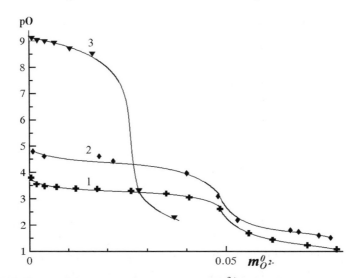

Fig. 3.7.16. Curves of the potentiometric titration of Ca^{2+} cations in molten KCl (1), KBr (2) and KI (3) at 800 °C.

solubility corresponds to the similar dependences in the KCl–NaCl and CsCl–KCl–NaCl melts at 600 and 700 °C. The study makes it possible to estimate the dissociation constants of PbO and CdO, which were close, and therefore, it can be assumed that the dissociation constants of oxides possessing close solubility product values are practically the same. The obtained solubility parameters are close to those in the KCl–NaCl melt at 700 °C. The $pP_{MeO} - r_k^{-2}$ plot constructed using the obtained data is practically

TABLE 3.7.18
Thermodynamic parameters describing the behaviour of metal in molten KCl at 800 °C

Oxide	pP_{MeO}, molalities	pP_{MeO}, mol fractions
SrO	3.04 ± 0.3	5.30
CaO	4.58 ± 0.2	6.84
MgO	9.32 ± 0.1	11.74
CoO	7.78 ± 0.2	10.04
NiO	8.59 ± 0.1	10.85
CdO	4.68 ± 0.1	6.94
CdO(pK)	2.23 ± 0.1	3.36
PbO	4.66 ± 0.2	6.92
PbO(pK)	2.44 ± 0.1	3.57

linear ($r_{xy} = 0.992$) and can be approximated by the following equation:

$$pP_{MeO,N} = 1.81(\pm 0.35) + 0.0514(\pm 0.003)r_k^{-2} \qquad nm \qquad (3.7.48)$$

Both coefficients coincide with those for the KCl–NaCl melt (1.8 ± 0.9, and 0.053 ± 0.01, respectively), which means that the character of the interaction between the metal cations and chloride ions remains unchanged within the $100-200\,°C$ temperature range.

We have also investigated the solubilities of MgO and CaO in molten potassium halides at $800\,°C$ [197] to elucidate the effect of anion composition of the halide melt on metal-oxide solubility. The MgO was found to be practically insoluble in chloride and bromide melts, and the iodide-melt could not be investigated owing to intense iodine evolution from strongly acidic solutions. In contrast, CaO solubility products were determined successfully in all the potassium halide melts at $800\,°C$, by the potentiometric titration method. The corresponding potentiometric titration curves are shown in Fig. 3.7.16.

The potentiometric titration curve of Ca^{2+} in molten KCl (see Fig. 3.7.16, curve 1) contains an initial section with a sharp pO decrease at low titrant weights, whereas in molten KBr this section is practically imperceptible (Fig. 3.7.16, curve 1). The solubility of CaO becomes very low in molten KI. It should be noted that a considerable loss of Ca^{2+} ions was observed in this melt owing to the above-mentioned oxidation process with the evolution of iodine.

The obtained solubility data are presented in Table 3.7.19. The oxide-solubility changes in molten potassium halides can be explained successfully in the framework of the HSAB concept. Indeed, Ca^{2+} ion belongs to the hard acids, whereas chloride ion is a hard base, bromide ion is an intermediate base and iodide ion is a soft base. Therefore, the stability of calcium halide complexes should reduce in the "chloride–bromide–iodide" sequence. Since an increase in the halide complex's stability results in the elevation of the oxide solubility in the melts, one should expect a reduction in the solubility of

TABLE 3.7.19
Parameters of Lux acid–base reactions in molten potassium halides at $800\,°C$ (molar fractions)

Equilibrium	KCl	KBr	KI
pP_{CaO}	7.09 ± 0.07	7.83 ± 0.04	12.35 ± 0.18
pP_{MgO}	11.74 ± 0.09	12.49 ± 0.09	–

CaO and other oxides in the sequence "chloride–bromide–iodide", which actually takes place.

3.7.1.7 Other solvents based on alkali-metal halides

Inyushkina et al. investigated the solubility of MgO in individual molten alkali-metal chlorides: CsCl (700–850 °C), KCl (820–1000 °C), NaCl (850–1100 °C), and RbCl (750–900 °C) [377]. The study was performed by the method of isothermal saturation: MgO was added to the melt in the form of the ordinary commercial powder, or of pills pressed at $p = 80$ MPa and calcined at 1250 °C. Saturation of the solutions with respect to MgO solubility was shown in Ref. [377] to be achieved after keeping the oxide in contact with the melt for 4 h.

The solubility of the powdered MgO in molten KCl was 140–420 times as high as that for the oxide which had been subjected to pressing and preliminary thermal treatment. This was explained by the formation of suspensions in the solutions contacting with the powdered oxide. The slope of the dependence of the solubility of MgO pressed in pills, against the inverse temperature, is three times as large as that for the powdered oxide, in molten KCl.

The dependence of MgO solubility ($\log N_{MgO}$) on the free volume of the melts studied shows that the solubility of this oxide proceeds according to the physical mechanism. In other words, the dissolution process consists, mainly, of filling voids in the melt's bulk by the oxide molecules, and only then by the following process of oxide dissociation. On the basis of the solubility data, the authors determined the enthalpy change upon dissolution of MgO as the slope of the dependence of solubility on the inverse temperature:

$$\log N_{MgO} = -\left(\frac{\Delta H_s}{2.3R} \right) T^{-1} + C, \tag{3.7.49}$$

These dependences are

$$\log N_{MgO} = -2810 T^{-1} - 4.76 \ \text{for NaCl}, \tag{3.7.50}$$

$$\log N_{MgO} = -2916 T^{-1} - 1.49 \ \text{for KCl powder}, \tag{3.7.51}$$

$$\log N_{MgO} = -6269 T^{-1} - 0.95 \ \text{for KCl pills}, \tag{3.7.52}$$

$$\log N_{\text{MgO}} = -10,450 T^{-1} + 3.37 \quad \text{for RbCl}, \tag{3.7.53}$$

$$\log N_{\text{MgO}} = -11,700 T^{-1} + 5.21 \quad \text{for CsCl}. \tag{3.7.54}$$

The values of ΔH_s determined from the constructed plots are 54 (NaCl), 120 (KCl), 200 (RbCl) and 224 (CsCl) kJ mol^{-1}. The experimental data show that there is no appreciable interaction of the dissolved MgO with the ionic solvent, i.e. from the viewpoint of chemical thermodynamics, the formation of $Mg^{2+} + O^{2-}$ ions in the melt is less favourable than the dissolution of the solid in the form of MgO.

In order to estimate the effect of the cation composition of melts based on alkali-metal halides upon the metal-oxide solubilities, we investigated the solubility of MgO in the following melts: KCl and CsCl at 800 °C, NaCl (830 °C) and molten CsCl–KCl–NaCl and KCl–LiCl eutectic mixtures at 800 °C. The solubility parameters obtained are collected in Table 3.7.20, where the oxoacidic properties of the chloride melts reduce when moving from the top to the bottom. From the data cited here, it follows that the solubility products of MgO increase with the substitution of the low-acidic constituent cation by that possessing enhanced acidic properties. The same can be said for the increase in the concentration of the constituent acidic cation in the melt: there is an increase in the MgO solubility in the sequence, KCl–CsCl–KCl–NaCl(CsCl)–NaCl, where the concentration of Na$^+$ and hence the melt acidity, increases.

The solubility product index of MgO in the molten KCl–LiCl exceeds the corresponding values in the low-acidic chloride melts by approximately 3.5pP units. This is in good agreement with the data obtained at 600 and 700 °C. Moreover, this difference shows that the relative acidic properties of ionic melts do not undergo a sharp change with elevation of the melt's temperature.

TABLE 3.7.20
Solubility parameters of MgO (pP_{MgO}) in the melts based on alkali-metal halides at $\sim$800 °C (molar fractions)

Melt	pP_{MgO}
KCl	11.74 ± 0.09
CsCl	11.40 ± 0.12
CsCl–KCl–NaCl	11.44 ± 0.10
NaCl	11.22 ± 0.03
KCl–LiCl	7.87 ± 0.04

The investigations of MgO solubility in the molten eutectic mixtures make it possible to trace the thermal dependence of the MgO solubility product in ionic melts possessing quite different acidic properties. These dependences are presented in Fig. 3.7.17. They are practically linear, although in both dependences there is a bending towards the inverse-temperature axis near 700 °C. Nevertheless, both plots can be approximated by the following equations (on the molar-fraction scale):

$$pP_{MgO,KCl-LiCl} = 0.79(\pm 1.4) + 7495(\pm 1400)T^{-1}, \qquad (3.7.55)$$

$$pP_{MgO,CsCl-KCl-NaCl} = 5.81(\pm 1.8) + 5920(\pm 1700)T^{-1}. \qquad (3.7.56)$$

While discussing the obtained slope values it should be noted that the theoretical slope value calculated according the Le Chatelier–Shreder equation (3.7.19) considered above is approximately equal to 7300. This value is practically coincident with the slope in equation (3.7.55), whereas the slope in equation (3.7.56) deviates from the theoretical one by less than 20%. This seems to be explained by the fact that the chemical properties of magnesium (and Mg^{2+} ion) resemble those of lithium rather than the properties of the other alkaline-earth metals (it is the so-called diagonal periodicity). Therefore, the properties of MgO and Li_2O in the molten KCl–LiCl eutectic are close to ideal, and the Le Chatelier–Shreder equation is correct just for such solutions. At the same time, the chemical properties of magnesium and lithium cations differ essentially from those of Na, K and Cs.

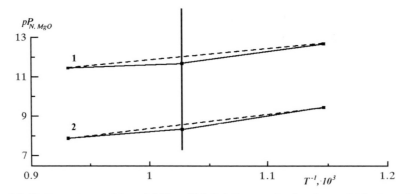

Fig. 3.7.17. Dependence of the solubility of MgO (pP_{MgO}, molar-fraction scale) in the molten CsCl–KCl–NaCl (1) and KCl–LiCl (2) eutectics against the inverse temperature.

Therefore, one should not expect ideal behaviour for MgO and Li$_2$O solutions in halide melts based on these cations.

We shall now consider some investigations, which cannot be classified among the systematic ones, but which nevertheless give some information about the metal-oxide solubilities in some ionic melts. Woskressenskaja and Kaschtschejev performed one of the first investigations of oxide solubilities in molten salts [327]. The solubilities of oxides of calcium, magnesium, zinc, chromium(III) and copper(II) in molten sodium and lithium as well as potassium sulfates and chlorides were determined in the 700–1200 °C temperature range. The solubilities of all the oxides studied were shown by the isothermal saturation method to increase with the melt temperature, although for the magnesium oxides this fact was not reliably confirmed. The solubilities of the metal-oxides decrease in the following sequence: CaO $\approx$ Cu$_{(2)}$O > ZnO > MgO > Cr$_2$O$_3$. Dissolution of copper(II) oxide in the chloride melt is accompanied with simultaneous oxidation of the latter with the reduction in Cu(II) to Cu$^+$. However, the total copper concentration in the melts did not depend on the initial degree of oxidation of the copper in the oxide, i.e. the addition of CuO or Cu$_2$O to the melt led to the same final distribution of copper between both of the oxidation degrees, and the solutions formed were nothing but the eutectic CuO–Cu$_2$O mixture.

The data on the solubility of CaO and ZnO in the chloride melts allow us to trace the effect of the acidic properties of the melt-constituent cation on the metal-oxide solubilities: the ratio of the solubilities in K$^+$, Na$^+$ and Li$^+$-based melts is 1:1.5:15. However, these are the total solubilities, and the data are distorted by the contribution of an appreciable concentration of non-dissociated oxide, which is not sensitive to changes in the melt-acidity. It should be emphasized that since we do not know the ratios of the ionized fraction to the non-dissociated one in all the melts studied, we cannot estimate quantitatively the oxoacidic properties of all the melts in question. This neglect of the metal-oxide's dissociation is because the said work was published in 1923, whereas the Lux definition was formulated only in 1939. Therefore, the authors of Ref. [327] did not take into account the dissociation of the oxides in the molten salts, which from our point of view, reduced the significance of this investigation.

Although Woskressenskaja and Kaschtschejev [327] constructed some correlations of the metal-oxide solubilities in their melts against the generalized ionic moments of the oxide and the molten salt, these dependences were found to be non-linear and were characterized by a considerable scattering of the points.

Grigorenko *et al.* investigated ZrO_2 solubility in molten KCl, KPO_3, and KCl + KPO_3 mixtures at 800 °C [378]. The ZrO_2 solubility was shown in this work to be 1.34 mass% in pure KPO_3, and in the equimolar KCl–KPO_3 mixture it achieves 3.25 mass%. In the opinion of the authors of Ref. [378], the main cause of the increase in the solubility of ZrO_2 while the KPO_3 concentration is decreasing in the mixed melt is the de-polymerization of potassium metaphosphate which results in the strengthening of the oxoacidic properties of the melt. If the concentration of KCl in the mixed melt exceeds 60 mol% then the solubility of ZrO_2 decreases sharply because the strengthening of potassium metaphosphate's dissociation under the said conditions cannot compensate for the decrease in the total concentration of the Lux acid.

The solubilities of nickel and thorium(IV) oxides in molten KF–LiF mixture at 550 °C were investigated by the potentiometric titration method using a membrane oxygen electrode [379, 380]. The solubility product of NiO in the fluoride melt was equal to 1.3×10^{-6}, which considerably exceeded the corresponding values in the chloride melts discussed above. The solubility of NiO seemingly increased owing to the fixation of Ni^{2+} cations in strong fluoride complexes.

Gorbunov and Novoselov reported an investigation of the interaction of LaF_3 and CeF_3 with CaO and La_2O_3 in a molten LiF–NaF (0.6:0.4) eutectic mixture [381]. In this melt, the solubilities of CaO and La_2O_3 were determined to be 0.1 and 0.05 mass%, respectively, at 800 °C. After determining these solubilities the authors added CaO and La_2O_3 to the solutions of the rare-earth fluorides in the melt. The stoichiometry of these interactions was found to be as follows by analysis of the lanthanide concentration after each CaO addition:

$$LnF_3 + CaO = LnOF + CaF_2. \tag{3.7.57}$$

The solubility of LaOF in the melt is equal to 0.8 mass% and that of CeOF is 1 mass%. The addition of CaO over the stoichiometry predicted by equation (3.7.57) leads to the following interaction according to the equation

$$2LnOF + CaO = Ln_2O_3 + CaF_2. \tag{3.7.58}$$

Addition of La_2O_3 to the fluoride melt containing LnF_3 results in the formation of LaOF and La_2O_3–LaOF solid solutions, and a mixture of composition $2LaOF + CeOF$ in the case of CeF_3. In the latter case, the phase-composition of the precipitate was not examined. Deanhardt and Stern reported investigations of the solubilities of CoO, NiO and Y_2O_3 in molten NaCl and Na_2SO_4 at 1100 K [244, 245]. The solubility products of the oxides were found by potentiometric titration to be $(0.99 \pm 3.3) \times 10^{-11}$ for Co_3O_4, $(1.9 \pm 4.9) \times 10^{-12}$ for NiO and $(1.4 \pm 2.2) \times 10^{-36}$ for Y_2O_3 in the chloride melt (molar fractions). The corresponding values in the sulfate melt were as follows: $(0.8 \pm 2.2) \times 10^{-10}$, $(1.6 \pm 8) \times 10^{-11}$ and $(4.5 \pm 6.4) \times 10^{-31}$. With excess of the titrant all the oxides formed showed appreciable oxoacidic properties and fixed the oxide ions into oxo-complexes such as NiO_2^-, CoO_2^- and YO_2^-. The considerable increase observed in the calculated solubility products of the metal-oxides in the sulfate melt as compared with chloride melts confirms the assumption that the acidic properties of the studied cations are levelled to the acidity of SO_3.

Combes and Koeller studied the solubility of CaO in molten NaCl at 1100, 1150 and 1200 K by the SAMs [382]. Elevation of the temperature in the temperature range studied led to a small increase in the solubility product of CaO from $pP_{CaO} = 4.3 \pm 0.3$ at 1100 K to 4.1 ± 0.3 at 1150 K and to 3.9 ± 0.3 at 1200 K. The first value is in good agreement with the data obtained in our work on CaO solubility in molten KCl at 800 °C, 4.8 ± 0.1, taking into account the fact that the oxobasicity index of NaCl is approximately equal to 0.5 of that of KCl. The dependences of e.m.f. against the initial concentration of CaO were characterized by the existence of an inclined section corresponding to the unsaturated solution, and the plateau which testified to the formation of the saturated solution of CaO. The initial sections of the plateau, in two cases from the three, contained a minimum connected with the partial dissolution of the oxide powder. Besides the CaO solubility, the dissociation constants of hydroxide ion and the hydrolysis constant of NaCl were determined at the same temperatures; their values at 1100 K were equal to 11.0 ± 0.3 and 1.6 ± 0.3, respectively. This shows that the strength of OH^- as a strong Lux base increases with the temperature and with the strengthening of the melt acidity.

A similar study of CaO solubility performed in Refs. [168, 383] allows us to refine the data of Ref. [382]. Indeed, the CaO solubility in Ref. [382] was investigated without additional calibration of the potentiometric cell with strong Lux bases, and therefore, the solubility product of CaO was calculated using the initial molality of CaO corresponding to the onset of the plateau. Our investigations show that the inclined section of the SAM plot is located above the E–pO plot obtained with the use of NaOH as strong Lux base. This means that CaO does not undergo complete dissociation in its unsaturated solutions in molten NaCl. The value of pK_{CaO} at 830 °C is equal to 2.59 ± 0.2. The solubility product of CaO in the chloride melt is estimated to be $pP_{CaO} = 3.77 \pm 0.05$, i.e. the solubility of CaO is somewhat higher than that obtained in Ref. [382]. The full solubility of CaO in molten NaCl at 830 °C is 0.079 mol kg^{-1}, where the Ca^{2+} molality is 0.013 mol kg^{-1} and the CaO molality is 0.066 mol kg^{-1}. The degree of CaO dissociation in the saturated solutions is equal to 16%. The total molality of CaO corresponds to a value of $-\log m^0_{CaO} = 1.1$ where the plateau begins.

Watson and Perry determined the solubility products of ZnO and K$_2$ZnO$_2$ in molten KCl at 800 °C as 2.3×10^{-8} (ZnO) and 3.2×10^{-12} (K$_2$ZnO$_2$) [172].

3.7.2 OXIDE SOLUBILITIES IN MELTS BASED ON ALKALI- AND ALKALINE-EARTH METAL HALIDES

Few investigations have been performed using such halide systems as solvents for oxides, although they are of great interest for practical purposes, as media for sodium and magnesium electrodeposition.

Bocage *et al.* reported the study of solubility of MgO in a BaCl$_2$–CaCl$_2$–NaCl (0.235:0.245:0.52) melt at 600 °C [185]. According to their data, the solubility product index of MgO is $pP_{MgO} = 6.2 \pm 0.1$. The initial section of the potentiometric titration curve includes a sharp pO decrease at small quantities of the base-titrant. This points to the fact that the MgO solution is unsaturated under these conditions. Treatment of the data in Ref. [185], taking into account the formation of the unsaturated solution, allows us to estimate the dissociation constant of MgO in the melts as $pK_{MgO} \approx 4.6$. This melt was shown by the authors to possess the upper limit of basicity connected with the precipitation of CaO from the saturated solution. The solubility of CaO in the molten BaCl$_2$–CaCl$_2$–NaCl mixture is $S_{CaO} = 0.12 \pm 0.07$ mol kg^{-1}.

Castrillejo *et al.* studied the solubility of magnesium oxide in the molten CaCl$_2$–NaCl equimolar mixture at 575 °C [178]. The solubility product index was found to be $pP_{MgO} = 5.3 \pm 0.1$, on the molality scale. Nevertheless, the information presented in Ref. [178] makes it possible to obtain additional data on the behaviour of MgO in the studied melt. To realize this let us consider Fig. 3.7.18. The first two points in the potentiometric curve can be established, on the basis of the treatment method in Part 6, to belong to the unsaturated

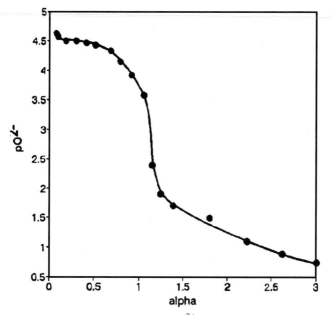

Fig. 3.7.18. Potentiometric titration curve of Mg^{2+} in the molten $CaCl_2$–$NaCl$ equimolar mixture at 575 °C ($m = 0.107$ mol kg^{-1}). The abscissa axis is the ratio of the initial molality of the base-titrant to that of the metal cation (by Castrillejo *et al.* [178]).

solution section. These data allow us to calculate the dissociation constant of MgO on the background in the $CaCl_2$–$NaCl$ melt as $pK_{MgO} = 3.5$–3.6; the molality of the non-dissociated MgO in this melt is equal to 0.016 mol kg^{-1}.

Kosnyrev *et al.*, investigated the solubility of calcium oxide in the molten $CaCl_2$–KCl–$NaCl$ mixture at 650–750 °C [384] and found that the solubility of CaO (mass%) rose with the $CaCl_2$ concentration in the studied melt. At 700 °C and within the range of 10–50 mass% of $CaCl_2$ the solubility increases in the following sequence: 0.26 mass% (at 10% of $CaCl_2$) $\rightarrow$ 0.68 mass% (30% of $CaCl_2$) $\rightarrow$ 1.45 mass% (50% of $CaCl_2$). The obtained solubility values are considerably lower than those in the $CaCl_2$–CaO system, where the temperature elevation from 970 to 1030 K causes the increase in CaO solubility from 2 to 10 mass% [385]. Threadill reported an investigation of the solubility of CaO in molten calcium chloride, and the examination of the process of electrochemical deposition of metallic Ca from the said melt [386]. An increase in the concentration of oxygen-containing admixtures in the melt was found to retard the process of formation of Ca powder by electrolysis.

Strelets and Bondarenko studied the interactions between $MgCl_2$ and CaO in carnallite melt ($KMgCl_3$ with the melting point at 488 °C) [387]:

$$MgCl_2 + CaO \downarrow \rightleftharpoons CaCl_2 + MgO \downarrow, \qquad (3.7.59)$$

using a radiochemical method which consisted in the addition of the Ca^{45}-isotope to ordinary CaO. The investigation of different layers of the carnallite melt after the interaction showed that the distribution of Ca^{45}, and, consequently of Ca^{2+}, was uniform in all the layers, whereas the Mg concentration was zero in the upper layers, and was found only in the precipitate, which was pure MgO. As estimated from the thermodynamic calculations, the equilibrium constant of equation (3.7.59) varies within the range $10^6 - 10^7$ at 700–800 °C, and the complete transformation of $MgCl_2$ into MgO was achieved under the experimental conditions (i.e. the action of a stronger Lux base). Besides the $KMgCl_3$ melt, equilibrium (3.7.59) was studied in the so-called calcium electrolyte: 10 mass% of $MgCl_2$ + 34 mass% of $CaCl_2$ + 50 mass% of $NaCl$ + 6 mass% of KCl, used for the deposition of metallic calcium by electrolysis of the melt. The addition of 1–3% of CaO to the said melt led to its complete transformation into $CaCl_2$ with the precipitation of an equivalent quantity of MgO. This meant that $MgCl_2$ suppressed the formation of CaO owing to the pyrohydrolysis of the said melt.

Another interesting result of the work mentioned above is that the rate of interaction of $MgCl_2$ with oxygen-containing admixtures considerably affects the sizes of the MgO crystals which are formed. Thus, upon interaction between the said melts and moisture, or on the addition of CaO (a slow interaction), finely dispersed particles of MgO are formed (with sizes in the order of 1 μm), whereas pyrohydrolysis of $MgCl_2$-containing melts without calcium results in the formation of relatively large (~ 20–30 μm) MgO crystals. In the authors' opinion, this means that finely dispersed crystals are formed owing to the intermediate formation of CaO in the process of pyrohydrolysis. There is another explanation for this effect: the presence of acidic cations of calcium in the melt retards the formation of the solid oxide phase, and results in the formation of smaller crystals.

Recently, we have studied the solubility of some metal-oxides in chloride melts based on alkali-metal salts with the addition of approximately 25 mol% of alkaline-earth metal chlorides ($CaCl_2$, $SrCl_2$ or $BaCl_2$). The solubility of MgO and NiO in the molten $CaCl_2$–KCl (0.235:0.765) mixture at 700 °C was determined in Ref. [191]. The range of oxides available for potentiometric investigations in this melt is very narrow owing to the levelling of the acidic properties of the metal cations: the acidic properties of Cd^{2+}, Sr^{2+}, Ba^{2+} and Pb^{2+} are undoubtedly levelled, and the solubilities of ZnO and MnO should be very high. Both of the oxides studied are practically insoluble in the chloride melts, even of relatively high acidity. Their solubility product indices (on the molality scale) are the following: $pP_{MgO} = 5.34 \pm 0.2$ and $pP_{NiO} = 5.24 \pm 0.2$. The calibration $E-pO$ plot in this melt, which is shown in Fig. 2.4.15, dependence 3, allows us to estimate the upper limit of the melt basicity connected with the formation of the saturated solution of CaO, which is accompanied by a plateau at low pO values. Such a feature of the calibration plot makes it possible to determine the solubility product of CaO in the $CaCl_2-KCl$ melt (see Table 3.7.21) using the SAM. However, the high solubility of CaO in the melt excludes the minimum in the plateau section, which can be expected in this case.

The average value of pP_{CaO} is equal to 0.72 ± 0.2. However, this averaged constant cannot reflect correctly the actual situation in the saturated melt. The point is that the potentiometric cell is calibrated with KOH and not with CaO. Therefore, the addition of KOH to the melt causes the precipitation of CaO from the melt and a persistent change in the melt composition: it is enriched with K^+, and the molar fraction of Ca^{2+} decreases. This means that each value of the solubility product corresponds to the specific composition of the $CaCl_2-KCl$ melt. When CaO was used for the calibration we could not observe the said changes of pO after achieving the saturation point, since subsequent additions of CaO did not dissolve in the melt and its composition did not change.

Thus, the first point from Table 3.7.21 is referred to the $CaCl_2-KCl$ melt of composition 0.233:0.767, saturated with CaO; the second point corresponds to the saturated solution of calcium oxide in the melt with the ratio of Ca to K cations equal to 0.229:0.771; for other points these

TABLE 3.7.21
Results of investigations of CaO solubility in the molten $CaCl_2-KCl$ eutectic at 700 °C

$m^0_{O^{2-}}$	E (V)	pO	$\bar{n}$	pP_{MeO}
0.082	0.290	1.24	0.009	0.79
0.132	0.285	1.18	0.024	0.74
0.184	0.280	1.13	0.039	0.69
0.305	0.270	1.01	0.073	0.59
0.805	0.241	0.68	0.210	0.33

ratios are 0.226:0.774 (third point), 0.218:0.792 (fourth point) and 0.186:0.814 (fifth point). Consequently, the obtained values for the CaO solubility products actually belong to chloride melts of different cation compositions, and the corresponding pO values are nothing but the upper limits of oxobasicity in these melts. The results presented in Table 3.7.21 show that an increase in Ca^{2+} cation concentration in the chloride melt results in a considerable increase in CaO solubility. This effect predominates over the other regularity, which consists in the decrease in the oxide solubility owing to weakening of the acidic properties of $CaCl_2$–KCl melts. The upper limit of oxobasicity in the said melts estimated, in Ω units, is near 5.3–5.4.

The dependence of the upper limit of basicity of $CaCl_2$–KCl melts on the Ca^{2+} concentration is presented in Fig. 3.7.19. As seen from this figure, there takes place a practically linear decrease in the upper pO values with the logarithm of the Ca^{2+} molar fraction in the melts at relatively small changes in the acidic ion concentration. However, if the Ca^{2+} concentration in the melt reduces by a factor of 1.3 then the increase in the oxobasicity limit becomes lower than the expected one. This is caused by the decrease in the melt acidity and by the reduction in Ca^{2+} cation concentration, which results in a reduction in the Ca_2O^{2+} complexes' stability and, consequently, the solubility of CaO in these melts.

The solubility products of some metal-oxides in the molten $SrCl_2$–KCl–NaCl (0.22:0.42:0.36) mixture at 700 °C were determined by the potentiometric titration method [192]. The changes (increase) in the solubility products of the studied oxides, as compared with the molten KCl–NaCl

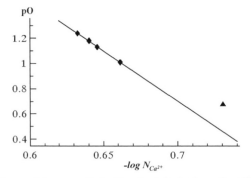

Fig. 3.7.19. Dependence of the upper limit of oxobasicity in the melts of $CaCl_2$–KCl system at 700 °C against the molar fraction of Ca^{2+} in the melts.

equimolar mixture were caused mainly by the strengthening of the Sr^{2+}-containing melt's acidity. As seen from the data of Table 3.7.22, they are practically the same. All the solubility products undergo small fluctuations about the average value of $+1.95pP_{MeO}$ units. The latter was discussed in Part 3 as being the oxobasicity index of the $SrCl_2$–KCl–$NaCl$ melt at 700 °C. The titration of Pb^{2+} cations with quantities of a strong Lux base did not result in the precipitation of PbO from the melt, and the attempt to determine the solubility product of PbO failed. Hence, the PbO solubility was high enough, and the solutions remained unsaturated at all the titrant weights. However, it was possible to estimate the dissociation constant value of PbO as $pK_{PbO} = 2.58 \pm 0.05$. The oxobasicity index found for the Sr^{2+}-containing chloride melt is in good agreement with the values of the dissociation constant indices of SrO in the molten KCl and CsCl–KCl–NaCl mixtures at 700 °C, which are equal to 2.30 and 2.14, respectively. The dependence of the solubility product values against the inverse-squared radius of metal cation in the KCl–NaCl–$SrCl_2$ melt is approximated by the equation

$$pP_{N,MeO} = -0.72(\pm 1.1) + 0.048(\pm 0.01)r_k^{-2}. \tag{3.7.60}$$

That is, in the Sr^{2+}-based chloride melt the slope of this dependence is close to that for the KCl–NaCl equimolar mixture (0.053), and this confirms again our assumption that the interactions of "multivalent metal cation–chloride ion" are subject to small changes upon change in the cation composition of the melt.

TABLE 3.7.22
Solubility products of some oxides in the molten $BaCl_2$–KCl–NaCl and KCl–NaCl mixtures at 700 °C

Oxide	pP_{MeO} (molalities)	pP_{MeO} (molar fractions)		ΔpP
		$BaCl_2$–KCl–NaCl	KCl–NaCl	
MgO	7.97 ± 0.12	9.76 ± 0.12	11.62	1.86
NiO	7.47 ± 0.11	9.26 ± 0.11	11.38	2.12
CoO	6.33 ± 0.12	8.12 ± 0.12	10.24	2.12
MnO	5.59 ± 0.06	7.38 ± 0.06	9.13	1.75
ZnO	5.18 ± 0.06	6.97 ± 0.06	9.28	2.31
CdO	3.64 ± 0.13	5.43 ± 0.13	7.54	1.92
CaO	$pK = 2.65 \pm 0.12$		–	–

The solubilities of metal-oxides in molten chloride mixtures with addition of Ba^{2+} cations were determined by the potentiometric titration technique in Refs. [193, 194]. The solubility products of MgO, NiO, CoO, MnO, ZnO and CdO in the $BaCl_2$–KCl–NaCl (0.43:0.29:0.28) melt were found to be considerably higher than those in the molten KCl–NaCl equimolar mixture. The titration of Ca^{2+} cations with KOH in the said melt did not result in CaO precipitation, although according to the titration data Ca^{2+} cations possessed appreciable oxoacidic properties.

The data on the oxide solubility in the molten $BaCl_2$–KCl–NaCl mixture are collected in Table 3.7.22. Similar to the melts based on alkali-metal chlorides, a tendency of the oxide solubilities to increase with the growth of the radius of the oxide cation is also observed in the Ba^{2+}-based melt. The dependence $pP_{MeO} = f(r_k^{-2})$ in the $BaCl_2$–KCl–NaCl eutectic melt is expressed by such an equation:

$$pP_{N,MeO} = -0.04(\pm 1) + 0.051(\pm 0.02)r_k^{-2}. \tag{3.7.61}$$

It can be seen that the slope of this dependence is close to that of the similar plot obtained in the molten KCl–NaCl equimolar mixture (0.053 ± 0.01); the intersects of these plots differs by 1.84. Such a behaviour of the $pP_{MeO} = f(r_k^{-2})$ plot points to the fact that the differences in metal-oxide solubilities in chloride melts of various cation compositions are mainly caused by the difference in the constituent cation acidities. The increase in the metal-oxide solubilities in the Ba^{2+}-based melt as compared with the KCl–NaCl melt (see Table 3.7.23) demonstrates that the oxoacidic properties of Ba^{2+} cation are considerably stronger than those of K^+ or Na^+ cations. It should be

TABLE 3.7.23
Solubility products of some oxides in the molten $SrCl_2$–KCl–NaCl and KCl–NaCl mixtures at 700 °C

Oxide	pP_{MeO} (molalities)	pP_{MeO} (molar fractions)		ΔpP
		$SrCl_2$–KCl–NaCl	KCl–NaCl	
MgO	7.61 ± 0.06	9.63	11.62	1.99
NiO	7.44 ± 0.09	9.56	11.38	1.82
CoO	6.25 ± 0.12	8.37	10.24	1.87
MnO	5.20 ± 0.17	7.32	9.13	1.81
ZnO	4.97 ± 0.07	7.09	9.28	2.18
CdO	3.21 ± 0.07	5.33	7.54	2.02

added that the value of the dissociation constant of BaO in molten KCl—NaCl, determined in Ref. [343] is 81. The logarithm of this value is equal to 1.91, which is very close to 1.84, and to the oxobasicity index of the $BaCl_2$–KCl–NaCl eutectic, which is 2.01, as has been said above.

Comparison of the obtained data with the results for Sr^{2+}-based melt shows that the oxobasicity indices of $SrCl_2$–KCl–NaCl (0.22:0.42:0.36) and $BaCl_2$–KCl–NaCl (0.43:0.29:0.28) melts are practically the same: $pI_{SrCl_2-KCl-NaCl} = 1.95 \pm 0.3$ and $pI_{BaCl_2-KCl-NaCl} = 2.01 \pm 0.3$; although in the latter case the concentration of alkaline-earth metal cation is approximately twice as high as in the former case. From this, an obvious conclusion may be made: that the acidic properties of Ba^{2+} cation, are to a certain extent, weaker than those of Sr^{2+}. To assert this without any doubts, we shall consider the study in the following Ba^{2+}-based melt.

We reported the investigation of the solubility of ZnO and CdO in the molten $BaCl_2$–KCl (0.26:0.74) eutectic [194]. The former oxide was shown by the potentiometric titration method to be practically insoluble in the said melt, whereas the latter oxide was not only appreciably soluble, but it did not show acidic properties in unsaturated solutions on the background of the Ba^{2+}-based melt. So, the oxoacidic properties of Cd^{2+} were found to be comparable with those of Ba^{2+} oxide. The results of the oxide solubility determinations, which are presented in Table 3.7.24, allow us to estimate the relative oxoacidic properties of both Ba^{2+}-containing melts: the corresponding oxobasicity indices are equal to 2.01 for the molten $BaCl_2$–KCl–NaCl eutectic, and 1.83 for the $BaCl_2$–KCl (0.26:0.74) melt. This shows that the increasing concentration of the acidic cation in the melt gives rise to the strengthening of the melt acidity, which in turn leads to the increase in metal-oxide solubilities.

The potentiometric curves obtained in the titration of Mg^{2+} cations ($MgCl_2$) in all the chloride melts studied at 700 °C may be considered to illustrate the effect of ionic melt acidity on the solubility of metal-oxides

TABLE 3.7.24
Solubility products of ZnO and CdO in some $BaCl_2$-based melts at 700 °C

Oxide	pP_m	pP_N		
	$BaCl_2$–KCl	$BaCl_2$–KCl	$BaCl_2$–KCl–NaCl	KCl–NaCl
ZnO	5.14 ± 0.1	7.14	6.97	9.28
CdO	3.62 ± 0.1	5.62	5.43	7.35

(see Fig. 3.7.20). With the strengthening of the oxoacidic properties of the melt cation there is observed a shift in the acidic section of the potentiometric titration curve to lower pO values: similarly, the potentiometric curves in the melts possessing close oxoacidic properties are practically coincident.

Of course, the acidic sections of all the presented potentiometric curves reconstructed on the scale of the common oxoacidity function Ω should coincide, and the sections corresponding to an excess of the base-titrant should be shifted upwards with respect to the KCl–NaCl equimolar mixture as the oxobasicity indices of the said melt increase (Fig. 3.7.21). As seen from this figure the titration of Mg^{2+} cations in chloride melts is characterized by Ω values in the order of eight at a considerable excess of the titrant. Further, they are similar to the titration curves obtained using weak Lux bases. This means that the basicity of oxides formed by the most acidic cation of the chloride melts weakens in the following sequence: Na_2O–$BaO(SrO)$–Li_2O–CaO. This latter result testifies to the fact that an increase in the oxoacidic properties of the ionic melts signifies simultaneous weakening of the "absolute" basicity of the solution of strong bases in these melts.

In concluding our consideration of oxide solubility studies in the molten $BaCl_2$-based chloride mixtures, we should mention the results of Ivenko $et~al.$ [388, 389], who studied the solubilities of BaO, CuO and Y_2O_3 in these melts

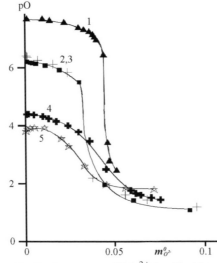

Fig. 3.7.20. Potentiometric titration curves of Mg^{2+} in chloride melts at 700 °C: (1) KCl–NaCl, (2) $SrCl_2$–KCl–NaCl, (3) $BaCl_2$–KCl–NaCl, (4) KCl–LiCl, (5) $CaCl_2$–KCl.

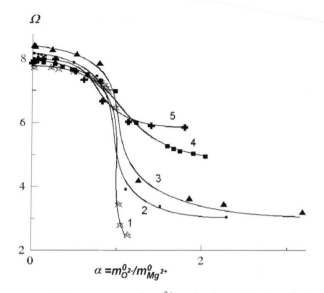

Fig. 3.7.21. Potentiometric titration curves of Mg^{2+} in chloride melts at 700 °C expressed via the Ω common oxoacidity function: (1) KCl–NaCl, (2) $SrCl_2$–KCl–NaCl, (3) $BaCl_2$–KCl–NaCl, (4) KCl–LiCl, (5) $CaCl_2$–KCl.

in the temperature range of 920–1200 K. The choice of the objects of investigation was caused by the development of superconductor materials, which was one of the most fashionable areas at that time. $YBa_2Cu_3O_{7-\delta}$ was considered among the most promising superconductor compounds [390, 391] and the methods of its synthesis (and co-precipitation from the melts) were developed extensively. The solubilities of the mentioned oxides in the Ba^{2+}-based melt were found to be appreciably higher than those in the molten KCl–NaCl eutectic, e.g. CuO solubility was 1.3 mol% at 1073 K and 1.6 mol% at 1123 K. Also, BaO was more soluble in the specified melt; its solubility increased with the temperature in the manner, 13–19 mol% at 973 K, 16–20 mol% at 1073 K and 20–22 mol% at 1173 K. Among the mentioned oxides, Y_2O_3 possessed the smallest solubility; it was only in the order of 0.01 mol%. It should be added that the investigations of CuO have no relation to the equilibrium

$$CuO \rightleftharpoons Cu^{2+} + O^{2-}, \tag{3.7.62}$$

since Cu^{2+} ions dissolved in the melt are subject to reduction by chloride ions, which results in the formation of Cu_2^{2+} ions (or Cu^+) and chlorine evolution from the melt studied. This means that actually the system of composition $(Cu_2O + CuO)–(Cu^I + Cu^{II})$ ions was studied. As for the solubility of BaO in the molten $BaCl_2–KCl–NaCl$ eutectic, this allows us to determine the upper limit of oxobasicity in this melt, which is equal to $pO = -0.3$. If we recalculate this value with respect to the common oxoacidity function Ω then it will be near $\Omega = 2$. After the onset of the solid-phase precipitation there is observed a small change in oxide-ion concentration in the melt, which the authors connect with the formation of barium oxochloride owing to the reaction of BaO powder with the melt:

$$BaO \downarrow + BaCl_2 \rightleftharpoons Ba_2OCl_2 \downarrow. \tag{3.7.63}$$

To conclude our consideration of metal-oxide solubility in molten ionic halides we should mention the investigation of cerium(III) oxide solubility in melts of $KCl–NaCl–CeCl_3$ mixtures, performed by Combes et al. [169, 392]. The obtained data are of interest for obtaining new scintillation materials LnX_3:Ce (La, Gd, Lu) and Me_nLnX_{3+n} (La, Gd, Lu) discovered by Van Eijk's group [280, 393]. The reason is that the raw materials used for the synthesis and growth from the melt of the said scintillators are exposed to pyrohydrolysis with the formation of the corresponding oxide deposits or inclusions (Ln_2O_3, or CeO_2 in the case of cerium). These impurities can exist both in the liquid and in the solid state, and worsen the optical and scintillation properties of lanthanide–halide-based materials. The data of Ref. [169] allow one to plot the solubility of the oxide against the concentration of LnX_3 in the melt, as shown in Fig. 3.7.22.

The reactions between cerium trichloride and oxide ions were initially studied in the pure $KCl–NaCl$ equimolar mixture at 1000 K by the method of potentiometric titration using a calcium-stabilized zirconia membrane electrode. The titration curves clearly demonstrated the existence of the soluble cerium oxychloride CeO^+ and precipitated cerium oxide:

$$Ce^{3+} + O^{2-} = CeO^+, \tag{3.7.64}$$

(Ce_2O_3) with respective dissociation constants 10^{-11} and 10^{-30} on the molality scale.

The increase in the LnX_3 concentration results in a rise in the solubility of the corresponding oxide in the melts; the solubility values exceed 0.5 mass% at a concentration of $CeCl_3$ in the order of 30 mol%. It shows that the homogeneous melt used for the growth of the scintillation crystals may

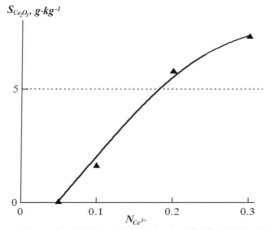

Fig. 3.7.22. Dependence of solubility of Ce_2O_3 in the $(CeCl_3)_x(KCl-NaCl)_{1-x}$ melts from the concentration of $CeCl_3$ plotted according to the data of Ref. [169].

contain appreciable concentrations of oxides of lanthanides (near 1 mass%). Cooling of this melt can result in the precipitation of the oxide or oxochloride, and creates a number of crystallization centres which favour the formation of polycrystalline materials instead of a single crystal. Therefore, a detailed treatment of the said compositions by chlorination agents, such as NH_4Cl (in the solid state, prior to melting) and CCl_4 (in the molten state) is recommended.

3.7.3 SOLUBILITIES OF ALKALI EARTH METAL CARBONATES IN KCl–NaCl EUTECTIC

Carbonates are often used for studies in molten halides, which is connected with their basic character, i.e. they are donors of O^{2-}. Since alkaline carbonates are suitable for Lux acid neutralization they are often used as Lux bases (oxide ion donors) in molten halide melts to study the acid–base equilibria [165, 178, 185]. Quantitative estimations of the carbonate ion's basicity in different melts are performed on the assumption of its dissociation according to equation (2.3).

Known investigations of the reaction (2.3) in molten chlorides were performed both under equilibrium (i.e. with a constant CO_2 pressure over the melt in the process of the experiment) [310, 312] and non-equilibrium conditions (without CO_2 pressure over the melts) [141]. These studies resulted in the conclusion that carbonate was a weak base under CO_2 pressure, and its dissociation was incomplete even in the absence of carbon dioxide in the atmosphere over the melts possessing a weak acidity, e.g. KCl–NaCl [139]. Complete dissociation in an inert gas atmosphere was assumed in $CaCl_2$-containing mixtures [178, 185] possessing stronger acidities than the KCl–NaCl melt.

All the above-said concerns alkali-metal carbonates dissolved in melts based on alkaline- and alkaline-earth metal halides. It should be mentioned that there is a lack of data on the behaviour of alkaline-earth carbonates, in high-temperature alkaline halide melts. Nevertheless, these data are necessary, in particular, for the purification of halide melts. For example, it is known that alkaline-earth oxides possess appreciable solubilities in molten alkaline halides, and that they can be used for precipitation of less soluble transition metal-oxides. After such a precipitation, the excess of alkaline-earth oxide remains in the dissolved state. The question arises, therefore, whether it can be removed from the halide melt by passing CO_2 according to the conventional reaction

$$MeO + CO_2 \uparrow \rightleftharpoons MeCO_3 \downarrow, \qquad (3.7.65)$$

where Me is an alkaline-earth metal. To answer the above question one should determine the solubility products of alkaline-earth carbonates in molten halide media, e.g. in the molten

$$MeCO_3 \downarrow \rightleftharpoons Me^{2+} + CO_3^{2-} \qquad (3.7.66)$$

KCl–NaCl equimolar mixture.

Equilibrium (3.7.66) must be investigated in the molten KCl–NaCl equimolar mixture at 700 °C for $CaCO_3$, $SrCO_3$ and $BaCO_3$, which are suitable for such studies, since their decomposition temperatures under a CO_2 pressure of 1 atm are close to 825, 1340 and 1450 °C, respectively, i.e. they appreciably exceed 700 °C. The decomposition of $MgCO_3$ under the said CO_2 pressure occurs at 650 °C, and therefore, it is inappropriate for the solubility studies, since $MgCO_3$ does not exist at 700 °C, even under CO_2 pressure in the order of an atmosphere.

Naturally, the studies of equilibrium (3.7.66) require data on the equilibrium constant of equation (2.3). Although in the literature there are similar data, $pK = 4.91$ [312], $pK = 5.17$ [310], we determined this value at

700 °C using our experimental technique as $pK = 5.14 \pm 0.2$. The equilibria mentioned were studied using the potentiometric cell with the liquid junction

$$Pt|Ag|Ag^+, \; 0.1 \; mol \; kg^{-1}, \; KCl–NaCl\|KCl–NaCl$$

$$+ CO_3^{2-} |YSZ|Pt(O_2) \qquad\qquad (3.7.67)$$

which was previously calibrated with known weights of KOH in a CO_2 atmosphere. Such a calibration allowed a plot to be made of $E–p(CO_3^{2-})$ which was used to analyse the results of $Me^{2+}–CO_3^{2-}$ interactions.

The solubility products of alkaline-earth metal carbonates according to reaction (3.7.66) were determined by a SAM [333], which consisted in the introduction of the weight of carbonate into the solvent, up to its saturation. The latter was detected by measurements of the e.m.f. of the potentiometric cell (3.7.67). The calibration plot $E–p(CO_3^{2-})$:

$$E = 0.239(\pm 0.006) + 0.086 \pm (0.008)p(CO_3^{2-}) \qquad V \qquad (3.7.68)$$

allowed us to estimate the degree of alkaline-earth metal carbonate dissociation in their unsaturated solutions. Passing CO_2 through the melt leads to a shift in the reaction (2.3) to the left, and CO_3^{2-} remains as the main base in the melt. The slope of the plot is close to $1.15RT/F$ (0.096 V), which is characteristic of this e.m.f. range [317].

To determine the solubilities of alkaline-earth carbonates in molten KCl–NaCl we used the calibration-like method [350], which consisted in the addition of weights of the substance studied to the solvent with subsequent measurement of the equilibrium e.m.f. Let us consider the applicability of such a technique to our case. The increase in the total initial concentration of alkaline-earth metal carbonate in the solution leads to the growth of ion concentrations on the right-hand side of equation (3.7.66); quite naturally, their product should also increase. Under the constant pressure of CO_2 (1 atm) the concentration of carbonate ion is proportional to that of O^{2-} according to the simple relationship

$$m_{CO_3^{2-}} = m_{O^{2-}} \times 1.38 \times 10^5. \qquad\qquad (3.7.69)$$

In particular, this means that the alkaline-earth carbonate solution is contaminated with only 0.01% of the corresponding oxide. Strict observance of the above-said proportion allows us to conclude that in chloride melts saturated with CO_2 (under a constant pressure) the oxygen electrode serves simultaneously as a carbonate one. So, in the region of the unsaturated solution, the slope of the E–$p(MeCO_3)$ plot should coincide with that calibrated with weights of carbonate (Fig. 3.7.23, sloping line). Indeed, we observed a good agreement of the plots obtained in unsaturated solutions with that obtained for the calibration of the cell with KOH additions under CO_2 pressure (equation (3.7.67)). This confirms the complete dissociation of the alkaline-earth carbonates in the unsaturated-solution regions. When the solubility product value is achieved, there arises a plateau in the $p(CO_3^{2-})$–$p(MeCO_3)$ plot, since the carbonate and oxide ion concentrations remain constant (Fig. 3.7.23, sections 1–3).

As is well known, alkaline-earth metal-oxides possess a limited solubility in molten chlorides at ~ 1000 K. In the saturated solutions of these oxides equilibrium (3.6.1) takes place. Therefore, there are two simultaneous reactions in the carbonate solutions, namely, equations (3.6.1) and (3.7.66), which may result in the precipitation of a solid phase. For equation (3.6.1) all we have said above for the carbonate remains true, i.e. the plateau (Fig. 3.7.23, sections 1–3) may also be a result of alkaline-earth metal-oxide precipitation.

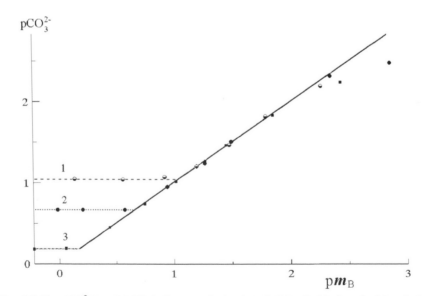

Fig. 3.7.23. $p(CO_3^{2-})$–$p(MeCO_3)$ diagram obtained at $CaCO_3$ (1, black-and-white circles), $SrCO_3$ (2, black circles) and $BaCO_3$ (3, quadrates) additions in the molten KCl–NaCl equimolar mixture, 700 °C, CO_2 flow.

In order to attribute the obtained results either to carbonate or oxide properties (solubility), one should know which of the substances—oxide or carbonate—precipitates. It may be assumed that these precipitates are pure carbonates.

Indeed, the first argument confirming the formation of pure carbonate precipitate consists of the following. The system considered is under equilibrium conditions, and therefore, it is clear that the solid phase is not only under equilibrium conditions with the melt, but it also contacts with the gas atmosphere. As seen from the discussion above, the equilibrium CO_2 pressures over all alkaline-earth metal carbonates are efficiently less than 1 atm, and therefore, the formation of the oxides under these conditions is practically excluded.

We shall now consider whether our results are in agreement with this prediction, or are inconsistent with it. As seen from Fig. 3.7.23, the coordinate of the intersection point is bound up with the solubility product of the precipitate. If carbonate precipitates, then

$$pP_{MeCO_3} = 2p(CO_3^{2-}) \qquad (3.7.70)$$

if, on the contrary, there is precipitation of the alkaline-earth metal-oxide in the solid phase, then

$$pP_{MeO} = 2p(CO_3^{2-}) + pK_{(1.2.3)} \qquad (3.7.71)$$

Taking into account the fact that, under a CO_2 pressure of 1 atm, the interrelation between the concentrations of oxide ion and carbonate ion can be expressed in the logarithmic form as

$$pO = p(CO_3^{2-}) + pK_{(1.2.3)}. \qquad (3.7.72)$$

From the known data on BaO (obtained by a method close to that described above) [239] and on CaO [238], we can compare $m_{Me^{2+}} m_{O^{2-}}$ at the inflection point with the solubility product taken from literature data. The SrO solubility (pP_{SrO}) obtained by the IST method (analysis of cooled sample of saturated solution) is 5.84 [328]. The estimations are presented in Table 3.7.25.

As is seen, the precipitation begins at rather high concentrations of the carbonates added. The comparison of oxide solubility products taken from the literature with the products $m_{Me^{2+}} m_{O^{2-}}$ in the saturated solutions considered in the present work, shows that in all cases the solutions are unsaturated with respect to the corresponding oxides. Hence, we can conclude that in molten KCl–NaCl saturated with CO_2, alkaline-earth metal carbonate precipitation

TABLE 3.7.25
Data on oxide and carbonate solubilities (in molalities) in molten KCl–NaCl equimolar mixture at 700 °C

Carbonate	$p(CO_3^{2-})$ saturation	The calculated parameters	
		pP_{MeCO_3}	$m_{Me^{2+}} m_{O^{2-}}$
BaCO$_3$	0.18	0.36	5.50 vs 2.31
SrCO$_3$	0.67	1.34	6.48 vs 5.84
CaCO$_3$	1.02	2.04	7.18 vs 5

starts at lower equilibrium concentrations of oxide (and, hence, carbonate) ions than that of the corresponding oxide. Therefore, both arguments lead to the conclusion that the solids precipitated under the experimental conditions are pure carbonates.

Such high solubilities of alkaline-earth metal carbonates mean that their precipitation according to equation (3.7.66) is practically impossible. However, there exists an opportunity to dissolve alkaline-earth oxides possessing lower solubilities, and to hold them in solution as carbonates by passing CO_2 from the saturated solutions of metal-oxides. The observed effect can be used for the growth of crystals of mixed metal-oxides, and is especially convenient for metal-oxides which form carbonates which are unstable at high temperatures. To explain the essence of the synthesis routine, we shall consider the following scheme of the assumed technological process:

- addition of the oxide mixture to a low-acidic chloride melt,
- passing CO_2 through the melt,
- removal of CO_2 from the melt,
- separation of the obtained material from the melt.

At the first stage, the mixture of oxides is placed into chloride melts: the system is heterogeneous since all the oxides are slightly soluble in the ionic liquid.

Passing carbon dioxide through the melt results in a considerable increase in pO (acidity), which, in turn, causes an increase in the metal-oxide's solubility. At this stage the oxides dissolve in the melt, and at a certain "metal-oxide/melt" mass-ratio the mixture can be completely dissolved in the melt.

The third stage can be realized in two ways. One consists of a slow decrease in the partial pressure of CO_2 over the melt, which causes a gradual reduction in the melt's acidity and results in small supersaturation values. Hence, the rate of the oxide crystal's growth from the melt will be low, and the crystals will be relatively large. This routine may be very promising for obtaining oxide single crystals from the halide melts.

Another way is the fast removal of carbon dioxide, e.g. by bubbling inert gas through the melt, which favours high supersaturation providing faster crystallization processes than in the first case. The oxide crystals obtained will be smaller, resulting in the formation of fine-size mixtures, which can be used for subsequent thermal treatment. The fourth stage consists of the separation of the oxide crystals from the melts, which can be performed in various ways, which need not be discussed here.

Thus, treatment of metal-oxide solutions in molten salts with carbon dioxide may be proposed as a promising method to monitor the crystal-size and composition of complex oxide materials.

3.7.4 CONCLUSION

Our consideration of the data on metal-oxide solubilities in ionic melts based on alkali-metal halides shows that there is much experimental material, and an attempt at systematizing it is made in this book. Some regularities of metal-oxide solubilities in different melts and over a wide temperature range are obtained.

As mentioned in Section 3.7.1.2, there is a considerable scatter of solubility product values obtained in the molten KCl–NaCl eutectic using different methods of solubility determination. This disagreement in the solubility parameters may be explained by differences in the sizes of oxide particles whose solubility is to be determined. The difference in size causes the scatter of the solubility data according to the Ostwald–Freundlich equation; and the employment of the isothermal saturation method, which implies the use of commercial powders (often pressed and sintered), leads to values which are considerably lower than those obtained by the potentiometric titration technique where the metal-oxides are formed *in situ*. Owing to this fact, the regularities connected with the effect of physico-chemical parameters of the oxides or the oxide cations should be derived only from solubility data obtained under the same or similar experimental conditions. However, this does not concern the dissociation constants of the oxides, since homogeneous acid–base equilibria are not sensitive to the properties of the solid phase of the oxide. It should be noted that estimation of the parameters connected with the surface properties of metal-oxides results in values of the surface-energies at the "metal-oxide–chloride melt" interface boundary in the order of some decades of $J\,m^{-2}$. This value is surprisingly high, although it successfully explains a considerable scattering of the oxide-solubility data obtained by different methods. Possibly, these values are true, but they may be caused by unknown features of the behaviour of finely dispersed systems in ionic liquids. The action of other unknown factors is also possible: in this case the calculated "surface energy" values are actually the sum of the true surface energy and the energetic equivalent of this unknown action.

As a rule, elevation of the temperature of ionic melts results in an increase in the metal-oxide solubility in these melts, which occurs according to the Le Chatelier–Shreder equation. In any case, the deviation from this rule is not so considerable and this allows us to calculate the metal-oxide solubilities near the temperatures of the experiment. The thermal coefficient of solubility was found in our investigations to be in good agreement with that predicted based on the Le Chatelier–Shreder equation. Appreciable deviations from this regularity are observed only for soluble oxides (BaO, SrO, PbO), whose dissolution leads to the appearance of foreign particles—both charged (Ba^{2+}, Sr^{2+}) and uncharged (PbO)—in the melt.

The dependence of the metal-oxide solubility in all the alkali-metal halides studied against the melting points of the oxides are found to consist of two plots: the alkali-metal-oxides belong to one of them, and the second group consists of the oxides of the 3d-elements. This may be caused by different degrees of the shift of electron density from the metal to oxygen for oxides belonging to the different groups. It has been demonstrated that for chemical analogues (Ca–Sr–Ba, Zn–Cd) there is an obvious dependence of the solubilities of the corresponding oxides against their melting points, and the slopes of these dependences are found to be coincident with those predicted on the basis of the Le Chatelier–Shreder equation.

For oxides of the type MeO, there exists a linear dependence of the solubility product index of the oxide against the inverse-squared radius of the metal cation. The slopes of these plots in the melts based on alkali-metal halides and alkaline-earth metal halides are stated to be approximately the same. This can give evidence that, at high temperatures in the order of 1000 K, the changes in solvation ability of the ionic melts, proceeding from one melt to another, are close for different cations with radii in the order of 0.1 nm (from 0.74 nm for Mg^{2+} to 1.38 nm for Ba^{2+}). An increase in the melt temperature makes the $pP-r_k^{-2}$ dependence more rigorous, which demonstrates a weakening of the solvation processes in the melts with the temperature elevation, and an increase in the contribution of the electrostatic component in the interactions between the ions in the oxide solutions. The dependences found can be proposed for estimations of the solubility products of metal-oxides in the corresponding halide melts, on the basis of the cation radius.

As with the dependences discussed above, the correlation of the metal-oxide solubility with the electronegativity by Allred and Rochow consists of two plots characterized by close slopes. The observed shift of the correlation dependences for s^2- and 3d-elements can be connected with the systematic difference of the calculated effective charge of the metal cations belonging to the mentioned groups.

The solubilities of the metal-oxides in alkali-metal halide melts are stated to increase with increase in the parameter a, of the crystalline lattice of the oxides. This linear regularity applies for all the studied oxides possessing the cubic lattice, and can be observed for two oxides, ZnO and PbO, characterized by a hexagonal lattice.

The solubility products of metal-oxides in alkali-metal halide melts containing such acidic cations as Li^+, Ba^{2+}, Sr^{2+} and Ca^{2+} are considerably higher than those in low-acidic melts—in particular, KCl–NaCl and CsCl–KCl–NaCl mixtures—and increase in the mentioned sequence of the alkali- and alkaline-earth cations. Changes in the melt composition cause practically the same changes in the solubility products, i.e. the changes of the oxoacidic properties of the melts are the main reason for the oxide solubility changes.

The solubilities of metal-oxides (on the molar-fraction scale) in chloride melts possessing practically the same oxobasicity-index values are close. This shows the negligibly small effect of low-acidic constituent cations of the melt on its acidity and the solubility product values owing to the levelling of their acidity by that of the most acidic cation of the melt. The solubilities of the oxides in bromide and iodide-melts are considerably lower than those in chloride melts. This may be explained by the stronger association of bromide and iodide ionic melts, and by weakening of the stability of the complexes formed by the hard bases which most cations belong to, with the intermediate (Br^-) and soft (I^-) bases.

Afterword

We have concluded our consideration of papers on oxoacidity and related topics. This book not only presents "classic" works on oxoacidity—which are cited in most published works devoted to this topic—it also contains relatively new and extensive scientific material published a few decades ago but remaining on bookshelves without any treatment. This mainly concerns the works on molten salt chemistry published in some journals of the former Soviet Union, whose authors did not relate their results to the oxoacidity branch by using the corresponding keywords. Therefore, these papers did not attract the attention of appropriate researchers and remained unknown for a long period of time.

Does this mean that all the directions considered have been studied exhaustively and that further extensive development of the presented and new work is not possible? No, it is not so! On the contrary, the material presented in this book reveals many questions which seem to be solved but actually are not being developed. There is much "blank space" in oxoacidity, which has not received attention so far and remains unstudied. Let us dwell on this briefly.

Concerning the homogeneous Lux acid–base reactions, it should be noted that these investigations in molten nitrates need thorough verification and, undoubtedly, further experimental studies. The point is the demonstrated effect of the levelling of acidic properties by these solvents. Therefore, the range of the Lux acids weaker than NO_2^+ is to be determined and the corresponding equilibrium constants are to be estimated. Moreover, recently there have been reported some metal-oxide electrodes which are promising for potentiometric investigations in these media. However, the authors who introduced these electrodes into practice made nothing but the classic, but doubtful, "acidity rows". Studies of acid–base reactions require more reliable methods, with computer-aided processing of the obtained results, to allow confirmation of the reaction stoichiometry. This, in its turn, makes it possible to obtain the actual equilibrium constants. The problem of correctly processing the results of complicated acid–base equilibria in molten salts has not been solved up to now, although it does not, I think, contain any additional features as compared with similar routines for aqueous solutions.

The instability of the solvent acids with temperature rise, and narrowing of the acid–base ranges in oxygen-containing ionic solvents, make it possible to develop new methods for the synthesis of complex oxide materials. The principles of this synthesis are considered in this book and may be used as a starting point for future applied works.

Further, the acid–base ranges in molten oxygen-containing systems at different temperatures are practically unstudied. For example, we know the ionic product of water at 25 °C (10^{-14}) and at 100 °C (5.5×10^{-13}) and this allows us to predict some processes connected with the use of this solvent at different temperatures. While reading the paper by Afanasiev and Kerridge [113], we may assume that a temperature rise considerably narrows the intrinsic acid–base range of nitrate melts; however, we have no quantitative data on this problem. As for sulfate melts, they are only slightly studied owing to the relatively high melting points both of the individual salts and their eutectics. Nevertheless, besides the obvious routine of performing the oxoacidity studies at elevated temperatures, there is another way: to prepare mixed chloride–sulfate solvents possessing lower melting points and to study the behaviour of SO_4^{2-} ions there. Similarly, there are no works devoted to the problem of the arrangement of oxygen-containing melts in the general oxoacidity scale, although such work will be very interesting.

As for halide melts, it should be emphasized that most results presented in this book are obtained from studies of chloride systems, whereas other molten halides are practically uninvestigated. In my opinion, this gap should be filled, although I cannot propose another way apart from the conventional potentiometric titration using a membrane or the newly found metal-oxide electrodes. This way is obvious enough, but is too laborious.

The study of fluoride melts basically requires the development of new experimental equipment, since the use of Al_2O_3-based ceramics leads to strong pollution of the studied fluorides by aluminium derivatives and glass–carbon crucibles are not suitable for oxide-solubility studies since carbon reduces the oxides formed from transition metals.

From the viewpoint of the general oxoacidity scale, it is of interest to continue investigations of phenomena such as the levelling of acidic and basic properties by ionic solvents and the effect of temperature on the reciprocal arrangement of ionic melts in the oxoacidity scale.

Concerning investigations of the oxygen-electrode's reversibility and studies of the dissociation of different Lux bases, it should be mentioned that we recently found a two-slope structure of the calibration E–pO plots for a $Pt(O_2)|YSZ$ membrane electrode. I think this feature should attract the attention of experienced scientists who investigate the electrochemical aspects of the usage of such electrodes. There should be examined some aspects connected with the working of these electrodes at low pressures of oxygen in the inner space and the behaviour of the calibration dependences at low pO values within 0.5–2.

It is also expedient to study alternative kinds of oxygen electrodes. For example, the behaviour of a number of metal-oxide electrodes, such as $Zr|ZrO_2$, $Nb|Nb_2O_5$ and $Ta|Ta_2O_5$ in high-temperature melts (1000 K and higher), has not been studied yet. One may expect that the former electrode will be the most convenient of the electrodes mentioned owing to the weakest oxoacidic properties of ZrO_2. However, this statement requires experimental verification.

As for studies of the dissociation of oxide-ion donors in molten salts, this is a very wide field. The behaviour of the strongest bases, such as peroxides and hydroxides, has not been studied in halide melts other than chlorides, and sulfate melts are not an exception. Data on the dissociation of carbonate ion in sulfate melts are not known. Therefore, if there are no data even for particular temperatures, to talk about the polythermal dependences of the corresponding dissociation constant is senseless.

Considering the situation for oxide-solubility studies in molten salts, it should be noted that most of the data are devoted to the behaviour of MeO-type oxides in ionic melts. Nevertheless, the practical significance of the studies of the solubility products for oxides such as Al_2O_3, ZrO_2 and Fe_2O_3 in the melts can hardly be overestimated. The corresponding investigations require the development of new experimental routines providing detection by the measurements, or calculations of the concentrations of the various ions formed as a result of the oxide dissociation: AlO^+ and Al^{3+} in the case of Al_2O_3, ZrO^{2+} in the case of ZrO_2, for example.

Comparison of the solubility data obtained by different methods shows that there is a considerable, but well-reproducible difference between two groups divided by the method of the investigation. This difference is caused by the different ways of obtaining the saturated solutions of the oxide studied: the lowest solubility values are obtained for pressed and sintered pellets and powders of the oxide. In contrast, if the oxide is deposited from the corresponding cation solutions, it has a higher solubility value. This difference is explained by the surface phenomena: an increase of the average particle size and, hence, of the molar surface area results in an increase of the equilibrium solubility of the metal oxide. The estimation yields values of the corresponding surface energies to be of the order of $20–30\,\mathrm{J\,m^{-2}}$. This is very good but, as can be found in the handbooks, the corresponding values of the surface energies in the "solid–liquid" systems are considerably lower. What does it mean? Maybe such a difference in the solubilities is explained by the action of other factors, and in this case these factors should be discovered. Naturally, these values may be real and this does not require finding out additional reasons. Nevertheless, the investigations of metal-oxide solubilities by the SAM method introduced in this book should give much interesting new material, both on the solubility and dissociation parameters and the effect of the grain-size of the oxide on its solubility in molten salts. This method is suitable for solving the following problem: the data obtained by the isothermal saturation method are often used by different authors to estimate the activity coefficients of the oxide by comparing the values of the solubility obtained experimentally and those calculated from the Le Chatelier–Shreder equation. Nevertheless, the obtained activity coefficients are referred to the mixture of particles possessing quite different properties (charges, sizes, etc.), and, therefore, the speculations based on these data are often incorrect. The SAM method allows one to divide the total solubility into the ionized and the non-dissociated components, which should facilitate the processing of these data from the viewpoint of chemical thermodynamics.

Naturally, this is not the complete list of tasks waiting for their solution. Besides the above-mentioned, there exist many problems in this area of chemistry: some of them are not yet discovered, the others are beyond my understanding.

Nevertheless, these concluding remarks will be interesting both for experienced investigators and for those scientists who are going to start their investigations in this field of molten salts chemistry.

REFERENCES

[1] Yu.Ya. Fialkov, Solvent as a Mean of Monitoring Chemical Processes (Khimiya, Leningrad, 1990) p. 240.

[2] G. Wypych (ed), Handbook of Solvents (Chemtec Publishing, Toronto, NY, 2001) p. 1675.

[3] Yu.K. Delimarskii and L.P. Barchuk, Applied Chemistry of Ionic Melts (Naukova Dumka, Kiev, 1988) p. 192.

[4] T. Ogawa and K. Minato, Dissolution and Formation of Nuclear Materials in Molten Media, Pure Appl. Chem. 73 (2001) 799–806.

[5] C.A. Zell, W. Freyland and F. Endres, In Situ STM Study of Ni Electrodeposition on Au(111) from the Room Temperature Molten Salt MBIC–AlCl$_3$, Molten Salt Forum 7 (2000) 597–606.

[6] V.A. Oliveira, M.E. Elias and A.M. Elias, Air and Water Stable 2-Methylpyridinium Tetrafluoroborate and +LiF$_4$B Room Temperature Ionic Liquids, Molten Salt Forum 7 (2000) 391–397.

[7] B.G. Zaslavsky, Automated Pulling of Large-Diameter Alkali Halide Scintillation Single Crystals from the Melt, J. Cryst. Growth 200 (1999) 476–482.

[8] B.G. Zaslavsky, Distinctive Features of Automated Pulling of Large Scintillation Alkali Iodides Single Crystals without Oxygen-Containing Impurities, J. Cryst. Growth 218 (2000) 277–281.

[9] Yu.K. Delimarsky, Chemistry of Ionic Melts (Naukova Dumka, Kiev, 1980) p. 372.

[10] A.I. Shatenschtein, Theories of Acids and Bases. History and the Modern Situation (Goskhimizdat, Moscow, Leningrad, 1949) p. 316.

[11] S.P. Miskidgyan and A.P. Garnovskii, Introduction in the Modern Theory of Acids and Bases (Vyscha Shkola, Kiev, 1979) p. 153.

[12] M.I. Kabachnik, The New in the Theory of Acids and Bases, Uspekhi Khimii 48 (1979) 1523–1547.

[13] G.N. Lewis, Valence and the Structure of Atoms and Molecules (Dover Publications, NY, 1966) p. 254.

[14] M.I. Usanovitch, On acids and bases, Zh. Obschch. Khim. 9 (1939) 180–192.

[15] M.I. Usanovitch, What are Acids and Bases, Izv. AN Kaz. SSR (1953) 71, Alma-Ata.

[16] J.E. Huheey, Inorganic Chemistry. Principles of Structure and Reactivity, 3rd edn. (Harper and Row Publishers, New York, 1983).

[17] H. Lux, "Säuren" und "Basen" im Schmeltzfluss: die Bestimmung der Sauerstoffionen-Konzentration, Z. Elektrochem. 45 (1939) 303–309.

[18] H. Flood and T. Forland, The Acidic and Basic Properties of Oxides. I. Acid–Base Definitions, Acta Chem. Scand. 1 (1947) 592–604.

[19] H. Flood and T. Forland, The Acidic and Basic Properties of Oxides. II. The Thermal Decomposition of Pyrosulphates, Acta Chem. Scand. 1 (1947) 781–789.

[20] H. Flood, T. Forland and B. Roald, The Acidic and Basic Properties of Oxides. III. Relative Acid–Base Strengths of Some Polyacids, Acta Chem. Scand. 1 (1947) 790–798.

[21] F.A. Cotton and J. Wilkinson, Principles of Inorganic Chemistry (Mir, Moscow, 1979) p. 678.

[22] R.W. Suter, H.C. Khachel, V.P. Petro, J.H. Howatsou and S.G. Shore, Nature of Phosphorus(V) Chloride in Ionizing and Non-ionizing Solvents, J. Am. Chem. Soc. 95 (1973) 1474–1479.

[23] G. Jander and J. Weis, Geschmolzenes Antimon(III)-Bromid als Ionisierendes Lösungsmittel und Das Verhalten Darin Geloster Substanzen, Z. Elektrochem. 61 (1957) 1275–1283.

[24] G. Jander and J. Weis, Geschmolzenes Antimon(III) chlorid als ionisierendes wasserahnliches Losungmittel. II. III, Z. Anorg. Allg. Chem. (1959) 54–79, see also 80–86.

[25] E.D. Whitney, R.O. MacLaren, C.E. Fogle and T.J. Hurley, Solvolysis Reactions in Chlorine Trifluoride and Bromine Pentafluoride: Preparation of the Tetrafluorochlorates and Hexafluorobromates of Potassium, Rubidium and Cerium, J. Am. Chem. Soc. 86 (1964) 2583–2586.

[26] T.G. Waddington, Liquid Sulfurous Anhydride, in: Non-aqueous Solvent Systems (Khimiya, Moscow, 1971) pp. 238–267.

[27] D.E. Burge and T.H. Norris, Isotopic Exchange Reactions in Liquid Sulfur Dioxide. V. The Acid Halide Catalyzed S^{35} Exchange between Thionyl Chloride and Sulfur Dioxide, J. Am. Chem. Soc. 81 (1959) 2324–2329.

[28] I.G. Dioum, J. Vedel and B. Tremillon, Acido-Basic Properties of Molten Potassium Tetrahalogenogallates (X = Cl, Br), J. Electroanal. Chem. 137 (1982) 219–226.

[29] G. Torsi and G. Mamantov, Acid–Base Properties of the System $AlCl_3$–MCl (M = Li, K,Na,K,Cs), Inorg. Chem. 11 (1972) 1439.

[30] J.F Wagner, Electrochemical Study in Molten Sodium Fluoroborate at 420°C, Report, CEA-N-2350 15, 2 (1983) Abstr. N. 15:005221.

[31] M. Brigaudeau, A. Gadri and J. Wagner, Acid–Base Concept in Molten Fluorides, Proceedings of the First International Symposium on Molten Salts Chemical Technology, 1983) pp. 273–275.

[32] F. Taulelle, B. Tremillon and T. Gilbert, On the Behaviour of Barium Carbonate and Barium Chloride in Molten Sodium Chloroaluminate at 175 °C, J. Electroanal. Chem. 134 (1982) 141–146.

[33] R.G. Pearson, Hard and Soft Acids and Bases, J. Am. Chem. Soc. 85 (1963) 3533–3539.

[34] R.G. Pearson, Hard and Soft Acids and Bases, HSAB. Part I. Fundamental Principles, J. Chem. Educ. 45 (1968) 581–587.

[35] R.G. Pearson, Hard and Soft Acids and Bases, HSAB. Part II. Underlying Theories, J. Chem. Educ. 45 (1968) 643–648.

[36] V.L. Cherginets, Acid–Base Equilibria in Ionic Solvents (Ionic Melts), in: Handbook of Solvents, ed G. Wypych (Chemtec Publishing, Toronto, NY, 2001) pp. 616–638.

[37] V.L. Cherginets and T.P. Rebrova, On the Effect of Ionic Melt Compositions on their Acid–Base Properties, Phys. Chem. Liq. 39 (2001) 367–381.

[38] B. Tremillon and G. Letisse, Propriétés en Solution dans le tétrachloroaluminate de Sodium Fondu. II. Comportement des Protons et des Ions Oxyde, J. Electroanal. Chem. 17 (1968) 387–394.

[39] G. Torsi and G. Mamantov, Electroreduction of Bi^{3+} Ions in Molten $AlCl_3$ + NaCl, J. Electroanal. Chem. 30 (1971) 193–201.

[40] D. Lambertin, J. Lacquement, S. Sanchez and G.S. Picard, Dismutation of Divalent Americium Induced by the Addition of Fluoride Anion to a LiCl–KCl Eutectic at 743 K, Electrochem. Commun. 3 (2001) 519–523.

[41] R.V. Tchenov, E.G. Ivanova and L.D. Dyubova, Anion Exchange between Halides of Silver and Potassium in Melt, Zh. Neorg. Khim. 30 (1985) 1326–1327.

[42] Barin and O. Knacke, Thermochemical Properties of Inorganic Substances (Springer, Berlin, 1973) p. 922.

[43] B.G. Korschunov and D.V. Drobot, Interaction of Chlorides of Gadolinium and Dysprosium with Sodium and Potassium Chlorides in Melts, Zh. Neorg. Khim. 10 (1965) 939–942.

[44] V.L. Cherginets, Generalised Definition of Acids and Bases for Solvent System, Rep. Natl. Acad. Sci. Ukraine N1 (1997) 158–161.

[45] V.L. Cherginets, Acid–Base Equilibria in Ionic Melts. The Generalized Definition of Acids and Bases in Liquid Media (Review), Funct. Mater. 3 (1996) 233–247.

[46] P.G. Zambonin, On the Presence of NO_2^+ in Molten Nitrates, J. Electroanal. Chem. 32 (1971) App. 1–4.

[47] R.N. Kust and F.R. Duke, A Study of the Nitrate Ion Dissociation in Fused Nitrates, J. Am. Chem. Soc. 85 (1963) 3338–3340.

[48] R.G. Bates and G. Schwarzenbach, Über pH-werte nichtwässeriger Lösungen, Helv. Chim. Acta 38 (1955) 699–716.

[49] N.A. Izmailov, Acidity of Non-aqueous Solutions, in: Modern Methods and Devices for Determination of Compositions, Properties and States of Substances, Issue 1. pH- and rH-Metry (CINTI, Moscow, 1962) p. 16.

[50] N.A. Izmailov, Electrochemistry of Solutions (Khimiya, Moscow, 1976) p. 488.

[51] N.A. Izmailov, Common Acidity Scale, Doklady Akad. Nauk. SSSR 127 (1959) 104–107.

[52] W.L. Jolly and C.J. Hallada, Liquid Ammonia, in: Non-aqueous Solvent Systems (Khimiya, Moscow, 1971) pp. 7–49.

[53] R. Lyonnet, C. Ciaravino, R. Marote, J.P. Scharff, B. Duran and J.P. Deloume, Reactivity in Molten Oxonitrates of Lanthanum and Manganese Salts. Synthesis of $La_{1-x}MnO_3$, in: Advances in Molten Salts. From Structural Aspects to Waste Processing, ed M. Gaunr-Escard (Begell House, NY, 1999) pp. 169–177.

[54] G. Matei, I. Jitaru, E. Andronescu and A. Novac, Preparation of Lanthanum Oxocuprate in Alkali Metal Nitrate Melts, in: Advances in Molten Salts. From Structural Aspects to Waste Processing, ed M. Gaunr-Escard (Begell House, NY, 1999) pp. 400–406.

[55] G. Matei, I. Jitaru, E. Andronescu and A. Novac, Preparation of Copper Ferrite by Reaction of Various Precursors with the Molten $NaNO_3$–KNO_3 Medium, in: Advances in Molten Salts. From Structural Aspects to Waste Processing, ed M. Gaunr-Escard (Begell House, NY, 1999) pp. 242–248.

[56] Y. Wang, W.Y. Huang, Z. Wu, Y. Chun and J.H. Zhu, Superbase Derived from Zirconia-Supported Potassium Nitrate, Mater. Lett. 46 (2000) 198–204.

[57] Y.S. Hong, C.H. Han and K. Kim, Preparation of Polycrystalline HT–LiCoO$_2$ using Molten Salt Synthesis Method at 280 °C, Chem. Lett. 12 (2000) 1384–1385.

[58] F. Lantelme, A. Barhoun and K. Kuroda, Role of the Oxoacidity and Ligand Effect in the Electrowinning of Titanium in Fused Salts, in: Refractory Metals in Molten Salts (Their Chemistry, Electrochemistry and Technology), eds D.H. Kerridge and E.G. Polyakov 3/53 (Kluwer Academic Publishers, Dordrecht, 1998) pp. 157–172.

[59] V.L. Cherginets, Oxoacidity in Ionic Melts, Uspekhi Khimii 66 (1997) 661–677.

[60] V.L. Cherginets, Acid–Base Properties by Lux–Flood in Different Kinds of Melts, Preprint of Inst. Single Cryst., ISC-89-6, Kharkov, 1989, p. 17.

[61] V.L. Cherginets, Acid–Base Reactions by Lux–Flood in Different Kinds of Melts, Dep. in ONIITEKHIM, Cherkassy, N740-хп 89 (1989) p. 16.

[62] Yu.F. Rybkin and A.S. Seredenko, Acidity Scale in Molten Potassium Chloride, Ukr. Khim. Zh. 36 (1970) 133–136.

[63] Yu.F. Rybkin and A.S. Seredenko, Acidity Scale in Molten Potassium Chloride and Caesium Iodide, Ukr. Khim. Zh. 40 (1974) 137–141.

[64] Yu.F. Rybkin and V.V. Banik, Acidity Scale in Molten Sodium Iodide, in: Single Crystals and Technics (Inst. Single Cryst., Kharkov, 1973) Issue 2 (9), pp. 153–158.

[65] N.N. Ovsyannikova and Yu.F. Rybkin, Effect of Cations and the Medium Basicity in Molten KCl–NaCl Eutectic, Ukr. Khim. Zh. 42 (1976) 151–155.

[66] M. Dratovsky and D. Havlichek, A Mixture of Nitrogen Dioxide and Oxygen as a Lux Acid in Nitrate Melts, Electrochim. Acta 28 (1983) 1761–1766.

[67] A.M. Shams El Din and A.A. El Hosary, Potentiometric Acid–Base Titrations in Molten Salts. The Acid Character of Group (VI) as Inferred from Their Reaction with Molten KNO$_3$, J. Electroanal. Chem. 9 (1965) 349–360.

[68] B.J. Brough, D.H. Kerridge and M. Mosley, Indicators in Fused Salts, J. Chem. Soc. (A) N11 (1966) 1556–1558.

[69] J.A. Duffy and M.D. Ingram, Establishment of an Optical Scale for Lewis Basicity in Inorganic Oxyacids, J. Am. Chem. Soc. 93 (1971) 64448–66454.

[70] J.A. Duffy and M.D. Ingram, Lewis Acid–Base Interactions in Inorganic Oxyacids, Molten Salts and Glasses. II. Ligand Competition Reactions of Lead(II) and Bismuth(III) with Various Anions in Molten Dimethylsulphone, J. Inorg. Nucl. Chem. 36 (1974) 39–42.

[71] J.A. Duffy and M.D. Ingram, Lewis Acid–Base Interactions in Inorganic Oxyacids, Molten Salts and Glasses. III. Co-ordination Studies of Thallium, Lead and Bismuth in Sulphate/Chloride/Bromide Glass Systems, J. Inorg. Nucl. Chem. 36 (1974) 43–47.

[72] J.A. Duffy and M.D. Ingram, Optical Basicity IV. Influence of Electronegativity on the Lewis Basicity and Solvent Properties of Molten Oxyanion Salts and Glasses, J. Inorg. Nucl. Chem. 37 (1975) 1203–1206.

[73] J.A. Duffy, M.D. Ingram and I.D. Sommerville, Acid–Base Properties of Molten Oxides and Metallurgical Slags, J. Chem. Soc. Faraday Trans. 74 (1978) 1410–1419.

[74] M.S. Slobodyanik, Synthesis of Inorganic Compounds in Molten Alkali Metal Phosphates, Abstract of 14th Ukrainian Conference on Inorganic Chemistry, Kiev (1996) 129.

[75] T.Y. Maekawa and T. Yokokawa, X-ray Fluorescence Spectroscopy of Inorganic Solids SiKα and AlKα Energies of Ternary Oxide Glasses, Nippon Kagaku N6 (1982) 900–904.

[76] A. Kato, R. Mishibashi, M. Nagano and I. Mochida, Relative Activity of Oxygen Ions in Molten Sodium Phosphate as Determined from the Solubility of Sulfur Trioxide, J. Am. Ceram. Soc. 55 (1972) 183–185.

[77] N. Iwamoto, Indicators for the Determination of Basicity in Molten Oxides, Youen N21 (1978) 287–312.

[78] R.K. Kurmaev and S.A. Amirova, Solubility of $VOCl_3$, $TiCl_4$ and $SiCl_4$ in Molten Chlorides of Potassium and Sodium, Zh. Neorg. Khim. 13 (1968) 2258–2261.

[79] F.R. Duke and S. Yamamoto, Acid–Base Reactions in Fused Salts. I. The Dichromate–Nitrate Reaction, J. Am. Chem. Soc. 80 (1958) 5061–5063.

[80] F.R. Duke and S. Yamamoto, Acid–Base Reactions in Fused Salts. The Absolute Concentration of NO_2^+ Ion in Fused Nitrate, J. Am. Chem. Soc. 81 (1959) 6378–6379.

[81] M.V. Smirnov, Electrode Potentials in Molten Chlorides (Nauka, Moscow, 1973) p. 248.

[82] E.A. Bordushkova, Thermal Stability and Decomposition Kinetics of Alkali Metal Nitrates and Nitrites, PhD thesis (Rostov State University, Rostov, 1984).

[83] J.M. Shlegel, Acid–Base Reactions in Fused Salts. The Dichromate Chlorate Reaction, J. Phys. Chem. 69 (1965) 3638–3640.

[84] J.M. Shlegel, The Decomposition of Chlorate in a Dichromate Solution in Fused Nitrates, J. Phys. Chem. 71 (1967) 1520–1521.

[85] F.R. Duke and J.M. Shlegel, Acid–Base Reactions in Fused Salts. The Dichromate–Bromate Reaction, J. Phys. Chem. 67 (1963) 2487–2488.

[86] I. Slama, Azidobasishe Reaktion von Kupfer(II)-ionen in Einer Alkalinitrat-Schmelze, Coll. Czechoslov. Chem. Commun. 28 (1963) 985–990.

[87] I. Slama, Reaktion von Kobalt(II)-ionen in Einer Alkalinitratschmelze, Coll. Czechoslov. Chem. Commun. 28 (1963) 1069–1072.

[88] A.M. Shams El Din and A.A. El Hosary, Potentiometric Acid–Base Titrations in Molten Salts. The Acid Character of Group (V) as Inferred From their Reaction with Molten KNO_3, J. Electroanal. Chem. 8 (1964) 312–323.

[89] A.M. Shams El Din and A.A.A. Gerges, Acid–Base Equilibrium Constants in Fused KNO_3, J. Inorg. Nucl. Chem. 26 (1963) 1537–1538.

[90] A.M. Shams El Din and A.A.A. Gerges, On Potentiometric Acid–Base Titrations in Molten Salts. II. The Titration of NaH_2PO_4, Na_2HPO_4, $NaPO_3$, $Na_4P_2O_7$ with Na_2O_2 in Molten KNO_3, Electrochim. Acta 9 (1964) 123–131.

[91] A.M. Shams El Din and A.A. El Hosary, A Potentiometric Study of the Reaction between V_2O_5 and Molten KNO_3, J. Electroanal. Chem. 7 (1964) 464–473.

[92] A.M. Shams El Din and A.A. El Hosary, The Formation of the Basic Arsenates in Molten KNO_3, J. Electroanal. Chem. 17 (1968) 238–240.

[93] A.M. Shams El Din, A.A. El Hosary and A.A.A. Gerges, Potentiometric Acid–Base Titrations in Molten Salts. The Neutralization of Sodium Metavanadate and Sodium Metaphosphate in Chloride and Nitrate Melts, J. Electroanal. Chem. 6 (1963) 131–140.

[94] A.M. Shams El Din, On Potentiometric Acid–Base Titrations in Molten Salts: The System $K_2Cr_2O_7$–K_2CrO_4–KOH, Electrochim. Acta 7 (1962) 285–292.

[95] A.M. Shams El Din and A.A.A. Gerges, Potentiometric Acid–Base Titrations in Molten Salts. The System $Cr_2O_7^{2-}$–CrO_4^{2-}–O^{2-}, J. Electroanal. Chem. 4 (1962) 309–313.

[96] A.M. Shams El Din and A.A. El Hosary, Potentiometric Acid–Base Titration in Molten Salts. The Acid Character of Li^+, Na^+, K^+, Ca^{2+}, Sr^{2+}, Ba^{2+} and Pb^{2+} as Inferred from the Reaction of Their Carbonates with $K_2Cr_2O_7$ in Molten KNO_3, J. Electroanal. Chem. 16 (1968) 551–562.

[97] A.M. Shams El Din, A.A. El Hosary and H.D. Taki El Din, Potentiometric Acid–Base Titrations in Molten Salts. A Comparative Study of the Acid Character of Some of Group I and Group II Cations, Electrochim. Acta 13 (1968) 407–416.

[98] A.A. El Hosary, M.E. Ibrahim and A.M. Shams El Din, Effect of Li^+ on Acidity of Metaphosphate–Alkali Nitrate Melts, Electrochim. Acta 24 (1979) 645–650.

[99] A.M. Shams El Din and A.A. El Hosary, Potentiometric Acid–Base Titration in Molten Salts. Basic Character of Some Oxide-Ion Donors in Molten KNO_3, Electrochim. Acta 13 (1968) 135–148.

[100] A.M. Shams El Din and A.A. El Hosary, On the Formation of Pyronitrate in Molten Salts, Electrochim. Acta 12 (1967) 1665–1675.

[101] J.D. Burke and D.H. Kerridge, Concerning Oxygen Anions in a Nitrate Melt, Electrochim. Acta 19 (1974) 251–256.

[102] A.M Shams El Din and A.A.A. Gerges, Potentiometric Acid–Base Titration in Fused Salts, Proceedings of the First Australian Conference on Electrochemistry (Pergamon Press, Oxford, 1964) pp. 562–577.

[103] A.M. Shams El Din and A.A.A. Gerges, Potentiometric Acid–Base Titrations in Molten Salts. The Use of a Metal/Metal-Oxide Electrode as an Indicator Electrode and the Determination of Acid–Base Equilibria in Fused KNO_3, Electrochim. Acta 9 (1964) 613–627.

[104] N. Coumert, M. Porthault and J.-C. Merlin, Analyse Potentiometrique de Phosphates Condenses en Solution dans les Nitrates Alcalins a 350 °C, Bull. Soc. Chim. France 33 (1965) 910.

[105] N. Coumert, M. Porthault and J.-C. Merlin, Reactions Acide–Base dans les Nitrates Alcalins Fondus Entre Phosphates et Molybdates, Bull. Soc. Chim. France 35 (1967) 332–338.

[106] M. Hassanein and N.S. Youssep, Cryoscopic Studies in Molten Salts. Dissociation of Some Alkali Isopolymolybdates and Some Related Molybdenum(VI) Compounds in Molten Potassium Dichromate and Potassium Nitrate, Indian J. Chem. A21 (1982) 72–74.

[107] C. Liteanu, E. Cordos and L. Margineanu, Automatic Potentiometric Titration in Molten Media. I. A New Automatic Procedure for Potentiometric Titration in Molten Media with Solid Standard, Rev. Roumaine Chim. 15 (1970) 583–587.

[108] J.M. Schlegel, Lewis Acid–Base Titration in Fused Salts, J. Chem. Educ. 43 (1966) 362–363.

[109] Y. Hoshino, T. Utsunomiya and O. Abe, On the Thermal Decomposition of Sodium Nitrate and the Effect of Several Oxides on the Decompositions, Bull. Chem. Soc. Jpn 54 (1981) 135–191.

[110] A. Baraka, A. Abdel-Razik and A.J. Abdel-Rohman, The Use of the Nb–Nb$_2$O$_5$ Electrode as an Indicator Electrode in Potentiometric Acid–Base Titration in Fused KNO$_3$, Surf. Technol. 25 (1985) 31–38.

[111] A. Baraka, A.J. Abdel-Rohman and E.A. El-Taher, The Use of Tantalum/Tantalum Oxide Electrode as an Indicator Electrode in Potentiometric Acid–Base Titration in Fused Potassium Nitrate, Mater. Chem. Phys. 9 (1983) 447–456.

[112] A. Baraka, A.J. Abdel-Rohman and E.A. El-Taher, The Use of Zirconium/Zirconium Oxide Electrode as an Indicator Electrode in Potentiometric Acid–Base Titration in Fused Potassium Nitrate, Mater. Chem. Phys. 9 (1983) 583–595.

[113] P. Afanasiev and D.H. Kerridge, Reactivity of V$_2$O$_5$, MoO$_3$ and WO$_3$ in Molten KNO$_3$, Studied by Mass Spectrometry, J. All. Comp. 32 (2001) 97–102.

[114] A.I. Efimov, L.P. Belorukova, I.V. Vasilkova and V.P. Chechev, Properties of Inorganic Compounds (Khimiya, Leningrad, 1983) p. 392.

[115] R.N. Kust, A Study of Dimolybdate and Dichromate Formation in Fused Equimolar Sodium Nitrate–Potassium Nitrate Solvent, Inorg. Chem. 6 (1967) 2239–2243.

[116] J.M. Schlegel and R. Bauer, A Potentiometric Study of Polymeric Anions of Molybdenum in Nitrate Melts, Inorg. Chem. 11 (1972) 909–911.

[117] P.P. Poppel and S.N. Khoroshevskaya, Interaction of Lead(II) Oxide with Molten Nitrates of Alkali Metals, Zh. Neorg. Khim. 30 (1985) 197–200.

[118] A.A. Attia, The Use of the Nb–Nb$_2$O$_5$ as an Oxide Ion Indicator Electrode in Fused Salts, Electrochim. Acta 47 (2002) 1241–1248.

[119] R. Combes, J. Vedel and B. Tremillon, Dissociation des Hydroxydes Alcalins en Solution dans le Melange NaCl–KCl Fondu, an Moyen d'une Electrode Indicatrice d'ions Oxyde, C.R. Acad. Sci. C273 (1971) 1740–1742.

[120] V.I. Posypaiko, E.A. Alekseeva, N.A. Vasina, E.S. Gryzlova, V.N. Afonova, N.N. Petrova, I.G. Popova, Z.L. Guseva and V.T. Shemyatenkova (eds), Fusion Diagrams of Salt Systems. Part I. Binary Systems with Common Anion [from AgBr–CsBr to In$_2$(WO$_4$)$_3$–Rb$_2$WO$_4$] (Metallurgiya, Moscow, 1977) p. 416.

[121] I.T. Goronovsky, Yu.P. Nazarenko and E.F. Nekryach, Short Handbook on Chemistry (Naukova Dumka, Kiev, 1987) p. 830.

[122] I.N. Belyaev and I.N. Tchikova, Ternary systems K$_2$SO$_4$–Li$_2$SO$_4$–Cs$_2$SO$_4$ and Li$_2$SO$_4$–Rb$_2$SO$_4$–Pb$_2$SO$_4$, Zh. Neorg. Khim. 8 (1963) 1442–1448.

[123] E.K. Akopov and A.G. Bergman, Fusion Diagrams for the Ternary System of the Sulfates of Lithium, Sodium and Potassium, Zh. Neorg. Khim. 4 (1959) 1146–1152.

[124] Y. Kaneko and H. Kojima, Dissociation Behaviour of Vanadium Dioxide and Basicity in Fused Alkali Metal Sulphate Mixtures, Proceedings of the First International Symposium on Molten Salts Chemical Technology, 1983) pp. 441–444.

[125] Y. Kaneko and H. Kojima, Basicity and Formation of Vanadium Bronze in Molten Alkaline Sulphate Mixtures, Youen 28 (1985) 109–126.

[126] A. Rahmel, Potentiometrische Saure-Base-Titrationen in Alkalisulfat-Schmelzen, J. Electroanal. Chem. 61 (1975) 333–341.

[127] O.A. Esin and P.V. Geld, Physical Chemistry of Metallurgical Processes (Metallurgizdat, Sverdlovsk, 1962) p. 672.

[128] H. Flood and W. Knapp, Acid–Base Equilibria in the System PbO–SiO$_2$, J. Am. Ceram. Soc. 46 (1963) 61–65.

[129] T. Forland and M. Tashiro, Acid–Base Relationship in Glass Systems. Development of an Oxygen Electrode for Measurements of Acid–Base Properties of Glasses, Glass Ind. 37 (1956) 381–385, see also 399, 402.

[130] T. Forland, Syre-Base Reaksjoner i Silikatsmelter, Glastekn. Tidskr. 7 (1962) 35–42.

[131] V.I. Minenko, S.M. Petrov and N.S. Ivanova, The Use of Reversible Oxygen Electrode in Oxygen-Containing Melts, Izv. Vysschikh Utchebn. Zav., Chern. Metallurgia N7 (1960) 10–13.

[132] P.A. Perkins, Oxygen Sensor and Method for Determining the Oxygen Activity in Molten Glass, US 4,313,799. C1204-1T; G01 N27 (58) 02 Feb 1982. Appl. 137,163 04 Apr. 1980, p. 8.

[133] H. Itoh and T. Yokokawa, Thermodynamic Activity of Sodium Oxide in Sodium Oxide–Silicon Dioxide–Aluminium Oxide Melt, Trans. Jpn. Inst. Met. 25 (1984) 879–884.

[134] T.T. Bobrova, V.V. Moiseev, T.V. Permyakova and G.E. Sheshukova, Interaction of Alkaline Aluminosilicate Glasses with Molten Calcium, Strontium and Barium Nitrates, Izv. AN SSSR, Neorg. Mater. 9 (1973) 1416–1419.

[135] R. Didtschenko and E.G. Rochow, Electrode Potentials in Molten Silicates, J. Am. Chem. Soc. 76 (1954) 3291–3294.

[136] G.I. Janz, Molten Salts Handbook (Academic Press, London, 1967) p. 588.

[137] L.I. Knuniantz (ed), Short Chemical Encyclopaedia, Vol. 5 (Sovetskaya Entsyklopedia, Moscow, 1967) p. 1184.

[138] L.I Knuniantz (ed), Short Chemical Encyclopaedia, Vol. 2 (Sovetskaya Entsyklopedia, Moscow, 1967) p. 1188.

[139] V.I. Shapoval and O.G. Tsiklauri, Acid–Base Titration by Lux in Molten KCl–NaCl Eutectic, in: Physical Chemistry and Electrochemistry of Molten Salts and Solid Electrolytes, Part 2, Sverdlovsk, 1973) pp. 32–33.

[140] V.I. Shapoval, A.S. Avaliani and O.G. Tsiklauri, Acid–Base Reactions of Formation of Chromates in Molten Chlorides, Soobsch. AN Gruz. SSR 72 (1973) 585–588.

[141] Yu.K. Delimarskii, V.I. Shapoval, O.G. Tsiklauri and V.A. Vasilenko, Potentiometric Study of Acid–Base Reactions by Lux in Molten KCl–NaCl Eutectic, Ukr. Khim. Zh. 40 (1974) 8–13.

[142] V.L. Cherginets and V.V. Banik, Acidic Properties of Dichromate Ion, Molybdenum and Tungsten Oxides in Molten KCl–NaCl Eutectic at 973 K, Rasplavy N2 (1991) 118–120.

[143] T.P. Boyarchuk, E.G. Khailova and V.L. Cherginets, Potentiometric Titration of Potassium Dichromate in Molten Equimolar KCl–NaCl Mixture at 700 °C, Ukr. Khim. Khurn. 58 (1992) 758–760.

[144] H. Remy, Lehrbuch der Anorganische Chemie (Akademische verlagsgessellschaft Geest & Portig K.G., Leipzig, 1960) Bd. 2.

[145] V.I. Shapoval, A.S. Avaliani and N.A. Gasviani, Acidic Properties of Molybdenum Oxide in Molten Chlorides, Soobsch. AN Gruz. SSR 72 (1973) 105–108.

[146] V.V. Zyryanov, Mechanochemical Equilibrium at Synthesis of Pb_2MoO_5, Izv. SO AN SSSR, Ser. Khim. 2 (1990) 101–106.

[147] S.I. Kutcheiko and I.Ya. Turova, On the Obtaining of Oxochlorides of Tungsten, Zh. Neorg. Khim. 30 (1985) 1327–1328.

[148] V.I. Shapoval, V.F. Grischchenko and L.I. Zarubitskaya, Behaviour of Platinum–Oxygen Electrode in KCl–NaCl Melt Containing, Tungstate, Ukr. Khim. Zh. 38 (1972) 1088–1091.

[149] V.I. Shapoval, V.F. Grischchenko and L.I. Zarubitskaya, Acidic Properties of Tungsten Oxide in Molten Chlorides, Ukr. Khim. Zh. 39 (1973) 867–869.

[150] V.I. Shapoval, Yu.K. Delimarskii and V.F. Grischchenko, Electrochemical Processes with Fast and Slow Acid–Base Reactions in Molten Electrolytes, in: Ionic melts, Issue 1 (Naukova Dumka, Kiev, 1974) pp. 222–241.

[151] O.G. Tsiklauri and N.A. Gasviani, On Kinetic Stage of Process of Dichromate Ion Electroreduction in Molten Equimolar Mixture, in: Materials of Conference of Young Scientists of Isct. Chemistry and Electrochemistry of Academic Science of Gruz. SSR (Institute of Inorganic Chemistry and Electrochemistry, Tbilisi, 1976) pp. 63–67.

[152] V.I. Shapoval, O.G. Tsiklauri and N.A. Gasviani, Significance of Acid–Base Properties of Molten Electrolyte in Kinetics of anion electroreduction, Soobsch. AN Gruz. SSR 88 (1977) 609–612.

[153] H.B. Kushkhov and V.I. Shapoval, High-temperature Electrochemical Synthesis of Refractory Compounds in Ionic Melts, Abstract of the 10th All-Union Conference on Physics Chemistry and Electrochemistry of Ionic Melts and Solid Electrolytes, Ekaterinbourg, Vol. 2. 1992) p. 3.

[154] R. Combes and B. Tremillon, Dissociation and Solubility Variation Versus pO^{2-} of Scheelite $CaWO_4$ in Molten NaCl–KCl (at 1000 K), J. Electroanal. Chem. 83 (1977) 297–308.

[155] M.E. Poloznikova, A.S. Kabaktchi, V.A. Balashov, O.I. Kondratov and V.V. Fomi-tchev, Vibrational Spectrum of Ba_2WO_5, Zh. Neorg. Khim. 35 (1990) 802–804.

[156] M. Tazika, S. Mizoe, M. Nagano and A. Kato, Acid–Base Equilibria of Phosphate Ions in Molten Alkali Chloride, Denki kagaku 46 (1978) 37–41.

[157] V.L. Cherginets and V.V. Banik, Potentiometric Investigation of Acidic Properties of Metaphosphate and Tetraborate Ions and Vanadium(V) Oxide in Molten KCl–NaCl Eutectic at 973 K, Rasplavy N6 (1990) 92–96.

[158] R. Combes, F. De Andrade and L. Carvalho, Proprietes Oxoacido-Basiques du Vanadium(V) en Solution dans NaCl–KCl a 727 °C, C.R. Acad. Sci C285 (1977) 137–140.

[159] P. Claes, F. Mernier, L. Wery and J. Glibert, Composition Dependence of the Oxoacidobasic Properties of Molten Hydroxides, Electrochim. Acta 44 (1999) 3999–4006.

[160] V.L. Cherginets, Potentiometric Investigation of Acidic Properties of Niobium(V) and Germanium(IV) Oxides in Chloride Melts (Dep. in ONIITEKHIM, Cherkassy, 1991) N241–91, p. 7.

[161] V.L. Cherginets, Acid–Base Interactions in Molten Chlorides, PhD thesis (Kharkov State University, Kharkov, 1991).

[162] V.L. Cherginets and T.P. Rebrova, Studies of Some Acid–Base Equilibria in the Molten Eutectic Mixture KCl–LiCl at 700°C, Electrochim. Acta 45 (1999) 469–476.

[163] V.L. Cherginets and T.P. Rebrova, Acidic Properties of Potassium Dichromate in Molten Alkali Metal Chlorides, Rasplavy N1 (2001) 81–84.

[164] Yu.F. Rybkin and V.V. Banik, Acid–Base Interactions in Molten Sodium Iodide, in: Methods of Obtaining and Investigations of Single Crystals and Scintillators, N5 (Institute for Single Crystals, Kharkov, 1980) pp. 121–125.

[165] Yu.F. Rybkin and V.V. Banik, Acid–Base Potentiometric Titration in Molten Sodium Iodide, in: Single Crystals and Technics (Inst. for Single Cryst., Kharkov, 1974) Issue 1 (10), pp. 111–114.

[166] V.L. Cherginets and V.V. Banik, Potentiometric Investigation of Acidic Properties of Sodium Vanadate and Sodium Tetraborate in Molten Sodium Iodide at 700 °C, Zh. Fiz. Khim. 68 (1994) 145–147.

[167] V.L. Cherginets and V.V Banik, Acidic Properties of Sodium Vanadate and Sodium Tetraborate in Molten Sodium Iodide (Abstract of the 10th All-Union Conference on Physics Chemistry and Electrochemistry of Ionic Melts and Solid Electrolytes, Ekaterinbourg, Vol. 1. 1992) p. 85.

[168] V.L. Cherginets and T.P. Rebrova, Equilibrium Constants of Some Acid–Base Reactions in Molten Sodium Chloride, Proceedings of the International Symposium on Ionic Liquids in honour of M. Gaune-Escard (2003) 449–458.

[169] R. Combes, M.N. Levelut and B. Tremillon, Oxoacidity and Its Influence on the Electrochemical Properties in Molten Mixtures of $CeCl_3$ and Equimolar NaCl–KCl at 1000 K, Electrochim. Acta 23 (1978) 1291–1295.

[170] R. Combes, B. Tremillon, F. De Andrade, M. Lopes and H. Ferreira, Definition and Use of an Oxoacidity Function Ω for the Comparison of Acidity Levels at 1000 K of Some Chloride Melts Involved in Electrometallurgical Processes, Anal. Lett. 15A (1982) 1585–1600.

[171] R. Combes, M. Cadi and B. Grindre, Oxoacidity and its Influence on Electrochemical Properties of Molten Tin(II) Chloride, Electrochim. Acta 25 (1980) 867–870.

[172] R.F. Watson and G.S. Perry, Solubility of ZnO and Hydrolysis of $ZnCl_2$ in KCl Melt, J. Chem. Soc. Faraday Trans. 87 (1991) 2955–2960.

[173] D. Ferry, Y. Castrillejo and G.S. Picard, Acidity and the Purification of Mixed Melt $ZnCl_2$ (33.4 mol%)–NaCl (66.6 mol%), Electrochim. Acta 34 (1989) 313–316.

[174] R.G. Bates, Determination of pH. Theory and Practice (Wiley, London, 1964).

[175] V.L. Cherginets, V.V. Banik and A.B. Blank, Results of Oxide Solubility Studies in Molten Equimolar Mixture of Sodium and Potassium Chlorides, Preprint of Inst. Single Cryst., IMK-90-3 (Kharkov, 1990) p. 15.

[176] V.I. Shapoval, V.F. Makogon and O.F. Pertchik, Electrochemical Investigation of Solubility of Cu_2O in the Molten LiCl–KCl, Ukr. Khim. Zh. 45 (1979) 7–10.

[177] V.I. Shapoval and V.F. Makogon, Features of Metal Oxide Electroreduction in Chloride Melt, Ukr. Khim. Zh. 45 (1979) 201–205.

[178] Y. Castrillejo, M. Martinez, G.M. Haarberg, B. Borresen, K.S. Osen and R. Tunold, Oxoacidity Reactions in Equimolar Molten $CaCl_2$–NaCl mixture at 575 °C, Electrochim. Acta 42 (1997) 1489–1494.

[179] W.J. Burkhardt and J.D. Corbett, The Solubility of Water in Molten Mixtures of LiCl and KCl, J. Am. Chem. Soc. 79 (1957) 6361–6363.

[180] R. Combes, J. Vedel and B. Tremillon, Realisation d'une Electrode Indicatrice d'ions Oxyde dans sels Fondus au Moyen d'Electrolytes Solides, Anal. Lett. 3 (1970) 523–529.

[181] V.L. Cherginets, Oxide Solubilities in Ionic Melts, in: Handbook of Solvents, ed G. Wypych (Chemtec Publishing, Toronto, NY, 2001) pp. 1484–1495.

[182] V.I. Posypaiko and E.A. Alekseeva (eds), Fusion Diagram of Salt System. Ternary Systems (Khimiya, Moscow, 1977) p. 328.

[183] V.L. Cherginets, O.V. Demirskaya and T.P. Rebrova, Oxoacidic Properties of Potassium Halide Melts at 800 °C, Electrochem. Commun. 2 (2000) 762–764.

[184] V.L. Cherginets, O.V. Demirskaya and T.P. Rebrova, Oxobasicity Index as a Measure of Acidic Properties of Ionic Melts: Definition and Methods of Acidity Estimation, Molten Salts Forum 7 (2000) 163–166.

[185] P. Bocage, D. Ferry and G.S. Picard, Proprietes Oxydoreductrices et Oxoacides du Magnesium dans le Melange $BaCl_2-CaCl_2-NaCl$ (23.5%–24.5%–52% Molaire) Fondu a 600 °C, Electrochim. Acta 36 (1991) 155–162.

[186] T.P. Boyarchuk, E.G. Khailova and V.L. Cherginets, Potentiometic Investigation of Solubilities of Oxides in Molten Eutectic Mixture of Caesium, Potassium and Sodium Chlorides, Preprint of Institute for Single Crystals, IMK-91-17 (Kharkov, 1991) p. 14.

[187] V.L. Cherginets, T.P. Rebrova and O.V. Demirskaya, Acidic Properties of Chloride Melts at 600 °C, Zh. Fiz. Khim. 76 (2002) 1345–1347.

[188] T.A. Lysenko, E.G. Khailova and V.L. Cherginets, Solubilities of Oxides of Bivalent Metals in the Molten CsCl–KCl–NaCl Eutectic at 700 °C, Preprint of Inst. Single Cryst., IMK-92-19 (Kharkov, 1992) p. 23.

[189] V.L. Cherginets and T.P. Rebrova, Studies of Oxide Solubilities in Molten KCl–LiCl at 700 °C, in: Advances in Molten Salts. From Structural Aspects to Waste Processing, ed M. Gaune-Escard (Begell-House. Inc., NY, 1999) pp. 108–111.

[190] V.L. Cherginets and T.P. Rebrova, Acidic Properties of Cations and Oxide Solubilities in the KCl–LiCl Eutectic Melt at 700 °C, Zh. Fiz. Khim. 73 (1999) 687–689.

[191] V.L. Cherginets and T.P. Rebrova, Potentiometric Studies of Acidic Properties of Some Cations in Molten Eutectic $CaCl_2-KCl$ mixture at 700 °C, Electrochem. Commun. 2 (2000) 97–99.

[192] V.L. Cherginets and T.P. Rebrova, Acidic Properties of Cations and Oxide Solubilities in the $SrCl_2-KCl-NaCl$ Eutectic Melt at 700 °C, Zh. Fiz. Khim. 74 (2000) 244–246.

[193] V.L. Cherginets and T.P. Rebrova, Oxide Solubilities in $BaCl_2-KCl-NaCl$ (0.43:0.29:0.28) Melt at 700 °C, Zh. Neorg. Khim. 46 (2001) 344–346.

[194] V.L. Cherginets and T.P. Rebrova, Oxoacidic Properties of Some Ba-Based Chloride Melts, Electrochem. Commun. 1 (1999) 590–592.

[195] V.L. Cherginets and T.P. Rebrova, Studies of the Relative Acidities of Some Chloride Melts, Electrochim. Acta 46 (2000) 25–30.

[196] V.L. Cherginets, T.P. Rebrova and O.V. Demirskaya, Determination of the Oxobasicity Indices of Ionic Solvents by Carbonate Methods, Abstract of Komar' All-Ukrainian Conference on Analytical Chemistry, Kharkov, 2000) 15–19 May, 2000, p. 52.

[197] V.L. Cherginets, O.V. Demirskaya and T.P. Rebrova, On 'Carbonate' Stability in Molten Caesium and Potassium Halides, J. Electroanal. Chem. 512 (2001) 124–127.

[198] V.L. Cherginets and T.P. Rebrova, On Carbonate Ion Dissociation in Molten Alkali Metal Halides at ≈ 800 °C, J. Chem. Eng. Data 48 (2003) 463–467.

[199] G.J. Janz and F. Saegusa, The Oxygen Electrode in Fused Electrolytes, Electrochim. Acta 7 (1962) 393–398.

[200] A.G. Keenan and I.J. Ferrer-Vinent, Transition Metal–Metal Oxide–Carbonate–Carbon Dioxide–Oxygen Electrodes at 350 °C in Fused Potassium Nitrate, J. Phys. Chem. 83 (1979) 358–360.

[201] V.S. Bagotsky, Fundamentals of Electrochemistry (Khimiya, Moscow, 1988) p. 400.

[202] B.A. Shmatko, R.Ya. Popilskii, T.F. Baranova, Zh.I. Ievleva, R.G. Belyanina and N.E. Golovkina, Conductivity and Some Technical Properties of Solid Electrolyte of Composition of 0.84 La_2O_3 0.16 CaO, Izv. AN SSSR, Neorg. Mater. 10 (1974) 1649–1653.

[203] B. Eichler, D. Naumann and H. Ullmann, Wirkungsweise und Aufbraueiner Sauerstoff-Elektrode mit Oxidionenleitender Th0. 8Y0.2O1.9 Festelelektrolytmembran fur Potentiometrische Messungen in Salz-Schmelzen, Z. Phys. Chem. 236 (1967) 372–377.

[204] K. Kuikkola, Ionic Conductivity of ZrO_2–CaO, J. Electrochem. Soc. 104 (1958) 379.

[205] Y. Miyashita and M. Mugita, Development and Application of Probes for Electrochemical Measurements of Free Oxygen Content in Liquid Steel, Nippon Kokau Tech. Rep. Overseas 33 (1981) 47–50.

[206] Z.S. Voltchenkova, Conductivity of composition of ZrO_2–Nd_2O_3 System, Izv. AN SSSR. Neorg. Mater. 5 (1969) 1096–1100.

[207] B. Chappey, B. Auclair, M. Goet and M. Guillou, Sur la Conductivite du Systeme Zircone-Oxyde de Cerium, C.R. Acad. Sci. C280 (1975) 1053–1056.

[208] J. Besson, C. Deportes and M. Darey, Sur une Electrode de Comparaison Utilicable en Bains de sels Oxygenes a Haute Temperature, C.R. Acad. Sci. 251 (1960) 1630–1632.

[209] D.A. Nissen, Solid Electrolyte Oxide Ion Electrode for Molten Nitrates, Energy Res. Abstr. 7 (1982) 22, Abstr. No 4347, Report 1981, SAND-81-8013; Ord N DE82003878.

[210] J.F. McAleer, P.T. Moseley, D.E. Williams and A. Maignan, The Use of Oxide-Ion Conducting Ceramics for Probing Reactions Occurring at the Surface of Semiconducting Oxide, Sens. Actuators 17 (1989) 313–318.

[211] M.V. Perfil'ev and G.I. Fadeev, Determination of Limitedly Low Temperature of Reversible Work of Electrodesin Cells with Solid Oxide Electrodes, Coll. Inst. Electrochem. Ural Sci. Cent. Acad. Sci. USSR N25 (1977) 102–106.

[212] K. Doerfel, Statistics in Analytical Chemistry (Mir, Moscow, 1969) p. 248.

[213] V.S. Bagotskii, Principles of Electrochemistry (Khimiya, Moscow, 1988) p. 400.

[214] T.P. Boyarchuk, E.G. Khailova and V.L. Cherginets, Potentiometric Measurements in Molten Chlorides. Solubilities of Metal Oxides in the Molten Eutectic Mixture CsCl–KCl–NaCl at 600 °C, Electrochim. Acta 38 (1993) 1481–1485.

[215] M. Frederics, R.B. Temple and G.W. Trickett, The Oxygen Electrode in Molten Nitrates, J. Electroanal. Chem. 38 (1972) App. 5–8.

[216] M. Frederics and R.B. Temple, The Solubility of Metallic Oxides and the Free Energy of Solvation of the Oxide-Ion in Molten Alkali Metal Nitrates, Inorg. Chem. 11 (1972) 968–970.

[217] R.N. Kust, Carbonate Ion Dissociation in Fused Alkali Nitrates, Inorg. Chem. 3 (1964) 1035–1036.

[218] R.N. Kust, The Oxygen Electrode in Fused Alkali Nitrates, J. Phys. Chem. 69 (1965) 3662–3663.

[219] E. Desimoni, L. Sabbatini and P.G. Zambonin, Oxygen Electrodes in Fused Salts. Potentiometric Behaviour of the System $CO_2,O_2/CO_3^{2-}$ in Molten Alkali Nitrates, J. Electroanal. Chem. 71 (1976) 73–79.

[220] J.M. Schlegel and D. Uhr, The Platinum–Oxygen Electrode in Dichromate–Chromate Solutions of Alkali Nitrate Melts, Inorg. Chem. 12 (1973) 595–597.

[221] A.G. Keenan and T.K. Williamson, The Platinum|Platinumoxide|Carbonate|Carbon Dioxide Electrode at 350 °C in Fused Potassium and Sodium Nitrates, J. Phys. Chem. 81 (1978) 46–49.

[222] V.M. Mikhailets and A.A. Plyshevskii, Investigation of Processes on Platinum Anode in Molten Oxides, (Physico-Chemical Investigations of Metallurgical Processes, Issue 1, Sverdlovsk, 1991) pp. 95–97.

[223] F. Colom and A. De la Plaza, Oxide Film Formation on Iridium Electrodes in KCl–LiCl Melts Containing Alkaline Oxide, (Proceedings of the 27th Meeting of International Society of Electrochemistry, Zurich, 1978) Ext. Abstr., S.1, s.a. N161.

[224] C. Tsaofang, K.J. Walsh and P.S. Fedkin, Cyclic Voltammetric Study of the Electrochemical Formation of Platinum Oxide in a Pt/Yttria-Stabilized Zirconia Cell, Solid State Ionics 47 (1991) 277–285.

[225] P.G. Zambonin and G. Signorile, Irreversibility of the System Pt (or Au) (O_2) H_2O/OH^- in Molten Nitrate at Low Hydroxide Concentrations. Potentiometric Findings and Mechanistic Considerations, J. Electroanal. Chem. 35 (1972) 251–259.

[226] P.G. Zambonin, Concerning the Oxygen Electrodes in Nitrate Melts, J. Am. Chem. Soc. 97 (1975) 682–683.

[227] R. Littlewood, A Reference Electrode for Electrochemical Studies in Fused Alkali Chlorides at High Temperatures, Electrochim. Acta 3 (1961) 270–278.

[228] I.M. Zharskii and G.I. Novikov, Physical Methods of Investigations in Inorganic Chemistry (Vysshaya Shkola, Moscow, 1988) p. 271.

[229] N.S. Wrench and D. Inman, The Oxygen Electrode in Molten Salts. Potentiometric Measurements, J. Electroanal. Chem. 17 (1968) 319–325.

[230] H.A. Laitinen and B.B. Bhatia, Electrochemical Study of Metallic Oxides in Fused LiCl–KCl Eutectic, J. Electrochem. Soc. 107 (1960) 705–710.

[231] G. Landresse and G. Duyckaerts, Diagramme Potentiel—pO^{2-} de l'uranium dans l'eutectique LiCl–KCl a 660 °C, Anal. Chim. Acta 65 (1973) 245–247.

[232] V.L. Cherginets and V.V. Banik, Interaction of KCl–LiCl Eutectic Melts with Gaseous HCl and H_2O at 773 K, Rasplavy N4 (1991) 98–101.

[233] V.L. Cherginets, On the Reversibility of Some Oxygen Electrode in the Molten KCl–NaCl Eutectic at 973 K, Rasplavy N1 (1991) 62–65.

[234] V.L. Cherginets, Oxide Ion Electrodes and Oxide Ion Donors in Molten Salts. A Consideration of Potentiometric Studies, Electrochim. Acta 42 (1997) 1507–1514.

[235] R. Littlewood and E.J. Argent, The Effect of Aqueous Contaminants on the Redox Potential of Chloride Melts, Electrochim. Acta. 4 (1961) 114–128.

[236] Yu.K. Delimarsky, V.I. Shapoval and N.I. Ovsyannikova, Potentiometric Study of Reactions of Formation of Some Oxides in Melts, Ukr. Khim. Zh. 43 (1977) 115–119.

[237] N.I. Kukhtina, Use of Solid-Electrolyte Oxygen Electrode for Determination of Basicity of Ionic Media, (Abstract of the Second All-Union Symposium on 'Solid electrolytes and their analytical exploitation', Sverdlovsk, 1985) p. 70.

[238] F. De Andrade, R. Combes and B. Tremillon, Determination du Produit de Solubilite de la Chaux dans le Melange Equimolaire de NaCl et KCl Fondus, C.R. Acad. Sci C280 (1975) 945–947.

[239] R. Combes, F. De Andrade, A. de Barros and H. Ferreira, Dissociation and Solubility Variation Vs. pO^{2-} of some Alkaline Earth Oxides in Molten NaCl–KCl (at 1000 K), Electrochim. Acta 25 (1980) 371–374.

[240] R. Combes, J. Vedel and B. Tremillon, Mesure de pO^{2-} au Moyen d'une Electrode a Membrane de Zircone Stabilisee et Determination Potentiometrique de Constantes d'Equilibre d'Echarge de O^{2-} dans le Melange Equimolaire NaCl–KCl Fondu, Electrochim. Acta 20 (1975) 191–200.

[241] S. Pizzini and R. Morlotti, Oxygen and Hydrogen Electrodes in Molten Fluorides, Electrochim. Acta 10 (1965) 1033–1041.

[242] V.L. Cherginets, Features of Work of Gas Oxygen Electrodes in Chloride Melts, Elektrokhimiya 35 (1999) 569–572.

[243] G. Brauer (ed), Handbook on Inorganic Synthesis, Vol. 5 (Mir, Moscow, 1985) p. 360, 6 volumes.

[244] M.L. Deanhardt and K.H. Stern, Solubility Products of Metal Oxides in Molten Salts by Coulometric Titration of Oxide Ion through Zirconia Electrode, J. Electrochem. Soc. 128 (1981) 2577–2582.

[245] M.L. Deanhardt and K.H. Stern, Solubility of Yttrium Oxide in Na_2SO_4 and NaCl Melts, J. Electrochem. Soc. 129 (1982) 2228–2232.

[246] V.E. Komarov and V.E. Krotov, Solubility of Zirconium Dioxide in Molten Equimolar Mixture of Chlorides of Sodium and Potassium, Coll. Inst. Electrochem. Ural Sci. Cent. Acad. Sci. USSR N27 (1978) 61–64.

[247] Yu.F. Rybkin and A.S. Seredenko, Ceramic Electrode Reversible to Oxygen Ions in Melt, Ukr. Khim. Zh. 40 (1974) 489–492.

[248] Yu.F. Rybkin and V.V. Banik, Oxygen Electrode for Measurements of Basicity in Medium of Molten Sodium Iodide, Coll. Inst. Single Cryst., Kharkov N3 (1979) 118–123.

[249] V.L. Cherginets, O.V. Demirskaya and T.P. Rebrova, Potentiometric Investigations of Carbonate Ion Stability in Molten Cesium and Sodium Iodides, Inorg. Chem. 41 (2002) 1045–1047.

[250] G. Brauer (ed), Handbook on Inorganic Synthesis, Vol. 3 (Mir, Moscow, 1985) p. 392, 6 volumes.

[251] K. Grjotheim, C. Krohn and M. Malinovsky, Aluminium Electrolysis (Aluminium-Verlag, Dusseldorf, 1982) p. 484.

[252] R. Lysy and G. Duyckaerts, Diagramme Potentiel—pO^{2-} du Neptunium dans l'Eutectique LiCl–KCl a 660 °C, Anal. Chim. Acta 96 (1978) 125–132.

[253] A.V. Gektin, N.V. Shiran and V.N. Belskiy, Fast UV Scintillation in CsI Crystals, Nucl. Tracks Radiat. Meas. 21 (1993) 11–13.

[254] L.N. Trefilova, Role of Admixture CO_3^{2-} Centers in the Processes of Creation of Centers of Luminescence and Colouring in CsI Crystals, PhD thesis (KharZkov, 2000) p. 170.

[255] N.W. Hanf and M.J. Sole, High-Temperature Hydrolysis of Sodium Chloride, Trans. Faraday. Soc. 66 (1970) 3065–3074.

[256] Yu.F Rybkin and Yu.A. Nesterenko, Interaction of Crystalline and Molten Sodium Iodide with Water Vapour (Naukova Dumka, Kiev, 1974) Isssue 2, pp. 184–189.

[257] Yu.F. Rybkin and Yu.A. Nesterenko, High-Temperature Interaction of Crystalline and Molten Sodium Iodide with Water Vapour and Oxygen, Zh. Fiz. Khim. 50 (1976) 781–788.

[258] M.V. Smirnov, I.V. Korzun and V.A. Oleynikova, Hydrolysis of Molten Alkali Chlorides, Bromides and Iodides, Electrochim. Acta 33 (1988) 781–788.

[259] C. Butler, J.R. Russel, R.B. Quincy and D.E. la Valle, Growth and Evaluation of High Purity KCl Crystals, J. Chem. Phys. 45 (1966) 968–971.

[260] F. Rosenberger, Purification of Alkali Halides, Ultrapurity Methods and Techniques (Dekker, NY, 1972) pp. 3–70.

[261] D. Ecklin, Fonte Zonalli et Analyse de Purete dans les Iodures Alkalins, Helv. Chim. Acta 50 (1967) 1107–1119.

[262] J.M. Peech, D.A. Bower and R.O. Rohl, Preparation of Pure Alkali Halide Crystals and Some of Their Properties, J. Appl. Phys. 38 (1967) 2166–2171.

[263] W.S. Frederics, L.W. Schuerman and L.G. Lewis, Investigation of Crystal Growth Process (Oregon state University, Corvallis, OR, 1966).

[264] R. Capulletti, V. Fano and M. Scavlini, Purification and Growth of Alkalihalide Crystals, Ric. Sci. 38 (1968) 887–890.

[265] U. Gross, Zone Purification of Alkali Iodides, Mater. Res. Bull. 5 (1970) 117–128.

[266] L.A. Ploom, Radiation Colouring and Thermostimulating Luminescence of KCl of Enchanced Purity, Coll. Inst. Phys. Astron. Est. SSR, Tartu 43 (1975) 81–89.

[267] V.V. Banik, O.V. Demirskaya and L.N. Chervonnaya, Fractional Separation of Oxygen-Containing Impurities in Sodium Iodide at Vacuum Distillation, Single-Crystalline Materials, Vol. N11 (Institute for Single Crystals, Kharkov, 1983) pp. 87–90.

[268] A.L. Novozhilov, E.I. Gribova and V.N. Devyatkin, Investigation of the State of HCl in Molten Chlorides of Alkali and Alkaline Earth by IR Spectroscopy Method, Zh. Neorg. Khim. 17 (1972) 2078–2080.

[269] A.L. Novozhilov, E.I. Gribova and V.N. Devyatkin, Solubility of Hydrogen Chloride in Molten Chlorides of Alkaline Earth Metals, Zh. Neorg. Khim. 17 (1972) 2570–2572.

[270] D. Bratland, K. Grjotcheim, C. Krohn and K. Motzfeld, On the Solubility of Carbon Dioxide in Molten Alkali Halides, Acta. Chem. Scand. 20 (1966) 1811–1826.

[271] G.J. Kipouros and D.R. Sadoway, A Thermochemical Analysis of the Production of Anhydrous $MgCl_2$, J. Light Met. 1, 2 (2001) 111–117.

[272] A.B. Salyulev and I.V. Kaluzhnikova, Investigation of Dehydration of Mixtures of Carnallite and Bishofite with KCl at Pressures of HCl up to 2.5 MPa in Molten State,

(Abstract of the 10th All-Union Conference on Physics, Chemistry and Electrochemistry of Ionic Melts and Solid Electrolytes, Ekaterinbourg, Vol. 1. 1992) p. 99.

[273] A.L. Novozhilov, Solubility of Hydrogen Chloride in Melts of $KCl-MgCl_2$ System, Zh. Neorg. Khim. 29 (1984) 218–221.

[274] Yu.M. Ryabukhin and N.G. Bukun, Mechanism of Dissolution of Chlorine in Molten Alkali Metal Halides, Zh. Neorg. Khim. 13 (1968) 1141–1145.

[275] N.M. Sinitsyn, V.N. Pitchkov, A.S. Kozlov, G.G. Novitskii, A.A. Sidorov and I.A. Khartonik, Interaction of Iridium with Chlorine in Melts of Alkali Metal Chlorides, Zh. Neorg. Khim. 25 (1980) 2603–2609.

[276] R. Taylor, Preparation of Water-Free Lantanide Halides, Chem. Rev. 62 (1962) 503–511.

[277] L.A. Ploom, N.I. Gindina and G.A. Yaanson, Determination of Basicity and Obtaining Single Crystals of KCl and KBr with Decreased Contain of Oxygen-Containing Admixtures, Problems of Purity and Perfection of Ionic Crystals (IFA AN ESSR, Tartu, 1969) pp. 38–44.

[278] V.V. Banik, N.I. Davidenko and Yu.A. Nesterenko, Synthesis of CaI2 of High Purity and Growing Single Crystals, Physics and Chemistry of Solid State (Institute for Single Crystals, Kharkov, 1983) N10, pp. 139–141.

[279] M.J. Weber, Inorganic Scintillators: Today and Tomorrow, J. Luminescence 100 (2002) 35–45.

[280] E.V.D. van Loef, P. Dorenbos, C.W.E. van Eijk, K.W. Kramer and H.U. Dudel, Optical and Scintillation Properties of Pure and Ce^{3+} Doped $GdBr_3$, Opt. Commun. 189 (2001) 297–304.

[281] K.S. Shah, Y. Glodo, M. Klugerman, L. Cirignano, W.W. Moses, S.E. Derenzo and M.J. Weber, $LaCl_3$: Ce Scintillators for Gamma Ray Detection, Nucl. Instrum. Method A-505 (2002) 76–81.

[282] K.S. Shah, L. Cirignano, R. Grazioso, M. Klugerman, P.R. Bennett, T.K. Gupta, W.W. Moses, M.J. Weber and S.E. Derenzo, RbGd2Br7: Ce Scintillators for Gamma Ray and Thermal Neutron Detection, IEEE Trans. Nucl. Sci. S-49 (2002) 1655–1660.

[283] Yu.K. Tselinskii and N.M. Samokhvalova, Low-Temperature Synthesis of Copper(I) and (II) Halides in Melts (Abstract of 14th Ukranian Conference on Inorganic Chemistry, Kiev, 1996) p. 125.

[284] M. Lebl and J. Trnka, Entfernung von Sauerstoff-heltigen Anionen aus Alkalihalogen-iden, Z. Phys 186 (1965) 128–136.

[285] O.V. Demirskaya and Yu.A. Nesterenko, Estimation of Effectiveness of Different Ways of Potassium Chloride Purification from Oxygen-Containing Impurities, Physics and Chemistry of Crystals (Institute for Single Crystals, Kharkov, 1977) pp. 155–159.

[286] Yu.F. Rybkin and O.V. Demirskaya, Purification of Alkali Metal Chlorides and a Method of its Control with the Use of Electrolytic Cell with Solid Electrolyte, Single Crystals and Technique (Institute for Single Crystals, Kharkov, 1974) Issue 1 (10), pp. 115–118.

[287] R.C. Pastor and A.C. Pastor, Crystal Growth in a Reactive Atmosphere, Mater. Res. Bull. 10 (1975) 117–124.

[288] R.C. Pastor and A.C. Pastor, Solid Solutions of Metal Halides under a Reactive Atmosphere, Mater. Res. Bull. 11 (1976) 1043–1050.

[289] A.I. Agulyanskii and P.T. Stangrit, To the Question of Purification of Some Alkali Metal Halides, Zh. Prikl. Khim. 10 (1977) 1201–1204.

[290] V.L. Cherginets and V.V. Banik, Purification of Lithium Chloride and the LiCl–KCl Eutectic from Oxygen-Containing Admixtures (Abstract of Seventh All-Union Conference on Chemistry and Engineering of Rare Alkali Metal Halides, Apatity, 1988) pp. 116–117.

[291] R.C. Pastor and A.J. Timper, Purification of Alkali Chlorides and Bromides [Hughes Aircraft Co] USA Patent, (C01 D3/14, C01 D3/20) No 3969491, (1976).

[292] E.P. Aleksandrov, B.D. Vasin, S.V. Krivonogov and A.G. Tyustin, Chlorination of Oxides of Scandium, Yttrium, Lanthanum, Praseodymium, Neodymium and Europium by Carbon Tetrachloride (Abstract of the 10th All-Union Conference on Physics, Chemistry and Electrochemistry of Ionic Melts and Solid Electrolytes, Ekaterinbourg, Vol. 1. 1992) p. 88.

[293] J. Eckstein, U. Gross and G. Rubinova, Growth of Alkaline-Halide Crystals with a Lowered Concentration of Oxygenous Using SiX_4 as a Removing Agent, Kristall Technik 3 (1968) 583–587.

[294] N.N. Smirnov and V.R. Lyubinskii, Thermodynamic Analysis of Processes of Interactions of Oxygen-Containing Impurities in Sodium Iodide with Different De-oxidants, Single Crystals and Techniques (Institute for Single Crystals, Kharkov, 1971) Issue 5, pp. 95–101.

[295] D.R. Flinn and K.H. Stern, Chemical and Electrochemical Studies of the Oxide-Ion—Oxygen Reaction in Equimolar Sodium Nitrate + Potassium Nitrate, J. Electroanal. Chem. 63 (1975) 39–57.

[296] J.M. De Jong and G.H.J. Broers, On the $O_2-O_2^-O_2^{2-}$ Equilibrium in Molten Alkali Nitrates with a Graphite Electrode, J. Electroanal. Chem. 56 (1974) 321–324.

[297] P.G. Zambonin and J. Jordan, Redox Chemistry of the System $O_2-O_2^-O_2^{2-}O^{2-}$ in Fused Salts, J. Am. Chem. Soc. 91 (1969) 2225–2228.

[298] P.G. Zambonin, Preliminary Note. Mechanism of the Oxygen Electrode in Molten Salts. An Unitary Interpretation of Potentiometric Findings in Unbuffered Melts, J. Electroanal. Chem. 24 (1970) 365–377.

[299] D.A. Tkalenko and S.A. Kydrya, Chemical and Electrochemical Behaviour of Peroxide Compounds in $NaNO_3$ Melt, Elektrokhimiya 14 (1978) 1579–1581.

[300] D.A. Tkalenko and S.A. Kydrya, On Stability of Oxygen Ions in Molten Lithium Nitrate, Physico-Chemical Properties of Molten Salts and Solid Electrolytes (Naukova Dumka, Kiev, 1979) pp. 96–103.

[301] D.A. Tkalenko, Electrochemistry of Nitrate Melts (Naukova Dumka, Kiev, 1983) p. 224.

[302] M. Cassir, G. Moutiers and J. Devynck, Stability and Characterization of Oxygen Species in Alkali Molten Carbonates: A Thermodynamic and Electrochemical Approach, J. Electrochem. Soc. 140 (1993) 3114–3123.

[303] M.V. Smirnov, O.Yu. Tkacheva and V.A. Oleynikova, Interaction of Oxygen with Molten Potassium Chloride, Institute of Electrochemistry of Ural Branch of Acad. Sci. USSR, Dep. in VINITI, 13.11.1989, N6809-B89, p. 26.

[304] M.V. Smirnov and O.Yu. Tkacheva, Interaction of Molten Sodium Chloride with Oxygen, Institute of Electrochemistry of Ural Branch of Acad. Sci. USSR, Dep. in VINITI, 31.08.1990, N4846-B90, p. 17.

[305] O.Yu. Tkacheva and M.V. Smirnov, Interaction of Molten Caesium Chloride with Oxygen, Institute of Electrochemistry of Ural Branch of Acad. Sci. USSR, Dep. In VINITI, 31.08.1990, N4847-B90, p. 15.

[306] V. Khokhlov, E. Nikolaeva and M. Smirnov, Investigation of Oxygen Electroreduction at the Pt-(O_2) Electrode in Molten Alkali Chlorides, in: Advances in Molten Salts. From Structural Aspects to Waste Processing, ed M. Gaune-Escard (Begell House, NY, 1999) pp. 300–308.

[307] M.V. Smirnov, O.Yu. Tkacheva and V.A. Oleynikova, Electrode and Redox Potentials of Oxygen Electrode in Molten Chlorides of Alkali Metals, Institute of Electrochemistry of Ural Branch of Acad. Sci. USSR, Dep. in VINITI, 31.08.1990, N4845-B90, p. 9.

[308] P.G. Zambonin, On the Voltammetric Behaviour of the Carbon Dioxide–Oxygen–Carbonate System in Molten Alkali Nitrates, Anal. Chem. 44 (1972) 763–767.

[309] A.P. Kreshkov, Principles of Analytical Chemistry. Theoretical Bases, Quantitative analysis, Vol. 2 (Khimiya, Moscow, 1976) pp. 185–189.

[310] R. Combes, R. Feys and B. Tremillon, Dissociation of Carbonate in Molten NaCl–KCl, J. Electroanal. Chem. 83 (1977) 383–386.

[311] A.L. Novozhilov, Solubility of CO_2 in Molten Chlorides of Alkali Metals, Zh. Neorg. Khim. 29 (1984) 2971–2973.

[312] N.M. Barbin, A.P. Pekar, V.N. Nekrasov and L.E. Ivanovskii, Equilibrium Constant of Reaction $Na_2CO_3 = Na_2O + CO_2$ in Melts of NaCl–KCl System, Rasplavy N4 (1994) 48–51.

[313] V.L. Cherginets and T.P. Rebrova, Dissociation of Carbonate Ion in Molten Alkali Metal Chlorides, Zh. Fiz. Khim. 76 (2002) 134–136.

[314] H. Remy, Lehrbuch Der Anorganischen Chemie, Vol. 1 (Akademische Verlagsgesselschaft Geest & Portig K.-G., Leipzig, 1960).

[315] V.L. Cherginets, O.V. Demirskaya, T.P. Rebrova, Studies of Some Heterogeneous Acid–Base Reactions in Molten KCl–LiCl Eutectic in 600–800 °C Temperature Range, Chemistry Preprint Server, http://preprint.chemweb.com/physchem/0203001, Uploaded 1 March 2002, p. 4.

[316] V.L. Cherginets, O.V. Demirskaya and T.P. Rebrova, Potentiometric Study of Acid–Base Equilibria in the KCl–LiCl Eutectic Melt at Temperatures in the Range 873 to 1073 K, J. Chem. Thermodyn. 36 (2004) 115–120.

[317] V.L. Cherginets, T.P. Rebrova and O.V. Demirskaya, On Carbonate Stability in Caesium Iodide Melt, Phys. Chem. Chem. Phys. 3 (2001) 1479–1480.

[318] Wrench, N.S., PhD thesis (University of London, London, 1967).

[319] G. Delarue, Reaction Chimiques Metallant en jen les ions O^{2-} et S^{2-} dans l'Eutectique LiCl–KCl fondu, Bull. Soc. Chim. France N8 (1960) 1654–1659.

[320] G. Delarue, Oxydation Electrochimique des Ions HO^- en O^{2-} dans l'Eutectique LiCl–KCl Fondu, J. Electroanal. Chem. 1 (1959) 13–25.

[321] V.L. Cherginets and E.G. Khailova, Investigation of Dissociation of Some Lux–Flood Bases in Molten Alkali Metal Halides, Zh. Neorg. Khim. 41 (1996) 734–736.

[322] B.P. Nikol'sky, Physical Chemistry. Theoretical and Practical Handbook (Khimiya, Leningrad, 1987) p. 880.

[323] J. Greenberg and L.J. Hallgren, Infrared Spectra of NaOH above and below the Melting Point, J. Chem. Phys. 35 (1961) 180–182.

[324] A.I. Novozhylov and E.I. Ptchelina, Investigation of Interaction of Water Vapour with Molten Sodium Chloride by Means of IR Spectroscopy, Zh. Neorg. Khim. 22 (1977) 2057–2060.

[325] G.E. Gladyshev and E.S. Shakhetova, Kinetics of Aggregation and Dissolution of Impurity in Crystals of Alkali Metal Halides, Izv. SO AN SSSR. Neorg. Mater. 25 (1989) 488–491.

[326] Yu.K. Delimarsky and V.N. Andreeva, Potentiometric Determination of Solubilities of Oxides in Molten Sodium Metaphosphate, Zh. Neorg. Khim. 5 (1960) 1123–1125.

[327] N.K. Woskressenskaja and G.N. Kaschtschejev, Solubility of Metal Oxides in Molten Salts, Izv. sektora fiziko-khimicheskogo analiza 27 (1956) 255–267.

[328] A.V. Volkovitch, Interaction of Oxides of Alkaline Earth Metals with the Melt of Equimolar Mixture of Chlorides of Sodium and Potassium, Rasplavy N4 (1991) 24–30.

[329] D. Naumann and G. Reinhardt, Loslichkeit von Erdalkalioxiden in Alkalichlorid-Schmelzen, Z. Anorg. Allg. Chem. 343 (1966) 165–173.

[330] V.L. Cherginets and V.V. Banik, Acidic Properties of Cations and Solubilities of Oxides in the Molten KCl–NaCl Eutectic at 973 K, Rasplavy N1 (1991) 66–69.

[331] V.L. Cherginets, Studies of the Cation–Oxide Ion Interactions in Halide Melts. The Potentiometric Control of Saturation at Different Oxide Ion Concentrations, J. Electroanal. Chem. 493 (2000) 144–147.

[332] V.L. Cherginets and E.G. Khailova, Determination of Solubility Products of Oxides in Molten by Potentiometric Method with the Use of Metallic and Oxygen Electrodes, Zh. Neorg. Khim. 36 (1991) 1277–1280.

[333] V.L. Cherginets, T.G. Deineka, O.V. Demirskaya, T.P. Rebrova, Estimation of interface energy of 'metal oxide/ionic melt' system from data on lead oxide solubility in molten KCl–NaCl eutectic, Prepr., Chemistry Preprint Server, http://preprint.chemweb.com/physchem/0103006 Uploaded 2 March 2001, p. 3.

[334] G.A. Laitinen, Chemical Analysis (Khimiya, Moscow, 1966) p. 656.

[335] G. Picard, F. Seon and B. Tremillon, Reactions of Formation and Stability of Iron(II) and (III) Oxides in LiCl–KCl Eutectic Melt at 470 °C, Electrochim. Acta 22 (1978) 1291–1300.

[336] F. Seon, G. Picard and B. Tremillon, Stability of Ferrous Oxide in Molten LiCl–KCl Eutectic at 470 °C, J. Electroanal. Chem. 138 (1982) 315–324.

[337] G. Delarue, Proprietes Chimiques dans l'Eutectique LiCl–KCl Fondu, Bull. Soc. Chim. France N5 (1960) 906–910.

[338] V.I. Shapoval and O.F. Pertchik, Polarographic Behaviour of Bi in the Presence of Oxide Ions on Background of KCl–LiCl, Elektrokhimiya 19 (1974) 1241–1245.

[339] M.P. Vorobey, A.S. Bevz and O.V. Skiba, Chlorination of Oxides of Uranium in Molten Chlorides of Alkali Metals and their Mixtures, Zh. Neorg. Khim. 23 (1978) 1618–1621.

[340] V.L. Cherginets and T.P. Rebrova, Solubility of Magnesium Oxide in Eutectic Melts of KCl–LiCl and CsCl–KCl–NaCl Systems, Zh. Neorg. Khim. 49 (2004) 1571–1574.

[341] O.A. Esin and S.E. Lumkis, Behaviour of Heavy Metals in Molten Chlorides, Zh. Neorg. Khim. 2 (1957) 1145–1148.

[342] S.A. Amirova, V.V. Petchkovskii and R.K. Kurmaev, Solubility of Vanadium Trioxide in Molten Chlorides of Sodium and Potassium, Zh. Neorg. Khim. 9 (1964) 1229–1231.

[343] V.I. Shapoval, O.G. Tsiklauri and N.A. Gasviani, Acidic Properties of Some Oxides (WO_3 MoO_3, CrO_3) and Cations (Ca^{2+}, Li^+, Ba^{2+}) on the Background of Molten Salt Systems, Soobsch. AN Gruz. SSR 89 (1978) 101–104.

[344] N.M. Barbin, V.N. Nekrasov, L.E. Ivanovskii, P.N. Vinogradov and V.E. Petukhov, Solubility of Dilithium oxide in Molten Equimolar NaCl–KCl Mixture, Rasplavy N4 (1990) 117–120.

[345] N.M. Barbin and V.N. Nekrasov, The Physicochemical and Electrochemical Behavior of Lithium Oxide in Equimolar NaCl–KCl Melt, Electrochim. Acta 44 (1999) 4479–4488.

[346] Y. Kaneko and H. Kojima, Reaction of Metallic Oxide and Anhydrous Silicic Acid in the Fused NaCl–KCl Mixture, Denki Kagaku 42 (1974) 304–309.

[347] R.L. Combes and S.L. Koeller, Solubilidade do SrO em KCl–NaCl Fundido a 727 °C, Quimica Nova 23 (2000) 34–36.

[348] V.L. Cherginets and A.B. Blank, Solubility of Metal Oxides in Molten Equimolar Mixture of Sodium and Potassium Chlorides, Ukr. Khim. Zh. 57 (1991) 936–939.

[349] L.M. Witing, High-Temperature Solvents–Melts, Handbook for Students of Universities (Moscow State University, Moscow, 1991) p. 221.

[350] V.L. Cherginets, T.G. Deineka, O.V. Demirskaya and T.P. Rebrova, Potentiometric Investigation of Oxide Solubilities in Molten KCl–NaCl Eutectic. Effect of Surface Area of Solid Particles on the Solubilities, J. Electroanal. Chem. 593 (2002) 171–178.

[351] A.I. Rusanov and V.A. Prokhorov, Surface Tensimetry (Khimiya, Saint-Petersburg, 1994) p. 400.

[352] A.A. Khokhryakov and A.M. Khokhlova, IR Spectra of Emission of Oxyhalide Complex Groups of S-Elements in Molten Mixtures of Alkali Metal Halides, Rasplavy N6 (1989) 66–71.

[353] V.L. Cherginets and E.G. Khailova, Solubility of Oxides of Alkaline Earth Metals in the Molten CsCl–KCl–NaCl Eutectic Mixture at 600 °C, Zh. Fiz. Khim. 66 (1992) 1654–1655.

[354] E.G. Khailova, T.A. Lysenko and V.L. Cherginets, Solubility of Two-Valent Metal Oxides in CsCl–KCl–NaCl Eutectic Melt at 700 °C, Zh. Neorg. Khim. 38 (1993) 175–179.

[355] Shen Tsin Nan and Yu.K. Delimarskii, Solubility of Oxides of Titanium, Molybdenum and Tungsten in Molten Borax, Ukr. Khim. Zh. 27 (1961) 454–457.

[356] V.A. Kireev, The Course of Physical Chemistry (Khimiya, Moscow, 1975) p. 776.

[357] V.A. Rabinovitch and Z.Y. Khavin, Short Handbook on Chemistry, 2nd edn. (Khimiya, Leningrad, 1978) p. 392.

[358] I.N. Bronstein and K.A. Semendyaev, Handbook of Mathematics for Engineers and Students of Highest Technical Schools, 13th edn. (Nauka, Moscow, 1986) p. 544.

[359] V.L. Cherginets and E.G. Khailova, On Some Regularities of Oxide Solubility in Chloride Melts, Zh. Neorg. Khim. 38 (1993) 1281–1285.

[360] V.L. Cherginets and E.G. Khailova, On the Solubilities of Bivalent Metal Oxides in Molten Alkaline Chlorides, Electrochim. Acta 39 (1994) 823–829.

[361] L. Pauling, The Nature of the Chemical Bond (Cornell University Press, Ithaca, NY, 1960).

[362] O.V. Demirskaya, E.G. Khailova and V.L. Cherginets, Potentiometric Determination of Oxide Solubilities in a Bromide Melt at 700 °C, Zh. Fiz. Khim. 69 (1995) 1658–1660.

[363] V.L. Cherginets and E.G. Khailova, Effect of Anion of Halide Melt on Oxide Solubilities, Ukr. Khim. Zh. 62 (1996) 90–92.

[364] V.L. Cherginets, E.G. Khailova and O.V. Demirskaya, Solubility of Oxides in Molten CsI at 700 °C, Zh. Fiz. Khim. 71 (1997) 371–373.

[365] F.R. Duke and M.L. Iverson, Complex Ions in Fused Salts, J. Phys. Chem. 62 (1958) 417–419.

[366] J. Braunstein, M. Blander and R.M. Lindgren, The Evaluation of Thermodynamic Association Constants in Solutions with an Application to Molten Salt Solutions, J. Am. Chem. Soc. 84 (1962) 1529–1533.

[367] Yu.K. Delimarskii, V.F. Grishchenko and L.I. Zarubitskaya, Potentiometric Investigation of the Complexation of Lead Ions in Chloride Melts, Zh. Neorg. Khim. 20 (1975) 921–925.

[368] Yu.K. Delimarskii, L.I. Zarubitskaya and V.F. Grishchenko, Potentiometric Investigation of Processes of the Complex Formation of Tin(II) Ions in Chloride Melts, Zh. Neorg. Khim. 21 (1976) 413–417.

[369] D. Inman, Complex Ions in Molten Salts: A Potentiometric Study of the Halide Complexes of Cadmium in Molten Equimolar NaNO₃–KNO₃ at 250 °C, Electrochim. Acta 10 (1965) 11–21.

[370] M. Blander and M.-L. Saboungi, The Coordination Cluster Theory. Application to Ternary Reciprocal Molten Salt Systems, Acta Chem. Scand. 34 (1980) 671–676.

[371] H.G. Damle, S.S. Katti and H.B. Mathur, Thermodynamic of Activity of Molten Salt Mixtures of AgJ and CdJ₂, Indian J. Chem. 15 (1977) 573–578.

[372] M.V. Smirnov, V.A. Khokhlov and A.A. Antonov, Density and Viscosity of Molten Reciprocal Fluorides of Sodium and Caesium, Coll. Inst. Electrochem. Ural Sci. Cent. Acad. Sci. USSR N25 (1977) 3–7.

[373] J.E. Enderby, Molten salts: Recent Progress in Determining Their Structure, J. Electrochem. Soc. 127 (1980) 407.

[374] L.T. Guyseva and V.A. Khokhlov, Determination of Solubility Product for MgO in Molten KCl Containing MgCl2 by Electrochemical Method with the Oxygen Electrode (Fifth International Symposium on Solubility Phenomena, Moscow, Russia, 1992) p. 69.

[375] S.F. Belov and G.D. Seredina, Interaction of Metal Oxides with the Molten NaCl–KCl, Izv. vuzov. Tsvet. Metall. N4 (1990) 19–23.

[376] T.P. Rebrova, Acidic Properties of High-temperature Ionic Solvents Based on Alkali and Alkaline Earth Halides, PhD Thesis (Karkov, 2003).

[377] T.L. Inyushkina, L.I. Petukhova and V.T. Kornilova, On the Solubility of Magnesium Oxide in Melts of Alkali Metal Chlorides, Zh. Neorg. Khim. 20 (1975) 1058–1060.

[378] F.F. Grigorenko, A.V. Molodkina, V.M. Solomakha and M.S. Slobodyanik, Studies of Zirconium Dioxide Solubility in Molten Potassium Metaphosphate, Ser. Chem. N1 (1973) 38–40, Visnyk Kyivskogo Universitetu, see also p. 81.

[379] Y. Ito, H. Hayashi, Y. Itoh and S. Yoshizawa, The Thermodynamic Behaviour of Nickel in a Molten Lithium Fluoride–Potassium Fluoride–Oxide System, Bull. Chem. Soc. Jpn 58 (1985) 3172–3175.

[380] Y. Ito, Stabilized Zirconia as an Oxide Ion Indicator, Proceedings of Electrochem. Soc. N1 (1986) 445–449, Molten Salts.

[381] V.F. Gorbunov and G.P. Novoselov, Interaction of Fluorides of Lanthanum and Cerium with Metal Oxides in the Medium of Molten Fluoride Salts, Zh. Neorg. Khim. 19 (1974) 1734–1736.

[382] R.L. Combes and S.L. Koeller, Estudo de Algumas Reacoes Oxoacidobasicas no Solvente NaCl Fundido Entre 1100 K e 1200 K, Quimica Nova 25 (2002) 226–230.

[383] V.L. Cherginets and T.P. Rebrova, The Effect of Temperature on Characteristics of Acid–Base Equilibrium in Medium of Molten Sodium Chloride, Zh. Fiz. Khim. 71 (2004) 1886–1889.

[384] G.T. Kosnyrev, V.N. Desyatnik, N.A. Kern and E.N. Nosonova, Total Alkalinity and Solubility of CaO in $CaCl_2$–KCl–NaCl Melts, Rasplavy N2 (1990) 121–123.

[385] B. Neumann, C. Kröger and H. Yuttner, Die Systeme Erdalkalichlorid–Erdalkalioxid und Zersetzung der Erdalkalichlorid Durch Wasserdampf, Z. Elektrochem. 41 (1935) 725–736.

[386] W.D. Threadill, The Calcium Chloride–Calcium Oxide Fused Salt Electrolytic System: Solubilities, Metal Contents and Freezing Points, J. Electrochem. Soc. 112 (1965) 632–633.

[387] K.L. Strelets and N.V. Bondarenko, Interaction of Magnesium Chloride with Calcium Oxide in Chloride Melts, Zh. Neorg. Khim. 8 (1963) 1706–1709.

[388] V. Ivenko, A. Shurygin, O. Tkacheva and S. Dokashenko, The Solubility of Yttrium, Barium and Copper Oxides in Molten $BaCl_2$–KCl–NaCl Eutectic (Fifth International Symposium on Solubility Phenomena, Moscow, Russia, 1992) p. 71.

[389] V.M. Ivenko, Determination of Solubility of Barium Oxide in BaCl2–KCl–NaCl Eutectic by Potentiometric Method (Abstract of 10th All-Union Conference on Physics, Chemistry and Electrochemistry of Ionic Melts and Solid Electrolytes, Ekaterinbourg, Vol. 1. 1992) p. 74.

[390] D.L. Nelson, M.S. Whittingham and T.F. George (eds), Chemistry of High-Temperature Superconductors (Mir, Moscow, 1988) p. 400.

[391] G.K. Strukova, I.I. Zverkova, V.S. Chekhtman and V.P. Korzhov, Evolution of Phases During the Process of Synthesis of $YBa_2Cu_3O_{7-x}$ and $YBa_4Cu_3O_{8.5+\delta}$ in Ammonium Nitrate Melt, Superconductivity: Physics, Chemistry, Technology 4 (1991) 2225–2228.

[392] R. Combes, M.-N. Levelut and B. Tremillon, Conditional Solubility Versus pO_{2^-} of Cerium(III) Oxide in Molten Equimolar NaCl + KCl at 727 °C, J. Electroanal. Chem. 91 (1978) 125–131.

[393] E.V.D. Van Loef, P. Dorenbos, C.W.E. Van Ejik, K. Kramer and H.U. Gudel, High Energy Resolution Scintillator: Ce^{3+} Activated $LaBr_3$, Appl. Phys. Lett. 79 (2001) 1573–1575.

Formula Index

A

AgBr   15, 148
AgCl   14, 47, 138, 148, 175
AgF   47
AgI   15,148
Ag_2O 267
Al_2O_3 41, 42, 60, 68, 177, 268, 269
$Al_2(SO_4)_3$ 269, 278
$AlCl_4^-$ 10
Am^0, Am^{2+}, Am^{3+} 14
As_2O_5 52, 53
$AuCl_3$ 268

B

BF_4^- 10, 19
BH_3 19
BO_2^- 42, 89, 96, 101
BO_3^{3-} 89
B_2O_3 41, 87
$B_4O_7^{2-}$ 42, 101, 120
$BaCO_3$ 338, 340
$BaCl_2$ 121, 175, 275, 326, 328, 336
$BaCrO_4$ 50
BaO   137, 160, 201, 232, 235, 249, 252,
     254, 258, 267, 277, 280, 281, 282, 293,
     300, 306, 309, 312, 336
BaO_2 161
Ba_2OCl_2 336
$BaSO_4$ 279
BeO   268
Bi^{3+} 39
Bi_2O_3 69, 267, 270
BiOCl   270

BrF_5, BrF_6^- 8
BrO_2^+ 51
BrO_3^- 51

C

C   198
C_6H_6 29
CH_3COO^- 58
CH_3COOH 5, 29
CCl_4 29, 166, 197, 337
CN^- 269
CO   198
CO_2 43, 86, 98, 125, 141, 144, 175, 206,
     209, 214, 337
$(COO)_2^{2-}$ 58
CO_3^{2-} 35, 43, 58, 86, 104, 124, 175, 204,
     217, 338, 340
$COCl_2$ 198
$CaCO_3$ 338, 340
$CaCl_2$ 109, 123, 175, 268, 275, 326, 328
CaF_2 324
CaO   42, 67, 79, 122, 124, 176, 252, 261,
     267, 276, 281, 282, 286, 289, 290,
     293, 300, 306, 309, 312, 318, 323,
     325, 326, 328, 331
$CaWO_4$ 79
$CdCl_2$ 47
CdO   69, 123, 235, 261, 267, 282,
     286, 289, 293, 300, 306, 309,
     318, 331
CdS   268
$CeCl_3$ 109, 336
CeO^+ 336

Ce_2O_3 109, 336
Ce_2S_3 268
Cl_2 50, 71, 149, 194, 205
ClF_3, ClF_4^- 8
ClO_2^+ 50, 51
ClO_3^- 50
$Co(NO_3)_2 \cdot 6H_2O$ 51
CoO 268, 270, 271, 274, 275, 282, 293,
     306, 318, 325, 331
CoO_2^- 325
Co_3O_4 325
$CrCl_3$ 166
CrO_2 70
CrO_3 22, 55, 69, 91, 101
Cr_2O_3 166, 167, 323
$CrO_4^{2-} \cdot O^{2-}$ 73, 101
$Cr_2O_7^{2-}$ 49, 52, 54, 70, 104, 116,
     119, 144
$CsBr$ 127, 171, 216
Cs_2CO_3 99, 170, 214
$CsCl$ 9, 127, 171, 194, 210, 216, 320
CsI 98, 105, 127, 170, 171, 173, 215,
     216, 308, 312
Cs_2O 201, 205
Cs_2SO_4 279
$Cu(NO_3)_2 \cdot 3H_2O$ 51
$CuCl$ 268
$CuCl_2$ 268
CuO 180, 197, 334
Cu_2O 173, 253, 255, 256, 275, 282, 293,
     300, 312, 323
Cu_2S 269

D
$DyCl_3$ 16

F
$[Fe(CN)_6]^{4-}$ 269
$FeCl_2$ 166
$FeCl_3$ 166, 268
FeO 67, 267, 275
Fe_2O_3, Fe_3O_4 267
FeS 269

G
$GdCl_3$ 16
GeO_2 89, 101
$Ge_2O_5^{2-}$ 91, 101

H
$HAsO_4^{2-}$ 54, 104
$H_2AsO_4^-$ 54
HCO_3^- 3, 58, 206
HCl 1, 108, 111, 183, 187, 189, 192
$HClO_4$ 7, 22
$H_2Cr_2O_7$ 73
HI 94, 186
H_2O 1, 3, 7, 22, 34, 102, 108, 111, 145,
     183, 187, 189, 218, 222, 223
H_3PO_4 62
HPO_4^{2-} 54, 62, 104
$H_2PO_4^-$ 54, 62, 94, 104
H_2SO_4 6, 29
HfF_4 47
$HgCl_2$ 13
HgI_2 13
HgO 13

I
I_2 95, 186, 195
IO_3^- 195

K
KBr 127, 171, 210, 216, 318
K_2CO_3 61, 98, 214
KCl 13, 14, 44, 98, 105, 127, 171, 194,
     210, 212, 214, 216, 276, 316, 318,
     320, 324, 326
$K_2Cr_2O_7$ 57, 60, 61, 66, 70, 91
$KGaCl_4$ 14
$KGaI_4$ 14
K_2HPO_4 61
KI 13, 127, 171, 216, 318
KLn_3X_{10} 16
K_3LnCl_6 16
KNO_2 98
KNO_3 22, 36, 49, 58, 61, 98, 142

K_2O 63, 113, 172, 201, 205
K_2O_2 172
KOH   13, 99, 150, 122, 156, 162,
    168, 170, 179, 252, 254, 267,
    268, 279
KPO_3 98, 99, 324
KSCN   38
K_2SO_4 98, 279
$K_2S_2O_7$ 98
KVO_3 53
$K_2V_6O_{16}$ 53
$K_3V_5O_{14}$ 53
K_2ZnO_2 326

L
La_2O_3 324
$LiCoO_2$ 32
LiCl   194, 210
$LiFeO_2$ 267
$LiNO_3$ 49, 60
Li_2O 110, 123, 152, 278, 219, 227,
    277, 323, 334
LnF_3 324
LnOF   324

M
Me_2TiCl_6 45
$MgCl_2$ 47, 109, 190, 191, 316, 328
$MgCl_2 \cdot H_2O$ 196
MgO   42, 60, 67, 110, 111, 118,
    124, 191, 196, 268, 271,
    273, 279, 280, 282, 293, 300,
    306, 316, 318, 320,
    323, 326, 328, 331
Mg_2O^{2+} 280
MnO   267, 282, 293, 300, 306, 331
MoO_2^{2+} 74
MoO_2Cl_2 74, 76
MoO_3 55, 57, 66, 101
$MoO_4^{2-} \cdot O^{2-}$ 101
$Mo_2O_7^{2-}$ 75
$Mo_3O_{10}^{2-}$ 54

N
NH_3 29
NH_4Cl 7, 196
NH_4VO_3 61
NO_2 37
NO_2^+ 18, 22, 36, 45, 51, 52, 54, 64, 100
NO_2^- 58, 104, 203
N_2O_5 36
$N_2O_7^{4-}$ 59
$NaAlCl_4$ 11
$Na_2B_4O_7$ 87, 96, 98
NaBr   126, 127, 171, 216
Na_2CO_3 76, 80, 82, 85, 88, 96, 134, 137,
    143, 152, 155, 170, 175, 207
NaCl   16, 44, 100, 126, 127, 170, 171,
    183, 194, 204, 210, 216, 226,
    276, 320, 325
$Na_4CrO_5 \cdot 13H_2O$ 73
NaF   15, 176
NaH_2PO_4 61
NaI   93, 105, 121, 127, 170, 186,
    215, 216
$NaNO_3$ 18, 49
Na_2O 68, 92, 103, 113, 160, 227, 279
Na_2O_2 59, 61, 142, 157, 186, 202,
    204, 227
NaOH   1, 59, 76, 80, 82, 85, 90, 96, 122,
    152, 158, 160, 162, 167, 168, 179,
    183, 218, 227, 240, 254
$NaPO_3$ 40, 42, 57, 61, 66, 84, 98, 99
Na_3PO_4 99
$Na_4P_2O_7$ 61, 66, 100
$Na_5P_3O_{10}$ 82
$Na_6P_4O_{13}$ 82
Na_2S 268
Na_2SO_4 49, 64, 98, 204, 279, 325
$Na_2S_2O_8$ 98
$NaVO_3$ 85
Na_2WO_4 78, 98, 99
$Na_2W_2O_7$ 99
NbO_3^- 90
NbO_4^{3-} 90
$NbO_4^{3-} \cdot O^{2-}$ 90
Nb_2O_5 90

$Nb_2O_7^{4-}$ 90
$NiCl_2$ 182
NiO   151, 164, 173, 179, 226, 235, 255,
    268, 270, 271, 275, 279, 282, 287,
    293, 300, 306, 318, 325, 331
NiO_2^- 325
NpO_2 270

O
O_2 37, 49, 71, 75, 101, 130, 143, 145,
    151, 153, 154, 159, 174, 186,
    205, 306
O_2^- 146
O_2^{2-} 131, 143, 146, 154, 159, 201, 203
OH^- 3, 35, 102, 145, 146, 189, 218, 222

P
PCl_5 7
PO_3^- 62, 66, 81, 83, 101, 104, 139
P_2O_5 53, 62, 81
$P_2O_7^{4-}$ 62, 82, 83, 94, 101, 104
PO_4^{3-} 62, 66, 101, 104
$PO_4^{3-} \cdot O^{2-}$ 84, 101
$POCl_3$ 7
Pb^{2+} 39
$PbCl_2$ 47, 148, 239
PbF_2 47
PbO   60, 61, 68, 69, 123, 151, 239, 261,
    263, 267, 270, 282, 289, 293, 300,
    306, 307, 309, 312, 318
PbO_2^{2-} 237, 239
Pb_3O_4 61
$PbO \cdot Pb(NO_3)_2$ 60
$PbSiO_3$ 69
PdO   270
PtO   164, 270

R
RbCl   210, 320
Rb_2O 201, 205
Rb_2SO_4 279

S
SO_2 9, 37, 95
SO_3 37, 43, 651, 95, 103, 269, 278, 325
SO_3^{2-} 269
SO_4^{2-} 18, 34, 94
$S_2O_7^{2-}$ 4, 18, 45, 52
$SbBr_3$, $SbCl_3$ 8
Sb_2O_3 267
$SiCl_4$ 44, 199
SiO_2 41, 42, 60, 67, 68, 90, 199
SiX_4(X=Cl, Br, I)   200
SnO   293
$SrCO_3$ 338, 340
$SrCl_2$ 121, 175, 328
SrO   160, 232, 235, 249, 252, 276, 279, 282,
    293, 300, 306, 309, 312, 318, 334
$SrSO_4$ 279

T
$TiCl_4$ 44
TiO_2 60, 173
$Ti(SO_4)_2$ 278
Tl^+ 39
$TlCl_3$ 268
Tl_2O 69

U
UCl_3 47
UO_2, UO_3, U_3O_8 271

V
VO_2 61
VO_3^- 86, 101, 116, 120
VO_4^{3-} 86, 101
$VO_4^{3-} \cdot O^{2-}$ 101
V_2O_3 276
V_2O5 53, 61, 66, 85–87, 89, 90, 101, 139
$V_2O_7^{4-}$ 86, 101
$VOCl_3$ 44

W
WO_3 55, 78, 101
WO_4^{2-} 78, 101

$WO_4^{2-} \cdot O^{2-}$ 80, 101

WO_2Cl_2 78, 80

$W_2O_7^{2-}$ 78, 99

$W_3O_{10}^{2-}$ 79

Y

YO_2^- 325

Y_2O_3 134, 166, 179, 325, 334

Z

$ZnCl_2$ 41, 47, 109, 268

ZnO 69, 258, 261, 268, 279, 282, 288, 289, 290, 293, 300, 306, 323, 326, 329, 331

ZnS 268

ZrO_2 60, 134, 166, 168, 179, 281, 324

Subject Index

A

Acid-base
  Definitions see *Definition of acids and bases*

Acidity (see also *Oxoacids*)
  BrO_2^+, ClO_2^+, NO_2^+ 51
  Cations   51, 57, 60
  Oxides   41, 43, 60, 68
  Functions
    Duffy $\Lambda_{Me^{n+}}$ 39
    Hammett H_0, H_- 24
    Oxobasicity index, pI_L 112, 116, 118, 121, 124
    Tremillon, Ω 107

Acidity
  Row   61, 93
  Scale   119
    Protic solvents   28
    Ionic solvents   30
    Hammett   25
    Izmailov   26

B

Bases
  BaO   201
  Carbonate, CO_3^{2-} 206
  Hydroxide, OH^- 146, 218
  Li_2O 110
  Na_2O_2, peroxides   146, 202–205

Basicity
  Carbonate, oxide group   58
  Formal, of slags   68
  In NaI   94

Optical   39
Upper limit   122

C

Calibration of oxygen electrodes
  Direct   135
  Indirect   137

Chlorination, uranium oxides   270, 271

Constant
  Dissociation see *Dissociation constant*
  Stability   270, 277

D

Definition of acids and bases
  Arrhenius   1
  Brønsted–Lowry   2
  Franklin   5, 17
  Hard and soft (HSAB) 11
  Lewis   3
  Lux–Flood XII
  Pearson   11
  Solvent system classic   5
  Solvosystem generalized   17
  Usanovitch   4

Dissociation constant
  Carbonate   175, 207, 215, 338
    Formal   208
  Hydroxide   218
  Oxides   230, 241
    Li_2O 110
    CoO   272
    PbO   123, 239

E

Electrochemical cell
    Scheme   46, 52, 149, 150, 153, 154,
        159, 316, 339
    Construction   136
Electrodes
    Calibration see *Calibration of oxygen*
        electrodes
    Gas oxygen ones   130, 148, 157
        $Au(O_2)$ 147
        Limits of use   132
        $Pt(O_2)$ 66, 99, 148, 159
        $Rh(O_2)$ 66
    Membrane
        Ni|NiO|YSZ   150, 160, 186, 218
        $Pt(O_2)$|YSZ   152, 154, 157, 159,
            160, 171, 175, 177, 220, 253
        $Pt(O_2)$|$ZrO_2(CaO)$ 159
    Metal-oxide   132, 144, 147, 150,
        163–164
        Limits of use   133
        Nb|Nb_2O_5 147
        Ni|NiO   157, 159, 221
        Ta|Ta_2O_5 147
        Zr|ZrO_2 147
    Oxide films on   144
    Oxide ion reversible   61, 130
    Peroxide function   143
    $Pt(O_2)$ 142
    Reference
        Chlorine   149
        Lead   149, 150
        Silver   134, 149
Electronegativity   303
    Allred–Rochow   304
Enthalpy of melting   297
Equation
    Antonov   292
    Le Chatelier–Shreder   294, 296
    Clausius–Clapyeron   295
    Nernst   130, 133
    Ostwald–Freundlich   259
    Shreder see *Equation* → *Le Chatelier-*
        Shreder

G

Glasses   39, 41

H

Hammett functions see *Acidity functions*
Hydrolysis   34, 245
    Individual halides   191
    KCl–NaCl   186
    $MgCl_2$ 190
    NaCl   183
    NaI   186

I

Intrinsic acid–base product   18
Investigation methods see *Methods of*
    investigations
Ionic moment, halide ions   128
Ionic product
    Solvent   6
    Water   348

L

Lanthanide halides   15
Levelling of acidic (basic) properties   7, 22,
    63, 123–124, 279
Liquid junction, potential of   47

M

Melt
    Ba^{2+}-based   120, 121, 175, 201, 256,
        275, 329, 332, 333
    Ca^{2+}-based   112, 118, 120–122, 175,
        236, 277, 326, 329
    CsBr   126, 171, 216
    CsBr–KBr   235, 248, 249, 305, 308,
        310, 312
    CsCl   210, 216, 320
    CsCl–KCl–NaCl   141, 156, 168, 169,
        213, 215, 308, 310, 312, 321
    CsI   98, 105, 125, 170, 171, 173, 215,
        216, 310, 312
    KBr   126, 171, 210, 216, 318

KCl   44, 98, 171, 210, 212, 214, 216,
   316, 318, 320, 324
KCl–LiCl   40, 91, 109, 114, 123,
   142, 149, 186, 215, 218, 221,
   267, 321
KCl–MgCl$_2$ 191
KCl–NaCl   44, 69, 109, 113, 156, 159,
   186, 207, 211, 218, 227, 232, 239,
   254, 255, 258, 259, 263, 275, 310,
   331, 337
KCl–NaCl–NaF   176
KI   216, 318
KCl–NaCl–SrCl$_2$ 331
KNO$_3$ 49, 52, 58, 147
KNO$_3$–LiNO$_3$ 38
KNO$_3$–NaNO$_3$ 45, 49, 56, 143, 148,
   218
KPO$_3$ 324
KSCN   38
K$_2$SO$_4$–Na$_2$SO$_4$ 37, 65, 103, 146, 148
LiCl   210
LiF–NaF   324
LiNO$_3$ 60
NaBr   126, 171, 216
NaCl   16, 44, 100, 126, 127, 170, 171,
   183, 204, 210, 216, 226, 320, 325
NaCl–AlCl$_3$ 40
NaI   93, 170, 215, 216, 121
NaNO$_3$ 60
Na$_2$SO$_4$ 204
Na$_2$O–B$_2$O$_3$ 39
Nitrates   143, 206
Oxygen-containing   31
Oxygen-less   31
PbSiO$_3$ 69
RbCl   210, 320
Silicates   67
Sulfates   64
Zn^{2+}-based   40
Methods of investigations
  Acidity
    Carbonate   125
    Gas solubility   43
    Indicator   37

    Kinetic   45
    Metallo-indicator   38
    Potentiometric   46
  Hydrolysis
    Dynamic   183
    Potentiometric   187
    Van't Hoff, OH$^-$ dissociation   222
  Solubility
    Isothermal saturation   231
    Potentiometric titration   233
    Sequential addition   257, 287

N
Nernst equation see *Equation → Nernst*
Non-ionizing solvents   9
  SO$_2$ 9

O
Oxides
  Insoluble   235
  Slightly soluble   236
  Moderately soluble   236
  Soluble   236
  Solubility (see *Solubility*)
Oxoacids
  As^V 52
  B^{III} 88, 96
  Cr^{VI} 49, 55, 56, 69, 91
  Ge^{IV} 89
  Mo^{VI} 54, 74,
  Nb^V 89
  P^V 52, 66, 81, 100
  S^{VI} 45,
  V^V 52, 84, 95
  W^{VI} 54, 77
Oxobasicity index (see *Acidity → Functions*)

P
pO XII, 106, 133, 139, 141, 254
Polarization, halide ions   128
Pressure of dissociation, CO$_2$ 214
Primary medium effect   26
  H$^+$ 26
  O^{2-} 107

Potentiometric titration curve
   Cations, description   235
   Differential   240
$p\theta$ 241, 247
Pyrohydrolysis see *Hydrolysis*

R
Removal oxide ions using
   HX(X=Cl, Br, I) 188
   NH_4X(X=Cl, Br, I) 195
   Carbohalogenation   197
   SiX_4(X = Cl, Br, I) 200

S
Slags   67
Solubility
   Alkali earth carbonates   337
   Al_2O_3 177
   $BaCrO_4$ 50
   BaO   277
   CaO   79, 122, 290
   $CaWO_4$ 79
   CdO   123
   Ce_2O_3 109, 336
   Complete   230, 231, 301, 306, 312
   CoO   274
   Cr_2O_3, study   167
   Cu_2O 255, 256
   FeO, Fe_2O_3, Fe_3O_4 267
   Gases, in molten salts   43
      Cl_2 194
      CO_2 209
      HCl   189
   Li_2O 278
   MgO   110, 111, 118, 273, 322, 323, 334
   Na_2SO_4 49

NiO   255
NpO_2 270
PbO   123, 239
SrO   232
Sulfides   268–269
UO_3 270
V_2O_3 277
ZnO   289, 290
ZrO_2 168, 281
Solubility product
   Oxides   230, 244, 254, 283, 297, 314,
      331
   In CsBr–KBr   306, 308
      In CsCl–KCl–NaCl   293, 294, 300
      In CsI   309
      In KBr, KI   319
      In KCl   318, 326
      In KCl–LiCl   273
      In KCl–NaCl   282, 291
Solvents
   first kind   20, 31
   second kind   23, 31
   Ionic with complex anion
      $KGaCl_4$, $KGaI_4$ 14
      KNO_3 22
      M^IAlCl_4 13
   Ionizing   6
      BrF_5 8
      ClF_3 8
      $SbCl_3$, $SbBr_3$ 8
Surface area, oxides   261

T
Theorem
   Lagrange   299